AF616459

SP
LLC

Plant Bioassays

Series Editor

S.S. Narwal

International Allelopathy Foundation,
101, Sector 14, Rohtak–124 001, Haryana, India

Co-Editors

D.A. Sampietro, M.A. Vattuone

Instituto de Estudios Vegetales,
Facultad de Bioquimica, Química Y. Farmacia,
Universidad Nacional de Tucumán,
España 2903, CP 4000, Tucumán, Argentina

C.A.N. Catalán

Instituto de Quimica Organica,
Facultad de Bioquímica, Química Y. Farmacia,
Universidad Nacional de Tucumán,
Ayacucho 471, Tucumán (T4000INI), Argentina

B. Politycka

Department of Plant Physiology,
August Cieszkowski Agricultural University,
60–637 Poznan, Wolynska 35, Poland

2009

www.studiumpress.com

Plant Bioassays

***Series Editor*: Dr. S.S. Narwal**

ISBN: 1-933699-42-6

Published by:

STUDIUM PRESS, LLC
P.O. Box-722200, Houston, Texas-77072, USA
Tel. 713-541-9400; Fax:713-541-9401
E-mail: studiumpress@studiumpress.com

Printed at:

Thomson Press India Ltd.

Editors

Prof. S.S. Narwal is an internationally renowned Allelopathy Scientist. He is the First National Fellow (Allelopathy) of Indian Council of Agricultural Research, New Delhi, from 1995–2005. Currently, he is the President of Indian Society of Allelopathy and the Chief Editor of Allelopathy Journal. Till now, he has authored/edited 18 books on Allelopathy and many books are under preparation.

He has traveled to more than 35 countries in all the Six-Continents to deliver lectures in International Allelopathy Conferences. In 1989, when he was Visiting Professor, Department of Allelopathy, Ukrainian Academy of Sciences, Kiev, he attended All Soviet Allelopathy Conference and also read Allelopathy Reprints from Prof. A.M. Grodzinsky's Collection. Prof. Narwal established Indian Society of Allelopathy in 1990. Praising him Prof. E.L. Rice, USA, wrote *'Congratulations, Prof. Narwal for Establishing the First National Society in the World'*. He has also organized four International Allelopathy Conferences in India, 1992 (Hisar), 1994 (New Delhi), 1998 (Dharwad) and 2004 (Hisar). He also founded International Allelopathy Society (1994) during the II International Allelopathy Conference, New Delhi.

Diego A. Sampietro is the Assistant Professor (Phytochemistry and Plant Biotechnology), National University of Tucumán, Tucumán, Argentina. He did his Ph.D. (Biochemistry) in 2005 and possesses 10 years experience of teaching and research. Currently, he is Regional Editor of Allelopathy Journal. He has authored/co-authored more than 30 scientific publications, including nine book chapters. He is member of the research group of Dr. Catalán and is actively working on sugarcane allelopathy and secondary compounds involved in plant defence.

César A.N. Catalán is full Professor (Organic Chemistry) since 1983. He has been the Head (Organic Chemistry Group), Faculty of Chemistry, Biochemistry and Pharmacy, National University of Tucumán, Argentina since 1985. He completed his Ph.D. (Chemistry) in 1977 from National University of Tucumán and in 1983, he was Postdoc at Standford University (California, USA). He possesses 40 years teaching and research experience at National University of Tucumán, Tucumán, Argentina and is Senior Researcher, National Council of Scientific and Technological Research (CONICET–

Argentina). He is also the Director of CONICET, Dean of Faculty of Chemistry, Biochemistry and Pharmacy, National University of Tucumán and Rector of this University. He has received the award 'Dr. Venancio Deulofeu 2005' from Argentinian Chemical Association (AQA) for his major contributions in the field of organic chemistry. He has also authored/co-authored 150 journal articles, 6 Books and has supervised over 20 M.Sc. and Ph.D. students in the fields of Chemotaxonomy, Bioactive Metabolites and Organic Synthesis, using readily available natural products. Currently, his research group is actively working on natural products (sesquiterpene lactones from Argentine Asteraceae and bioactive constituents) from aromatic and medicinal plants from South America.

Marta A. Vattuone is full Professor of Phytochemistry. Her main research interests are Plant Biochemistry and Natural Products. She has been Head of the Phytochemistry Group of the Faculty of Chemistry, Biochemistry and Pharmacy (National University of Tucumán, Argentina) since 1999. She completed her Ph.D. (Chemistry) in 1980 and was Postdoc in France, Spain and Belgium. She has 40 years of teaching experience at the National University of Tucumán, Tucumán, Argentina, and is Senior Researcher from the National Council of Scientific and Technological Research (CONICET–Argentina). Dr. Vattuone is working on plant biochemistry, secondary metabolites involved in allelopathic processes and antimicrobial activity of propolis and medicinal plants. She has published more than 100 papers in international journals and wrote seven book chapters.

Prof. Barbara Politycka is full professor and Vice Dean at August Cieszkowski Agricultural University, Poznan, Poland, where she is Teaching Plant Physiology. She has 30 years of experience in teaching and scientific studies. Her major scientific interests constitute allelopathy in soil sickness phenomenon and physiological responses of plants to phenolic compounds released from plant residues. She has co-authored a monograph, some book chapters and numerous research papers on allelopathy. In 2004, she was granted Life Time Achievements Award by International Allelopathy Foundation. She is the member of both Editorial Committee of Agriculture University in Poznan and Regional Editor of Allelopathy Journal.

Contributors

András Takács: *Georgikon Faculty of Agricultural Sciences Institute for Plant Protection, University of Pannonia, Keszthely, Hungary.*

Anita Mukherjee: *Department of Botany, Center of Advance Study in Cell and Chromosome Research Center, University of Calcutta, 35 Ballygunge Circular Road, Kolkata 700 019, India.*

Arshaid Javaid: *Institute of Mycology and Plant Pathology, University of the Punjab, Quaid-e-Azam Campus, Lahore–54590, Pakistan.*

Cesar A.N. Catalán: *Instituto de Química Orgánica, Facultad de Bioquímica, Química y Farmacia, Universidad Nacional de Tucumán, Ayacucho 471, San Miguel de Tucumán (T4000INI), Argentina.*

D.A. Sampietro: *Instituto de Estudios Vegetales "Dr. Antonio R. Sampietro", Facultad de Bioquímica, Química y Farmacia, Universidad Nacional de Tucumán, España 2903, 4000, San Miguel de Tucumán, Tucumán, Argentina.*

Elena G. Orellano: *Molecular Biology Division, IBR (Instituto de Biología Molecular y Celular de Rosario), Consejo Nacional de Investigaciones Científicas y Técnicas (CONICET), Facultad de Ciencias Bioquímicas y Farmacéuticas, Universidad Nacional de Rosario, Suipacha 531, (S2002LRK) Rosario, Argentina.*

Emma N. Quiroga: *Instituto de Estudios Vegetales "Dr. Antonio R. Sampietro", Facultad de Bioquímica, Química y Farmacia, Universidad Nacional de Tucumán, España 2903, 4000, San Miguel de Tucumán, Tucumán, Argentina.*

Erzsébet Nádasy: *Georgikon Faculty of Agricultural Sciences Institute for Plant Protection, University of Pannonia, Keszthely, Hungary.*

Fabricio Cassán: *Plant Physiology Lab, University of Río Cuarto, Ruta 8 Km 601, 5800, Río Cuarto, Córdoba, Argentina.*

Gabriella Kazinczi: *Georgikon Faculty of Agricultural Sciences, Institute for Plant Protection, University of Pannonia, Keszthely, Hungary.*

Geraldo Stachetti Rodrigues: *Embrapa Meio Ambiente (Empresa Brasileira de Pesquisa Agropecuária), Jaguariúna-SP, Brazil.*

Germán Dunger: *Molecular Biology Division, IBR (Instituto de Biología Molecular y Celular de Rosario), Consejo Nacional de Investigaciones Científicas y Técnicas (CONICET), Facultad de Ciencias Bioquímicas y Farmacéuticas, Universidad Nacional de Rosario, Suipacha 531, (S2002LRK) Rosario, Argentina.*

Ghazala Nasim: *Institute of Mycology and Plant Pathology, University of the Punjab, Quaid-e-Azam Campus, Lahore–54590, Pakistan.*

Jorgelina Ottado: *Molecular Biology Division, IBR (Instituto de Biología Molecular y Celular de Rosario), Consejo Nacional de Investigaciones Científicas y Técnicas (CONICET), Facultad de Ciencias Bioquímicas y Farmacéuticas, Universidad Nacional de Rosario, Suipacha 531, (S2002LRK) Rosario, Argentina.*

José Marcello Salabert de Campos: *Departamento de Biologia, Universidade Federal de Lavras, Lavras-MG, Brazil.*

Jose R. Soberón: *Instituto de Estudios Vegetales "Dr. Antonio R. Sampietro", Facultad de Bioquímica, Química y Farmacia, Universidad Nacional de Tucumán, España 2903, 4000, San Miguel de Tucumán, Tucumán, Argentina.*

József Horváth: *Georgikon Faculty of Agricultural Sciences Institute for Plant Protection, University of Pannonia, Keszthely, Hungary.*

Larissa Fonseca Andrade: *Departamento de Biologia, Universidade Federal de Lavras, Lavras-MG, Brazil.*

Lisete Chamma Davide: *Departamento de Biologia, Universidade Federal de Lavras, Lavras-MG, Brazil.*

Lucas D. Daurelio: *Molecular Biology Division, IBR (Instituto de Biología Molecular y Celular de Rosario), Consejo Nacional de Investigaciones Científicas y Técnicas (CONICET), Facultad de Ciencias Bioquímicas y Farmacéuticas, Universidad Nacional de Rosario, Suipacha 531, (S2002LRK) Rosario, Argentina.*

Lyderson Facio Viccini: *Departamento de Biologia, Universidade Federal de Juiz de Fora, Juiz de Fora-MG, Brazil.*

María Laura Tondo: *Molecular Biology Division, IBR (Instituto de Biología Molecular y Celular de Rosario), Consejo Nacional de Investigaciones Científicas y Técnicas (CONICET), Facultad de Ciencias Bioquímicas y Farmacéuticas, Universidad Nacional de Rosario, Suipacha 531, (S2002LRK) Rosario, Argentina.*

Marta A. Vattuone: *Instituto de Estudios Vegetales "Dr. Antonio R. Sampietro", Facultad de Bioquímica, Química y Farmacia, Universidad Nacional de Tucumán, España 2903, 4000, San Miguel de Tucumán, Tucumán, Argentina.*

Melina A. Sgariglia: *Instituto de Estudios Vegetales "Dr. Antonio R. Sampietro", Facultad de Bioquímica, Química y Farmacia, Universidad Nacional de Tucumán, España 2903, 4000, San Miguel de Tucumán, Tucumán, Argentina.*

Natalia Gottig: *Molecular Biology Division, IBR (Instituto de Biología Molecular y Celular de Rosario), Consejo Nacional de Investigaciones Científicas y Técnicas (CONICET), Facultad de Ciencias Bioquímicas y Farmacéuticas, Universidad Nacional de Rosario, Suipacha 531, (S2002LRK) Rosario, Argentina.*

Oscar Masciarelli: *Plant Physiology Lab, University of Río Cuarto, Ruta 8 Km 601, 5800, Río Cuarto, Córdoba, Argentina.*

Rabia Haouala: *Department of Biology, Higher Institute of Biotechnology of Monastir, University of Monastir, Av. Tahar Haddad, BP 74, 5000 – Monastir, Tunisia.*

Richard Gáborjányi: *Georgikon Faculty of Agricultural Sciences Institute for Plant Protection, University of Pannonia, Keszthely, Hungary.*

Rukhsana Bajwa: *Institute of Mycology and Plant Pathology, University of the Punjab, Quaid-e-Azam Campus, Lahore–54590, Pakistan.*

S.S. Narwal: *Department of Agronomy, CCS Haryana Agricultural University, Hisar-125004, India.*

Sergio Molinari: *Institute of Plant Protection, National Council of Research, C.N.R., Via G. Amendola 122/D, 70126 Bari, Italy.*

Stephen D. Murphy: *Department of Environment and Resource Studies, University of Waterloo, 200 University Avenue West, Waterloo, Ontario, N2L 3G1, Canada.*

Susana B. Popich: *Departamento de Ciencias Básicas y Aplicadas, Universidad Nacional de Chilecito, Ruta Los Peregrinos S/N. CP: 5360 Dpto., Chilecito. La Rioja, Argentina.*

Tomáš Gichner: *Institute of Experimental Botany, Academy of Sciences of Czech Republic, Na Karlovce 1a, 160 00, Prague 6, Czech Republic.*

Virginia Luna: *Plant Physiology Lab, University of Río Cuarto, Ruta 8 Km 601, 5800, Río Cuarto, Córdoba, Argentina.*

Plant Bioassays

Foreword

Few would deny that the subject of allelopathy is strewn with controversy. The root of these problems, since the conception of allelopathy, has been the perceived difficulties in its methodology and consequently the reliability of the data. The allelopathy experienced a wave of enthusiasm during the period 1965-1985 due to leadership of C.H. Muller and E.L Rice in the United States and A.M. Grodzinskii, Soviet Union. In the 1990's, the study of allelopathy declined, as the relevance of methodologies employed in most allelopathic studies, and the relevance of consequent data were seriously questioned. Eminent critics, such as J.L. Harper, cast a long and lasting, but not necessarily undeserved, shadow on allelopathic research. Despite this nadir, interest in allelopathy was renewed largely due to the efforts of Shamsher S. Narwal, Professor of Agronomy at Haryana Agricultural University, India. Narwal was steadfastly resolute in wishing to see allelopathy legitimised in the ecological and agricultural arenas, and this has led, under his aegis, to the formation of the International Allelopathy Society (IAS), foundation of the *Allelopathy Journal* and publication of an extensive series of monographs devoted to the science of allelopathy. The study of allelopathy offers great returns in enabling a better understanding and management of sustainable agriculture and forestry, and in reducing reliance on chemical pest control. It is no accident that the science of allelopathy flourishes foremost now in India, China, Japan and neighbouring countries.

Mindful of the methodological problems associated with so many studies in allelopathy, Narwal in 2004 set the agenda for publication of a series of monographs devoted to bringing together the various protocols and methods that have been used successfully by leading investigators in allelopathy and allied disciplines. To date there have been six volumes published. Five have appeared under the series title *Research Methods in Plant Sciences: Allelopathy*, with volume subtitles: *Soil Analysis*, *Plant Protection*, *Plant Pathogens*, *Plant Analysis*, and *Plant Physiology*. The series is not meant to provide the impossible, that is a single set of protocols for the allelopathy researccher, nor do they demand extraordinary laboratory equipment; rather, they bring together the tried methods, applications and experiences, as used by allelopathy researchers based mainly at Haryana Agricultural University, who have worked with different plant species and pest organisms, under various environmental conditions for the past 25 years. This valuable series acknowledges that allelopathic research must draw upon numerous interrelated disciplines. Not only do these volumes give details of the practical methods used across allelopathy studies, they also provide an understanding of the theoretical bases of the methods involved in order to allow informed decisions regarding their use. A groundbreaking sixth volume, edited by V.V. Roshchina and Narwal has been published separately as *Cell Diagnostics: Images, Biophysical and Biochemical Processes in Allelopathy*. Three new, eagerly awaited volumes in the primary series are now available: *Plant Bioassays*, *Biochemistry* and *Natural Products*.

I am honoured to write the Foreword to a new volume in *Research Methods in Plant Sciences: Allelopathy – Plant Bioassays*. This long awaited book departs significantly from previous volumes, in that it draws on a truly international group of world experts in allelopathic research, who have wrestled intimately with the problems of the intelligent and appropriate use of plant bioassay in allelopathic research, with considerable emphasis on interactions with microorganisms.

I can only commend Prof. Narwal and his team of co-editors, who have tirelessly and selflessly progressed the series, which provides nothing less than a small library of methods and background information for the allelopathy researcher. I also wish to commend Studium Press LLC, Houston, TX77072, USA who have continued their support in the publication of scientific monographs on allelopathy, at a refreshingly affordable price.

October 5, 2008

R.J. Willis
Botany School,
University of Melbourne
Parkville, Victoria,
Australia

Preface

Allelopathy is a newly emerging multidisciplinary field of agricultural research. Till now much Allelopathy research work has been done in various fields of Agriculture and Plant Sciences. However, no standard methods are being used by various workers due to lack of compendium on the techniques, hence, the results obtained are not easily comparable with each other. This situation has caused a lot of problems to researchers working in underdeveloped/third world countries in small towns, where library and research facilities are not available. Therefore, to make available the standard methods for conducting allelopathy research work, these books have been planned. In all the conferences held since 1990 in India or in foreign countries, a need has always been felt for a Manual of Allelopathy Research Methods. Hence, Prof. S.S. Narwal has planned this multi-volume book ***Research Methods in Plant Sciences: Allelopathy.*** This book series aims to provide basic information about various methods to research workers, so that they can conduct research independently without the requirement of sophisticated equipments. The methods have been described in a simple way just like 'DO IT YOURSELF' book. Till now 6 Books (Soil Analysis, Plant Analysis, Plant Physiology, Cell Diagnostics, Plant Pathogens, Plant Protection) have been published. Three Books (Plant Bioassays, Biochemistry, Natural Products) are in press and likely to be released in 2009.

Although studies on Allelopathy have increased in the last 50 years, full understanding of this phenomenon has been not possible. One of the reasons is the use of inappropriate methodology for performance of plant bioassays. In fact, students and scientists interested in Allelopathy sometimes have no access to appropriate literature providing methodological information about plant bioassays to study allelopathic phenomena and to differentiate these interactions from other ones occurring in nature. Allelopathic processes often imply complex interactions not only among plants but also between plants and other living organisms. Hence, comprehensive characterization of Allelopathic processes requires detailed knowledge of such interactions, which need macroscopic, physiological and biochemical studies. These concerns require reliable protocols for such studies, which sometimes are not available worldwide. Till now, there is no comprehensive compilation/Book of Methods to study allelopathic interactions at so many complexity levels. Majority of reseachers in under developed/Third world countries, do not consider detailed characterization of such interactions, mainly because of non-compilation of such methods in book form.

This book is divided into four sections. Section I: Plant–Plant Interactions, includes six chapters (Laboratory Assays in Allelopathy, Seed Bioassays, Pollen Allelopathy, Assays for Allelopathic Interactions Among Terrestrial Plants, Genetic Toxicology and Environmental Mutagenesis in Allelopathic Interactions, Comet Assay in Higher Plants). Section II: Plant–Microorganism Interactions has six chapters (Mycorrhiza Plant Bioassays, Biochemical Aspects of Plant–pathogen Interactions, Plant Growth Promoting Rhizobacteria, Hypersensitive Response, Plant–Virus Interactions, Search for Antiphytopathogenic Compounds Involved in Plant Defence). Section III. Plant–Phytophagous Interactions consists of two chapters (search for phytochemicals to control plant–insect interactions and Bioassays on plant–nematode interactions). Section IV: Appendices (abbreviations, chemical formulae and molecular weight of solvents and reagents).

This book will serve as ready reference material in laboratory/ greenhouse/field studies and in class room. It helps to solve many problems of plant bioassays. It will be useful for UG and PG students pursuing allelopathic work at laboratory, greenhouse and field levels and will also help Biologists and Ecologists. We have tried to provide appropriate solutions to the problems of plant bioassays. The users of this book can select suitable methods according to the available facilities.

We are indebted to the contributors, who have actually used all these methods in their fields of specialization for the last 10–25 years, in accepting challenging task to prepare chapters for the present book. All of them have made sincere efforts in presenting the procedure of various methods in very simple and lucid language, which can be easily understood by the beginners.

We will appreciate to receive valuable suggestions from the students and researchers a like, to make further improvements in future editions of this book and accordingly make it more useful and meaningful.

August 31, 2008

S.S. Narwal,
D.A. Sampietro,
C.A.N. Catalán,
M.A. Vattuone,
B. Politycka

Contents

SECTION I

Plant – Plant Interactions

CHAPTER 1

Laboratory Bioassays in Allelopathy

S.S. NARWAL[1], DIEGO A. SAMPIETRO[2] AND CESAR A.N. CATALÁN[3]

1. INTRODUCTION

The science of Allelopathy as defined by Molisch (1937) includes inhibitory as well as stimulatory effects of one plant on another, including microorganisms. After Molisch and till 1970, Allelopathy was commonly referred to as negative interactions between the plants. Thereafter, the definition of Molisch has been followed. The stimulatory or inhibitory effects of one plant on another are mediated through release of allelochemicals. There is increasing concern for potential use of these substances in sustainable agriculture, especially in underdeveloped countries, because they may constitute an alternative to synthetic polluting pesticides currently employed in crop protection. Interest in Allelopathy unfortunately has been accompanied with inconsistencies in the methodology applied to identify and exploit this phenomenon in modern agriculture (Dhawan and Narwal, 1995). Establishing the presence of a chemical within a plant and use of this substance in a formulated form to control plant growth is not the same as the study of the possible involvement of a plant material in an allelopathic interaction. Although standard procedures for isolation and characterization of plant allelochemicals are available, major lacunae so far has been both laboratory bioassays and field procedures applied to establish the inhibitory or stimulatory effects of these allelochemicals in allelopathic processes.

Laboratory bioassays are procedures that assess potency of a compound or a mixture of compounds (*i.e.* a plant leachate or root exudates) via its/their application-induced response on a receptor organism (Macías *et al.*, 2000). The response of bioassays should be

1. Department of Agronomy, CCS Haryana Agricultural University, Hisar-125 004, India. E-mail: allelopathy_1947@yahoo.co.in
2. Instituto de Estudios Vegetales "*Dr. A.R. Sampietro*", Facultad de Bioquímica, Química y Farmacia, Universidad Nacional de Tucumán, España 2903, San Miguel de Tucumán (4000), Argentina. E-mail: sampietro@tucbbs.com.ar
3. Instituto de Química Orgánica, Facultad de Bioquímica, Química y Farmacia, Universidad Nacional de Tucumán, Ayacucho 471, San Miguel de Tucumán (T4000INI), Argentina. E-mail: ccatalan@fbqf.unt.edu.ar

observed on physiological processes affected by an allelochemical. Nevertheless, most laboratory bioassays are based on plant responses. Bioassays properly designed can be more sensitive than chemical methods to detect bioactive substances. In most instances, response of germinating seeds or seedling growth to allelochemicals found in leachates/plant extracts are the first assays in allelopathic research. Controversy has arisen from ecological meaning of such laboratory assays, because they often have provided evidences of Allelopathy not verifiable in the field suggesting that in most cases they cannot explain processes occurring in natural conditions (Stowe, 1979). Despite these limitations, laboratory assays are undoubtedly important in allelopathic research. They allow analyzing how different factors (*i.e.,* modes of allelochemical release, microbial activity, soil properties and special growth conditions) may affect the whole allelopathic process. The major problem with laboratory bioassays, however, is the absence of general accepted standard protocols and designs (Waller and Feng, 1995). In this chapter, general aspects of laboratory bioassays for allelopathy are provided including plant sampling, plant sample processing and assays for germination and seedling growth. We hope that criteria and procedures provided may help beginners to improve design of their laboratory assays in allelopathic studies.

2. PLANT SAMPLING

The following aspects should be considered before plant sampling:

i. Plant samples used in laboratory assays should be truly representative of the suspected allelopathic phenomenon. Preliminary observations should be carried out in the field to identify which plant materials may be relevant for your study (see Chapter 4). Leachates from whole plants are often involved in allelopathic interactions. Leachates from plant organs (leaves, flowers, shoots and roots) are also assayed, when their allelochemical contents/relative biological activities are unknown. Hence, the plant materials collected should either be the whole plant or specific organs. The rationale for using a specific plant part should be justified through field evidences, *i.e.* suspected role of such organ in chemical interference or previous knowledge about its chemical composition. During sampling, green and yellow/senescent plant parts should be collected separately.

ii. Aerial plant parts should be collected during the day, 2–3 h after sunrise to prevent wetting from fog, mist or dew drops. In perennial plant species, the month, year and season (winter, summer, spring and autumn) of sampling considerably influences the chemical composition of plants. Hence, these must be registered. Allelochemical content is generally highest in plants under extreme weather conditions (*i.e.*, during drought or after frost), while, it is always minimal after precipitation (rain, dew, fog or snow) particularly after rains due to leaching losses. The plant residues or stubbles may be collected immediately after harvest, to avoid losses/changes in allelochemical content by irrigation/precipitation and decomposition.

iii. The chemical composition in annual plants generally varies with various phenological phases (germination, seedling, tillering/branching, rapid growth phase, pre-flowering, flowering, post flowering, seed development and maturity/senescence). In perennials, such phases are vegetative, pre-flowering, flowering and seed maturity/ripening, and sampling should be repeated annually. Therefore, phenological stages should be specified. In perennials, age in years and state of plant vigor may also be registered. In dry samples, like plant residues or litter, month, year and season of collection should be registered properly.

Experiment 1: Collection of plant samples

Materials required

Steel scissors and shovel, paper sheets for annotations, pencil and rubber, polyethylene bags with labels, dry ice storage containers.

Procedure

i. Visually identify the plant material to be collected. Select source of plant material at random to ensure samples representative of field situation under analysis. Use the sharp scissors to cut aerial plant parts or the shovel to extract roots. Clearly separate plant samples from soil, tissues from other organisms or other potential contaminating materials. Decide sample size according to the scale expected for your laboratory assay. Register time of day or season, stage of plant growth and plant part to be sampled, according to preliminary observation of the suspected allelopathic interaction.

ii. Keep each sample into a previously labeled polyethylene bag. Close each bag immediately after sample collection. Samples should be transported quickly to the laboratory on dry ice to preserve original humidity and allelochemical contents/natures found in plant materials.

iii. Weigh samples and register other relevant characteristics (*i.e.* sample color or morphological details of plant material). If samples are not immediately processed, they can be stored at –20 °C for a short time until preparation. However, it is advisable to immediatly clean and use of plant material.

3. CLEANING OF COLLECTED SAMPLES

This process involves removal of contaminants from plant samples like soil and other strange particles. Cleaning prevent further influence of these contaminants on chemical composition of plant extracts/leachates and should be performed immediately after arrival of plant sample to the laboratory.

Experiment 2: Removal off from plant contaminants surface

Materials required

Soft brush, cheese cloth or filter paper, freezer.

Procedure

i. Remove traces of adhering soil and dust particles by dry wiping with a soft brush. If it is not possible, wash gently with tap water only for few seconds, followed by a quick rinsing bath in distilled water.

ii. Dry plant material should be shaked vigorously with hand and then again dried by keeping in between clean cheese cloth or towels of absorbent paper. The sample should be in water for a minimum period, to avoid leaching losses of water soluble allelochemicals (Martin-Prevel *et al.*, 1987).

iii. To avoid sample fermentation, place clean plant material in polyethylene bags and store in the freezer at –20 °C. Dry materials can be directly stored in permeable cheese cloth or nylon bags for further processing. If fresh material is used, the sample is just ready for leaching. Fresh materials that require drying are immediately processed to avoid losses due to enzymatic and respiratory activities.

4. DRYING OF PLANT SAMPLES

Drying is the second step during processing of plant material for use in extracts/leachates because (i) it reduces the risk of purification and other changes during storage, (ii) it makes storage easy and (iii) it helps to obtain homogeneous fine powdered samples. When fresh plant samples are used in bioassays, it is important to know dry matter of plant materials, because variations in sample water contents will change the biological activity of extracts/leachates. To do it, weigh a known amount of each collected plant sample (sample fresh weight, sfw). These sample amounts are separately weighed again after drying in a stainless steel oven at 100 °C for 8 h (sample dry weight, sdw). The water content of sample is calculated by the following formula:

$$\text{Water content (g)} = \text{sfw (g)} - \text{sdw (g)}$$

4.1. Traditional drying methods

Sun and/or shade drying are the simplest and commonly used traditional methods. However, they often lead to long drying periods and incomplete drying, particularly in coastal areas and in rainy/winter seasons. Sun drying is comparatively quicker than shade drying, but long exposure to sun results in bleaching of samples and volatilization of certain compounds. If samples are not turned frequently, drying may not be uniform. Shade drying is free of problems encountered in sun drying but continuation of physiological processes (respiration, enzymatic activities) for long times significantly change chemical composition of plant samples. There are also chances of fermentation under low aerated conditions or when samples are not turned upside down.

4.2. Modern drying methods

They allow more efficient and quicker drying than traditional methods, hence, are considered to be more acceptable in plant science research.

i. **Hot-air oven drying:** A mass of fresh plant samples will not reach the temperature of oven until the water evaporates and heat is absorbed to supply the latent heat of vaporization. Therefore, the first stage of drying in oven at 60 °C is called moist incubation. During this period, there may be so much autolysis that the chemical composition of the final material may be different from its original composition. Besides, there is also loss of volatile substances. Some tissues may significantly lose their carbohydrates through respiration. These difficulties could be partly overcome by increasing the speed with which the high temperature is attained, by sub-dividing the material in thin lots (Pirie, 1956). Green material should be dried as rapidly as possible to reduce chemical and biological changes to a minimum. If drying is too delayed, considerable loss in dry weight may occur due to respiration, while proteins are broken down to simpler nitrogenous compounds (Piper, 1942). Hence, green material should be spread out in layers on shallow trays, killed quickly and drying is completed at 60 °C in forced warm air (Pirie, 1956). Green material may be dried in an oven with an exhaust fan, so that water vapor is rapidly removed from the immediate vicinity of the sample. The material to be dried should be very loosely packed in the oven to allow free movement of moisture laden air. Failure to secure good ventilation leads to considerable decomposition of some organic constituents in the sample. Rapidly dried samples are always much brighter green in color than those slowly dried. Drying is most conveniently done at 60 °C till constant weight (Waller and Feng, 1995). To avoid contamination of plant samples, use of stainless steel ovens are suitable (Pirie, 1956). The drying temperature should not exceed 60 °C, as higher temperatures may cause losses in metabolites owing to volatilization (Martin-Prevel *et al.*, 1987; Waller and Feng, 1995) and thermal degradation of the sample (Anderson and Chapman, 1987).

ii. **Vacuum drying:** Evaporation of water from the sample surface can be increased at comparatively low temperatures (40–45 °C) using vacuum oven as compared to hot-air oven. Thus, exposure to high temperature is eliminated. It solves the problem of oxidation in plant samples, which occurs during sun, shade and oven drying. In vacuum drying, there is little or no oxidation and drying takes place at very low temperature. The major limitation is that it is suitable only for small samples.

iii. **Freeze drying or lyophilization:** Lyophilization or freeze drying is a process in which water is removed from a frozen plant sample, placed under vacuum. This allows ice to change directly from solid to vapor without passing through a liquid phase. The process consists of three separate, unique and interdependent processes: freezing, primary drying (sublimation), and secondary drying (desorption). After sufficient time we get a completely dry sample. Freeze drying overcomes the problems encountered in oven drying, *i.e.* loss of volatile compounds and thermal degradation of sample. The major limitation is that drying is very slow and can be used only for small samples.

Owing to limitations in vacuum drying and lyophilization, drying in stainless steel oven at 60 °C is suitable.

5. GRINDING PLANT SAMPLES

Fine powdered samples are homogeneous and ensure maximum extraction of allelochemicals, because they provide maximum exposure to extraction solvents (Anderson and Chapman, 1987), which is important for phytochemistry studies. Main objection to grinding is that it allows qualitative and quantitative release of components from plant materials, which does not occur in natural conditions. Hence, this step may be avoided during ecological studies. Before grinding, bulk plant samples are best reduced by quartering until a convenient amount of representative material is obtained. The material should be carefully handled to avoid excessive separation of plant parts. Thereafter, sub-sample is chopped and further reduced by quartering. In allelopathy research, both dry and wet grinding is used.

Experiment 3: Dry grinding

Principle

The aim of this experiment is to obtain dry ground plant materials with homogeneous particle size.

Materials and equipments

Stainless steel Wiley mill, 20-mesh (0.846 mm) screen stainless steel sieves (see *Observation* 1), brush, dry plant samples, scissors.

Procedure

i. Chop plant material before grinding. Pieces should be small enough to prevent heating of wiley mill (see *Observation*).

ii. Clean Wiley mill with a brush before use to avoid sample contamination from previously grinded samples. Grind each sample. After grinding, mix thoroughly all sub-samples ensuring homogenization of ground material. For example, in a leaf sample the tissues of the petiole, the midrib, veins and various parts of parenchyma of every leaf must be proportionately distributed throughout the sample (Martin-Prevel *et al.*, 1987).

iii. Sieve each sample and store sieved material until use.

Observations

i. Mesh size indicates number of holes per square inch (Waller and Feng, 1995).

ii. Heating is undesirable since it means over working of Wiley mill and can change chemical composition.

Experiment 4: Wet grinding

Principle

The aim of wet grinding is to get the material being studied into solution or suspension, either through soaking and shaking the dry ground material in water/organic solvent or alternatively grinding fresh/green material in these solvents as proposed in the following experiment.

Materials and equipments

Warring blenders or Maxicum grinder, scissors, cheesecloth.

Procedure

i. Chop plant material before grinding. Pieces should be small enough to prevent blender heating (see Experiment 3, *Observation* 1).

ii. Rinse blender with distilled water before use to avoid sample contamination from previously grinded samples. Grind each sample. After grinding, mix thoroughly all sub-samples ensuring homogenization of ground material. A thoroughly wet ground sample having sufficient diluent is called dispersion slurry or suspension depending on the rate of sedimentation (Pirie, 1956).

iii. Filter ground material using a cheese cloth. Collect the filtrate for further studies.

6. EXTRACT/LEACHATE PREPARATION

The properties of the plant material extracted/leached influence the techniques used for extraction/leaching and even the nature of grinding (dry or wet). In some cases, where plant or its parts like fruits are succulent/juicy and the juice already contains the material that we want to extract, then juice can be extracted using a juicer. In general, a solvent is added and left in contact with the solid for enough time for allelochemical to release. The terms 'Extract' and 'Leachate' are not synonyms. Extract implies destruction of plant cells (*i.e.*, grinding of plant materials) and release of a high amount of cellular constituents. In contrast, leachate is prepared using intact plant parts without artificial destruction of cells. It includes easily soluble/releasable low molecular weight compounds.

Water is the most common solvent that solubilizes the allelochemicals in nature. Therefore, distilled water should be used as solvent for extract/leachate preparation. Tap water can not be used because it contains salts. Extraction/leaching from plant materials should be done in cold water to simulate natural release of allelochemicals and use of boiling water or organic solvents should be discouraged.

Solutes from the plant sample are released into the solvent through plasmolysis/diffusion. In a narrow ratio of solvent: plant material (less than 7:1), the extraction may be incomplete because equilibrium between solute and solvent reaches soon, when only some part of solvent is extracted. Therefore, wide ratios are useful for complete extraction of solutes. On the other hand, too wide ratios create problems in filtration of large volumes of extractive solvent. The most frequently used ratio of solvent:plant material varies from 7:1 to 10:1. This ratio should be selected taking into account the specific natural conditions observed during preliminary observations in field conditions (Sampietro *et al.*, 2006).

Diffusion of allelochemicals from plant materials to water or other solvent is usually slow. Solubilization can be accelerated by shaking, because it increases repeated contact of individual particle/organ of plant sample with the solvent. Therefore, vigorous shaking improves extraction and reduces soaking time of plant sample. Sonicating is also used to accelerate release of allelochemicals from plant materials. Both shaking and sonicating are often performed in phytochemical studies, while they are usually avoided in ecological studies.

7. REMOVAL OF SUSPENDED PARTICLES

7.1. Paper filtration

Experiment 5: Filtration using filter paper discs

In paper filtration, the extract/leachate passes through a paper disc by gravity, retaining soil particles. Traditionaly, this is most used technique for filtration of plant extracts/leachates.

Materials and equipments

Filter paper discs, glass funnel, glass rod, beaker or Erlenmeyer, steel ring and support, clumps.

Procedure

i. Select the size of the filter paper disc considering that, when folded, it should be a few millimeters below the rim of the glass funnel. Fold the paper into a cone by first folding it in half and then in half again (Fig 1).

ii. Support the glass funnel in the neck of an Erlenmeyer flask or in a ring (Fig 2). Wet the filter paper with a few milliliters of extract/leachate. Wetting the paper holds it in place against the glass funnel. Pour in portions the leachate to be filtered through the funnel, using the glass rod as indicated in Fig 2 (see *Observation*).

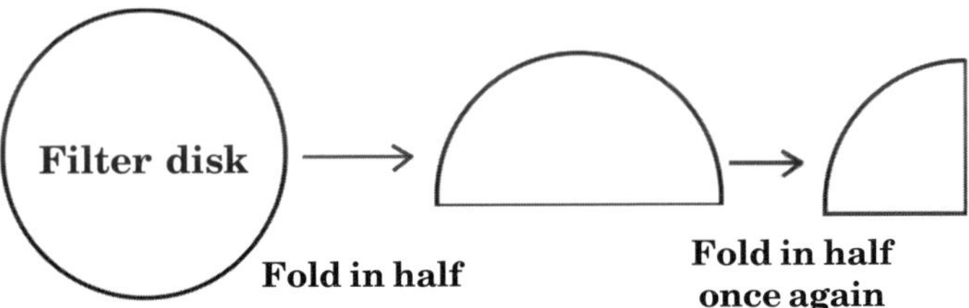

Fig 1. Preparing filter paper disc for use in paper filtration

Observations

Paper filtration is rarely satisfactory with finely ground plant materials owing to its open and hydrophilic texture, being a substantial part of extract components retained in the filter paper. Vacuum filtration (see Experiment 6) or filtration through a muslin cloth together with hand pressing can be alternatively performed in these cases (Ambler and Keith, 1996). Thereafter, the filtrate should be again filtered through filter paper, Whatman #1, to separate the suspended particles.

Experiment 6: Vacuum filtration

Vacuum filtration is faster than gravity filtration, because the plant extract/ leachate is forced to pass through the filter paper [usually placed in a Buchner

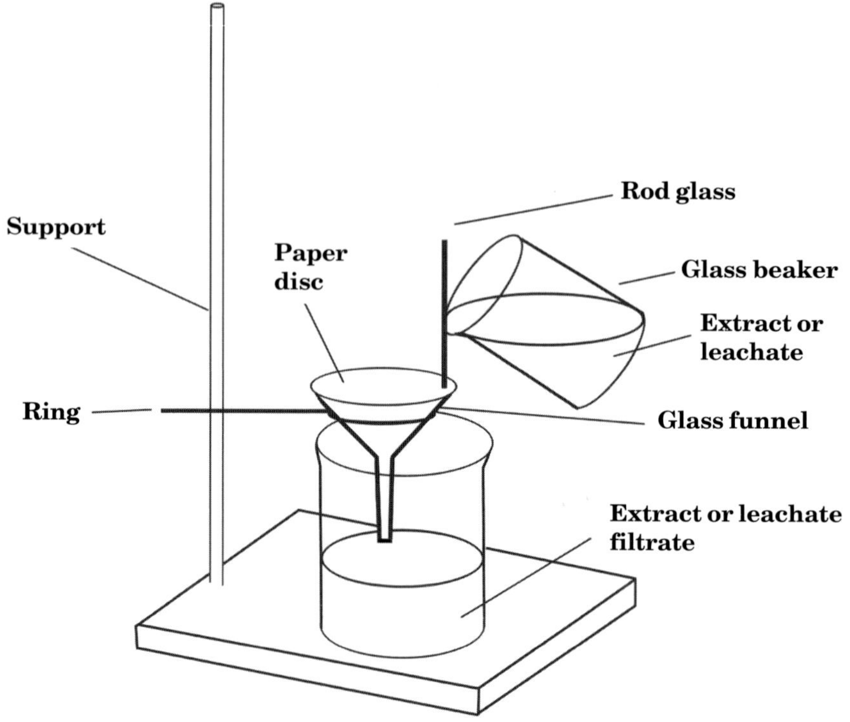

Fig 2. Paper filtration

funnel] by application of reduced pressure from a pump (Anderson and Chapman, 1987). Water jet pumps provide appropriate reduced pressures for filtration of plant extracts/leachates.

Materials and equipments

Heavy-walled Buchner funnel, side-arm filtering flask, heavy duty rubber tube, rubber adaptor or stopper to seal the funnel to the flask when under vacuum, water jet pump, clumps.

Procedure

i. Check the side-arm filtering flask carefully for cracks, since cracks could cause the flask to break when vacuum is applied. Clamp the flask securely to a ring stand. Add an adaptor and a Buchner funnel (Fig 3).

ii. Place a piece of filter paper in the funnel that must be small enough to remain flat but large enough to cover all the holes of the Buchner filter.

iii. Wet the paper with a small amount of the extract/leachate. Connect the jet water pump to the running water source. Then connect the pump to the side-arm filtering flask.

iv. Pour the mixture to be filtered onto the filter paper. The vacuum should rapidly pull the liquid through the funnel. Watch that particles do not creep under the edges of the paper. If this happens, start over and carefully pour portions of the solution onto the very center of the paper.

v. At the end of filtration process, disconnect the rubber tubing before turning off the jet water pump. Discard solid collected on paper disc and recover the filtrated extract/leachate.

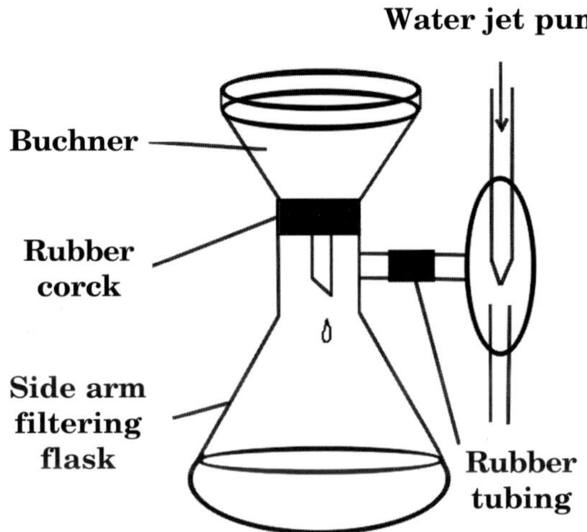

Fig 3. Vacuum filtration

7.2. Centrifugation

Centrifugation is a very useful technique to separate fine suspended solids from a liquid solvent, particularly when these are not easily filtered (Ambler and Keith, 1996). Under such conditions, it is the most common technique used, to make pellets of all suspended solid particles (Birnie and Rickwood, 1978). Centrifugation is recommended when filtrated extract/leachate is not completely free from small suspended solid particles, after paper filtration. Cold centrifugation (below 5 °C) in a refrigerated centrifuge at 10,000–15,000 rpm for 15–20 min. should be done. Do not use non-refrigerated centrifuge, because during centrifugation the rotor temperature rises quickly, which may adversely affect the allelochemicals present in the filtrate. Thereafter, all material is filtered through filter paper, the pellets are discarded and filtrated extract/leachate is used in further studies.

8. EXTRACT/LEACHATE CONCENTRATION

Concentrations of extract/leachate should be prepared considering the dry weight of plant material and volume of water used. For example, suppose 10 g plant material is soaked/mixed in 100 mL water and 60 mL of extract/

leachate are recovered after soaking, filtration and centrifugation. This filtrate is considered to be the 10 % concentration. The extract/leachate may be serially diluted for conducting bioassays, although some authors argued that each concentration should be separately prepared as indicated in the example above (Sampietro *et al.*, 2006).

9. GERMINATION AND SEEDLING GROWTH

Evaluation of seed germination and seedling growth under field conditions is often difficult, because receptor plants are exposed to several interacting environmental factors. Laboratory bioassays allow to control growth conditions, excluding several interference mechanisms (*e.g.,* resource competition, etc.) acting in the field. Assays may be performed as per the International Seed Testing Association, Switzerland Rules (ISTA, 1985), based on the standardized materials, methods and terminology for germination and seedling growth, providing reproducibility of results. However, these standard methods are not always appropriate as originally proposed, being necessary to adjust them for allelopathy research. The following experiments are examples of bioassays for germination and seedling growth adapted for assay allelopathic potential of plant extracts/ leachates.

Experiment 7: Seed germination assay

Principle

Seed germination is the resumption of growth by the embryo within the seed leading to the development of a new plant (Bewley, 1997). Germination comprises those events that begin with water uptake of the quiescent dry seed (imbibition) and finish with the elongation of the embryonic axis. Protrusion of the radicle through the testa allows visual detection of seed germination. In germination assays, seeds of a plant species are placed in contact with different concentrations of extracts/leachates or allelochemicals, including a control (distilled water). After a specific time interval under controlled conditions, the number of germinated seeds is counted and germination is expressed as one or more germination indices (Chiapusio *et al.*, 1997). Conditions needed for germination assays (*i.e.* dormancy breaking, temperature, photoperiod, volume of solution per Petri dish and pH) vary with the plant species and must be well identified (Leather and Einhellig, 1988; Macías *et al.*, 2000). The most common germination assay is performed in polystyrene or glass Petri dishes containing a proper support for germinating seeds and seedlings growth Whatman #1 filter paper, sand, agar, soil, vermiculite or cellulose sponge.

Materials required

Glass material: Petri dishes 9 cm, pipettes, 500 mL beakers, 500 mL Erlenmeyer's flasks, assay tubes, Whatman #1 paper discs (9 cm diameter), filter paper discs with 0.2-μm pore size membranes, tongs, stainless steel strainer (8 cm diameter or less), autoclave, laminar flow hood, lettuce (*Lactuca sativa* L.) seeds, balance with the precision of 0.01 mg, growth hamber, Parafilm.

Reagents

1% sodium hypochlorite; 1 N NaOH (dissolve 20g of NaOH in 500 mL of distilled water); 5 mM 2-[N-morpholino] ethanesulfonic acid (MES) buffer (dissolve 488 mg of MES in 500 mL of distilled water; adjust pH with 1 N NaOH. Final value should be closed to assayed plant extract/leachate); in a laminar flow hood, filter MES buffer and different concentrations of plant extract/leachate with a sterile 0.22-μm pore filter membrane; receive each sterile solution in a sterile 500 mL (Erlenmeyer flask).

Procedure

i. Put the Whatman #1 paper discs in a Petri dish (see *Observation* 1). Wrap glass material, tongs and the strainer with paper sheets or aluminium foil. Sterilize all materials in an autoclave at 120 ºC for 15 min before use.

ii. Weigh 100 seeds of lettuce. Calculate the seed amount to be used as follows:

$$\text{Amount of seeds (g)} = \frac{25 \times \text{DW}_{100}\,(\text{g}) \times N_R \times N_T \times 1.2}{100};$$

where DW_{100} is the dry weight of 100 lettuce seeds, N_R is the number of replications per treatment; N_T is the number of treatments. Weigh the calculated seed amount.

iii. Working in the laminar flow hood, place seeds in a 500 mL beaker containing 200 mL of 1 % sodium hypochlorite. After 15 min, drain seeds in the strainer. Dry the seeds using a sterile paper-disc towel.

iv. Transfer a filter paper disc with a tong to each Petri dish (see *Observation* 2). Then, add 4 mL of an extract/leachate solution or 5 mM buffer MES (control). Place 25 sterile seeds equidistant in each Petri dish (see *Observation* 3 and 4). Seal Petri dishes with a plastic bag or Parafilm.

v. Place Petri dishes in a growth chamber (25 °C, 16-h photoperiod at 400 μmol m^{-2} s^{-1} photosynthetically active radiation; see *Observation* 5). Count the number of germinated seeds every 6 h for 2 days, until no further seeds are germinated. A seed is considered germinated after rupture of seed coat and radicle emergence (see *Observation* 6).

Calculations

Simultaneous calculation of several germination indices has been proposed to provide a better interpretation of allelochemical activity on seed germination (Chiapusio *et al.*, 1997):

Total germination (G_T):

$$G_T = \frac{N_T \times 100}{N};$$

where N_T is the number of germinated seeds for each treatment at the end of assay and N is the total number of seeds.

Speed of germination (*S*):

$$S = (N_1 \times 1) + (N_2 - N_1) \times \tfrac{1}{2} + (N_3 - N_2) \times \tfrac{1}{3} + \ldots.. + (N_n - N_{n-1}) \times \frac{1}{\mathrm{n}};$$

where N_1, N_2, N_3.... N_n denote number of seeds germinated at 6 (1), 12 (2), 18 (3) .... and (n) hours after the assay beginning.

Speed of accumulated germination (A_S):

$$A_s = \frac{N_1}{1} + \frac{N_2}{2} + \frac{N_3}{3} + \ldots\ldots.. + \frac{N_n}{\mathrm{n}};$$

where N_1, N_2, N_3, ..., N_n denote cumulative number of seeds which germinated on time 1, 2, 3, ... or n, following set-up of the experiment.

Coefficient of rate of germination (CRG):

$$\mathrm{CRG} = \frac{(N_1 + N_2 + N_3 + \ldots\ldots.. + N_n)}{(N_1 \times T_1) + (N_2 \times T_2) + (N_3 \times T_3) + \ldots\ldots.. + (N_n \times T_n)};$$

where N_1 is the number of germinated seeds on time T_1, N_2 is the number of germinated seeds on time T_2 and N_n is the number of germinated seeds on time T_n.

Observations

1. Make at least three independent experiments (N = 3). Each treatment should have at least three replications.

2. Seeds may be germinated on filter paper, silica sand, quarts sand and soil representative of the allelopathic phenomenon under study. Prevent the use of compost/manures because usually they contain allelochemicals. Use of filter paper as substrate in germination and seedling growth assays has several disadvantages:

i. Natural substrates (*i.e.,* soil) adsorb plant allelochemicals, sometimes suppressing biological activity of plant extract/leachates, while filter paper only serves as an inert flat support, which in most cases cannot provide relevant ecological information.

ii. Non-uniform moisture conditions or swelling of the paper in localized areas can affect germination/root growth and must be avoided.

iii. Root curling often occurs onto filter paper, which implies less contact between root and test solutions leading to less uptake of allelochemicals. This is especially true for large-seeded-species.

iv. Paper often stuck to roots, making inaccurate measurements of root dry weight or elongations, etc. Roots also are often damaged during removal from filter paper, affecting measurement of root length.

3. The filter papers may be only kept moist, no excessive extract/leachate should be added to avoid water logging and anaerobic conditions.

4. Osmotic potential and pH of plant extracts/leachates sometimes inhibit *per se* germination and post-germination growth (Sampietro *et al.*, 2006). In these cases, osmotic potential should be measured with an osmometer and mannitol solutions with the same osmotic potentials as those determined in the assayed fractions should also be assayed on seed germination of the test plants.

5. Some researchers conduct germination/seedling growth assays in light, while, others do them in darkness. It is important to take into account that there is a negative relation between hypocotyl elongation and light intensity for several plant species. Thus, light may alter the relative growth of the hypocotyl, root, and shoot, altering assay sensitivity. Furthermore, biological activity of extract/leachate metabolites also may be modified by light.

6. Variations in germination response among plant species is often related with seed size. At the same concentration and with the same number of seeds per dish, the effect of an allelochemical may be less pronounced in large-seeded species than in small-seeded ones, because higher tissue volume may dilute the effect of a bioactive compound in a seed with large size (Dayan *et al.*, 2000).

Statistical analysis

Analyze data using one way Analysis of the Variance and Mann-Whitney U-test at the significant level of 0.05.

Experiment 8: Seedling growth assays

Principle

Most common seedling growth assay consists of exposing the pre-germinated seeds (with uniform radicle elongation) to different concentrations of an allelochemical/extracts/leachates, including a control (distilled water). Shoot and/or root elongation are measured after a specific time interval under controlled conditions, to prepare build a dose–response curve (Fig 4). Critical parameters such as the inhibition threshold (lowest concentration required to initiate inhibition), the IC_{50} (concentration required to obtain 50 % inhibition) and the LCIC (lowest complete inhibition concentration) are estimated from these curves, establishing the range of concentrations, where a plant extract/leachate exerts its biological activity (Dayan *et al.*, 2000). Allelochemical concentrations for further more expensive and/or time consuming assays can be selected according to these parameters.

Materials required and reagents

Use the same materials and reagents described for Experiment 7.

Procedure

i. Wrap glass material, tongs and the strainer with paper sheets or aluminium foil. Sterilize all materials in an autoclave at 120 °C for 15 min before use.

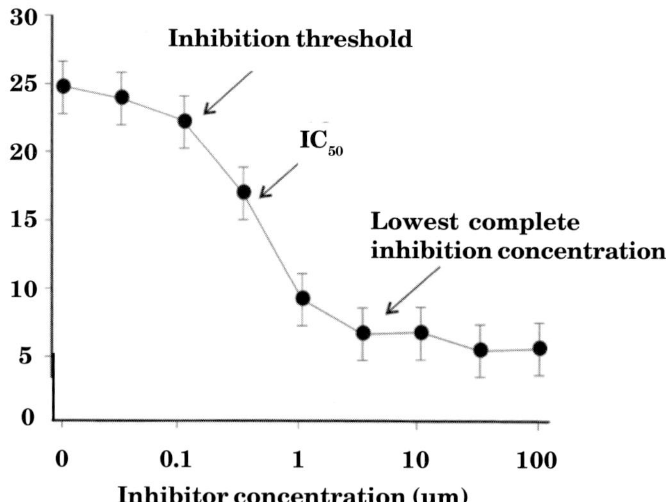

Fig 4. A typical dose–response curve illustrating various important parameters that can be estimated from the curve: inhibition threshold is the lowest concentration required to have an observable inhibition, IC_{50} is the concentration required for 50 % inhibition and LCIC is the lowest complete-inhibition concentration (*Source*: Dayan *et al.*, 2000)

ii. Calculate the seed amount needed as follows:

$$\text{Seed amount (g)} = \frac{100 \times DW_{100}\,(g) \times N_R \times N_T \times 1.2}{100};$$

where DW_{100} is the dry weight of 100 lettuce seeds, N_R is the number of replications per treatment; N_T is the number of treatments. Weigh the calculated seed amount. Prepare $N_R \times N_T$ Petri dishes. Work in a laminar flow hood.

iii. Place seeds in a 500 mL beaker containing 200 mL of 1 % sodium hypochlorite. After 15 min, drain seeds in the strainer. Dry the seeds using a sterile paper-disc towel. Work in a laminar flow hood.

iv. Place 100 seeds in each Petri dish on discs of Whatman #1 filter paper. Seed emerging radicles should be pointed downwards to the paper. Place the Petri dishes in a growth chamber (25 °C, 16 h photoperiod at 400 μmol m^{-2} s^{-1} photosynthetically active radiation) for 24 h.

v. In the laminar flow hood, transferring a sterile filter paper disc to each one with a tong. Then, add 4 mL of plant extract/leachate solutions or 5 mM buffer MES (control). Place 25 pregerminated seeds with uniform radicle length (up to 1 mm) in each Petri dish. Seal Petri dishes with Parafilm.

vi. Place Petri dishes in a growth chamber (25 °C, 16-h photoperiod at 400 μmol m^{-2} s^{-1} photosynthetically active radiation). After 4 days, measure root and shoot elongation with a ruler.

Calculations

1. Calculate root/shoot elongation (mm) for each treatment as mean ± S.D.

2. Plot root/shoot elongation *vs* plant extract/leachate concentration. Determine inhibition threshold, IC_{50}, and LCIC from the graph as indicated in Fig 1.

Observations

Observations are the same as indicated for Experiment 7.

Statistical analysis

Analyze variance of data using general lineal model (GLM) procedure and compare means using Dunnet T3 test at a significant level of 0.05. Sometimes dose–response curves of root/shoot elongation do not fit with linear regression functions, being needed other functions for statistical analysis (Inderjit *et al.*, 2002; Sampietro *et al.*, 2006; Finney, 1979).

Suggested Readings

Ambler, C.M. and Keith, F.W. (1966). *Techniques of Organic Chemistry*. Interscience Publishers, New York.

Anderson, R. and Chapman, N.B. (1987). *Sample Pretreatment and Separation*. John Wiley & Sons, New York.

Bewley, J.D. (1997). Seed germination and dormancy. *The Plant Cell* **9**: 1055–1066.

Birnie, G.D. and Rickwood, D. (1978). *Centrifugal Separations in Molecular and Cell Biology*. Butterworths, London.

Chiapusio, G., Sánchez, A.M., Reigosa, M.J., González, L. and Pellissier, F. (1997). Do germination indices adequately reflect allelochemical effects on the germination process? *Journal of Chemical Ecology* **23**: 2445–2453.

Dayan, F.E., Romagni, J.G. and Duke, S.O. (2000). Investigating the mode of action of natural phytotoxins. *Journal of Chemical Ecology* **26**: 2079–2094.

Dhawan, S.R and Narwal, S.S. (1995). Critical assessment of allelopathy bioassays in India. *In*: Proceedings of Abstracts International Symposium, *Allelopathy in Sustainable Agriculture, Forestry and Environment*. (*Eds*., S.S. Narwal, P. Tauro, G.S. Dhaliwal and J. Prakash). pp. 148. Hisar, India: Indian Society of Allelopathy, Department of Agronomy, CCS Haryana Agricultural University.

Finney, D.J. (1964). *Statistical Methods in Biological Assays*. Charles Griffin and Co., London.

ISTA (1985). International Rules of Seed Testing – 1985. *Seed Science and Technology* **13**: 297–513.

Leather, G.R. and Einhellig, F.A. (1988). Bioassay of naturally occurring allelochemicals for phytotoxicity. *Journal of Chemical Ecology* **14**: 1821–1828.

Macías, F.A., Castellano, D. and Molinillo, J.M. (2000). Search for a standard phytotoxic bioassay for allelochemicals. Selection of standard target species. *Journal of Agricultural and Food Chemistry* **48**: 2512–2521.

Martin-Prevel, P., Gagnard, J. and Gautier, P. (1987). *Plant Analysis*. SBA Publications, Kolkata.

Molisch, H. (1937). *Allelopathy: Influence of one Plant on Another*. Fisher, Jena (In German).

Piper, C.S. (1942). *Soil and Plant Analysis*. Hassel Press, Adelaide, Australia.

Pirie, N.W. (1956). Making and handling extracts. *In*: *Modern Methods of Plant Analysis* (*Eds*., K. Paech and M.V. Tracey). pp. 26–55. Springer Verlag, Berlin.

Purvis, M.J., Callier, D.C. and Walls, D. (1966). *Laboratory Techniques in Botany*. Butterworths, London.

Sampietro, D.A., Sgariglia, M.A., Soberón, J.R. and Vattuone, M.A. (2006). Effects of allelochemicals from sugarcane straw on growth and plant physiology of weeds. *Allelopathy Journal* **19**: 351–360.

Stowe, L.G. (1979). Allelopathy and its influence on the distribution of plants in an Illinois old field. *Journal of Ecology* **67**: 1065–1085.

Waller, G.R and Feng, M.C. (1995). Chemical analysis of allelopathic compounds. *In*: *Allelopathy: Field Observations and Methodology* (*Eds*., S.S. Narwal and P. Tauro). pp. 319–212. Scientific Publishers, Jodhpur, India.

Yongqin, M. (2005). Allelopathic studies of common wheat (*Triticum aestivum* L.). *Weed Biology and Management* **5**: 93–104.

CHAPTER 2

Seeds Bioassays

Rabia Haouala[1] and Shamsher S. Narwal[2]

1. INTRODUCTION

Bioassays are generally designed to test a hypothesis. Laboratory bioassays are important tools to understand a particular component of allelopathy, which cannot be replaced with any other analytical chemistry technique. In relatively simple and quick way, bioassays determine the allelopathic potential of crude extracts, exudates, volatile substances or isolated purified compounds derived from donor plants. Since the species show different sensitivity degrees to the allelochemicals, the choice of suitable target plants is a very significant element in the bioassays. Preliminary tests are thus necessary to evaluate the allelopathic potential of a plant: choice of target seeds, time of incubation, temperature, moisture and illumination. Although limited in their applications to field situations, laboratory bioassays can be designed to study the presence and subsequent release of allelochemicals, the growth of the target plant and interference of allelochemicals with their physiological processes, the fate the movement and the uptake by the target plant of allelochemicals in the soil environment and their detoxification when they are taken up by the target plant.

Bioassays are mostly conducted in Petri dishes. Substrate (filter paper, sand, soil or agar) is supplemented with a solution containing allelopathic substances. Seeds of target plants are placed on the medium surface. After defined time of incubation under controlled temperature, humidity and lighting, the number of germinated seeds, the length and dry matter of roots and aerial parts are determined. The seedling growth is more sensitive indicator to determine allelopathic potential than seed germination. Furthermore, seedling weight is more reliable indicator than root or shoot lengths.

1. Department of Biology, Higher Institute of Biotechnology of Monastir, University of Monastir, Av. Tahar Haddad, BP 74, 5000 – Monastir, Tunisia. E-mail: rabiahaouala@yahoo.fr

2. Department of Agronomy, CCS Haryana Agricultural University, Hisar–125 004, India. E-mail: allelopathy_1947@yahoo.com.ar

2. EXTRACTS PREPARATION

Water is commonly used as extraction solvent. Organic solvents are used in later stages, *i.e.,* during isolation and identification of molecules involved in the allelopathic activity. Dried or fresh plant material, litter as well as soil with plant residues are soaked into the solvent for a specific time interval. Extracts obtained are usually filtered or centrifuged. Obtained filtrate/supernatant or its different concentrations are used for bioassay.

Experiment 1: Aqueous extraction of fresh plant material

Materials required

Measuring scales, knife or scissors, mortar with pestle, beaker, graduated cylinder, distilled water, funnel and Whatman #1 filter paper, centrifuge.

Procedure

i. Weigh the sample of collected plant material, chop it into small pieces and homogenize in a mortar with a pestle.

ii. Add to homogenized sample distilled water in ratio 1–4:10 (w/v), according to the activity power of the plant material (preliminary tests are necessary to determine the adequate ratio), and mix thoroughly.

iii. Keep in dark at room temperature for 24 h.

iv. Filter the sample through Whatman #1 paper or centrifuge at 3000g for 30 min and use filtrate/supernatant immediately for bioassay. If needed, filtrate/supernatant could be diluted further to prepare its different concentrations.

Experiment 2: Aqueous extraction of dried plant material

Materials required

Oven with forced air circulation, grinding mill, sieve with 0.85 mm net, deep-freeze, scales, graduated cylinder, distilled water, centrifuge.

Procedure

i. Dry the collected plant material in an oven at 60 °C till constant weight.

ii. Grind to powder in a mill to pass a 0.85-mm net sieve and then mix thoroughly to obtain homogenous sample.

iii. Store the powdered material at –20 °C until extraction.

iv. Weigh the sample of powdered material and add the distilled water in ratio 1–4:10 (w/v) (preliminary tests are necessary to determine the adequate ratio).

v. Keep in dark at room temperature for 24 h.

vi. Filter the sample through Whatman #1 paper or centrifuge the sample at 3000g for 30 min and use filtrate/supernatant immediately for bioassay. If needed, filtrate/supernatant could be diluted further to prepare its different concentrations.

Experiment 3: Preparation of leachate solution

Materials required

Scales, graduated cylinder, distilled water, funnel and Whatman #1 filter paper, Erlenmeyer flask, refrigerator.

Procedure

i. Weigh the sample of collected intact plant parts and immerse them in distilled water in ratio of 1:10 (w/v) or more.

ii. Keep in a refrigerator at 8–10 °C for 24 h.

iii. Filter the leachate solution through Whatman #1 filter paper and use immediately to bioassay. If needed filtrate could be diluted further to prepare its different concentrations.

Experiment 4: Preparation of root exudate solution

Materials required

Plastic pots, mixture of soil with sand (3:1 ratio; w/w), seeds of donor plant.

Procedure

i. Fill plastic pots with equal quantity of soil and sow the seeds of donor plant. Keep some pots with the unseeded soil as control.

ii. After germination, leave the seedlings to grow and irrigate the pots with distilled water when needed, including soil control.

iii. When plants are grown up, harvest them and precisely remove the roots from the soil.

iv. Prepare the extract according to procedure described in Experiment 1. If needed, filtrate could be diluted further to prepare its different concentrations.

Experiment 5: Aqueous extraction of soil samples

Materials required

Pot with perforated bottom, filter paper, beaker, graduated cylinder, distilled water, Erlenmeyer flask, shaker, funnel and Whatman #1 filter paper.

Procedure

i. Line the perforated bottom of pot with a filter paper.

ii. Add soil into the pot and irrigate with distilled water. Level of water should reach up to 1/3 height of pot filled with soil.

iii. Keep at room temperature for 24 h and allow the soil to reach full saturation.

iv. Weigh the soil sample and mix with an equal volume of distilled water.

v. Shake the sample for 2 h on shaker or manually.

vi. Filter the soil extracts through Whatman #1 filter paper and use the filtrate immediately for bioassay. If needed, filtrate could be diluted to prepare different concentrations.

Experiment 6: Extraction with organic solvents

Materials required

Oven with forced air circulation; grinding mill; sieve with 0.85 mm net; deep-freeze; scales; graduated cylinder; distilled water; centrifuge; organic solvents with different polarity: petroleum ether; diethyl ether; ethyl acetate; acetone; ethanol; and methanol; Whatman #1 filter paper; rotary evaporator; DMSO.

Procedure

i. Dry the collected plant material in oven at 60 °C till constant weight.

ii. Grind to powder in the mill to pass 0.85-mm net sieve and then mix thoroughly to obtain homogenous sample.

iii. Store the powdered material at –20 °C until extraction.

iv. Weigh the powdered material and add the organic solvent in a ratio 1:9 (w/v).

v. Keep in dark at room temperature (20 °C) for 72 h.

vi. Filter the sample through Whatman #1 paper or centrifuge the sample at 3000g for 30 min.

vii. Evaporate the filtrate/supernatant to dryness under vacuum in a rotary evaporator at 40 °C.

viii. Determine the yield of the residue and store it at –20 °C until use.

ix. Dissolve each residue into 100 mL of 50 % aqueous acetone. Mix the mixtures with Tween 80 to obtain a concentration of 0.1 % and sonicate in an ultrasonic cleaner for 5 min for spreading and emulsification.

Use residue dissolved immediately for bioassay. If needed, extracts could be diluted further to prepare different concentrations.

3. PREPARATION OF AGAR MEDIUM FOR BIOASSAY

The most frequently used medium for bioassays is filter paper. However, if the seeds of test plant require a longer time for germination, filter paper becomes unsuitable medium, because the addition of large quantity of extract may lead to flooding of seeds. In this case, the alternative is agar medium, which prevents the occurrence of anaerobic conditions and simultaneously minimizes seed dehydratation.

Experiment 7: Preparation of agar medium

Materials and reagents

Scales, agar, graduated cylinder, distilled water, beaker, hot plate, glass rod, Petri dishes, automatic pippette.

Procedure

i. Add dry agar to water in the beaker and mix, the agar will not completely dissolve (when preparing 1 % agar medium, consider amount of extract to be added later).

ii. Heat the beaker on hot plate and boil the mixture to dissolve the agar.

iii. Pour into a flask and autoclave.

iv. Allow the medium to cool until it reaches to a temperature 60 °C, divide into parts as per experimental treatments. Add suitable amounts of sterile extract in different concentrations and mix. Add distilled water instead of extract to each medium. It serves as control medium.

v. When media are cool enough to touch, pour 15 mL into each Petri dish tilting them to cover the surface.

vi. After medium solidification, wrap up Petri dishes with aluminium foil.

Observations

The media may be prepared 1–2 days in advance and stored at 4 °C.

4. STERILIZATION OF MATERIALS AND DISINFECTION OF SEEDS

Bioassays should be done in semi-sterile conditions to avoid microbial infection. It is necessary to work in clean room without any air circulation, on the table and with surgical gloves disinfected with 70 % ethanol. The sterility of glassware and tools is very important. However, the largest danger to the performed bioassay are pathogens contaminating seeds.

The bioassay is conducted under conditions favorable to the development of these microorganisms, which may have a negative effect on both seed germination and seedling development, and thereby may change bioassay response. Sometimes microorganisms introduced into bioassays together with extracts, especially those prepared from soil and decomposing residues, may also pose problems.

4.1. Sterilization of glassware and tools

To eliminate bacterial and fungal contamination, the glassware should be sterilized. The best and most dependable method to sterilize glassware and tools is autoclaving and to kill fungal spores. Sterilization should be done at 120 °C 105 kPa for 20 min to kill fugal spores. Glassware and tools can also be sterilized with dry heating in an oven but it requires a higher temperature and longer exposure time.

Experiment 8: Sterilization with moist heat and pressure

Materials required

Petri dishes/glassware, pincettes, aluminium foil, autoclave.

Procedure

i. Separately wrap up Petri dishes/glassware and pincettes with aluminium foil.

ii. Prepare the liquids (*i.e.,* water or agar medium) in flasks that are filled about twice the volume of liquids and cover the flasks openings with aluminium foil.

iii. Sterilize them in an autoclave at 120 °C and 105 kPa for 20 min.

Observations

If autoclave is not available, pressure cooker can be used.

Caution: After sterilization, wear good protection on your hands and remove flasks containing liquids slowly and carefully, the liquid can boil over.

Experiment 9: Sterilization with dry heat

Materials required

Petri dishes/glassware, pincettes, aluminium foil, oven.

Procedure

i. Separately wrap up the Petri dishes and pincettes with aluminium foil.

ii. Sterilize them in an oven at 160 °C for at least 2 h.

Caution: Avoid the use of paper or other flammable materials, which can be fire hazard.

Experiment 10: Disinfection of seeds

Materials required

Erlenmeyer flask, 95 % ethanol, detergent, 0.1 % sodium hypochloride, filter paper sheet.

Procedure

i. Place seeds into an Erlenmeyer flask. Add an aqueous solution containing 95 % ethanol and 0.1 % detergent (v/v). Shake vigorously for 1 min.

ii. Allow seeds to settle and replace the ethanol solution with 0.1 % sodium hypochloride.

iii. After 30 min, rinse seeds several times with sterile distilled water to remove the sodium hypochloride residues and then dry in the sheet of filter paper.

4.2. Sterilization of extracts

Some allelochemicals present in the extracts may have thermolabile components, so instead of heat sterilization, the extracts should be filtered using millipore filters that are appropriate to the solvents.

Experiment 11: Sterilization of extracts

Materials required

0.22-mm pore size filter, set-up for vacuum filtration.

Procedure

Pass the extract through the 0.22 mm pore size filter.

5. BIOASSAYS

Temperature and light: Before beginning a bioassay, it is necessary to determine thermal and light conditions, under which it will be done. Temperature affects the rate and speed of germination. Seeds of different species germinate at a specific range of temperatures and the optimum temperature should be maintained during bioassay. Sometimes alternate temperatures are used during day and night periods. Most seeds germinate well in the light or dark, but seeds of many plant species require light for germination. This is particularly true for small-seeded plants such as annuals. When such species are test plants, the incubators should be equipped with low intensity cool-white fluorescent tubes that are switched on during the day. However, even the germination of light requiring seeds can be inhibited, if they are exposed to continuous high-intensity light.

Bioassay duration and quantity of extracts: It is very important to determine the duration of bioassay, the measurements should be done at the moment, when inhibitory or stimulatory effects are maximum. It is also important to determine the amount of extract or water to be supplied to seeds, to avoid growth inhibition either by (i) anaerobic conditions caused by an excessive amount of the applied solution or (ii) the medium dryness due to deficiency of solution. To avoid water losses by evaporation, high air humidity should be maintained in the incubator.

In bioassays, extracts should be assayed at various concentrations (*i.e.*, 0.1, 1, 10, 100 %). Experiments are conducted in completely randomized design. Every treatment should have at least three replications. If there are great differences among replications, more replications per treatment should be incorporated.

5.1. Bioassay for seed germination and early growth

Experiment 12: Bioassay with aqueous extract or leachate

Materials required

Sterile Petri dishes (9–10 cm diameter) lined with filter paper, Petri dishes with agar medium mixed with the extract or distilled water (control), automatic pipette, pincette, incubator, seeds of test plant.

Procedure

i. Add to each Petri dish lined with filter paper a known volume of extract or water (control). Use sterile Petri dishes with agar medium and use extract or water (control).

ii. Using a pincette, place a known number of sterilized seeds of the test plant on the top of a filter paper.

iii. Put Petri dishes with seeds into an incubator with a temperature suitable for seed germination.

iv. After a defined time, record the number of germinated seeds and estimate seedling lengths (root and shoot lengths) and dry weights.

Observations

Seeds are considered as germinated, when the length of the emerged radicle is equal to the diameter of the seed.

Experiment 13: Bioassay with organic extract

Materials required

Vacuum rotary evaporator, sterile Petri dishes (9–10 cm diameter) lined with filter paper, automatic pipette, pincette, seeds of test plant, incubator.

Procedure

i. Evaporate the organic solvent extract to dryness at 40 °C using vacuum rotary evaporator.

ii. Dissolve the residue in 96 % ethanol to obtain solution with required concentration.

iii. Pipette 2 mL of ethanol solution or 96 % ethanol (control) into a Petri dish lined with filter paper.

iv. Wait until complete evaporation of ethanol and then add a defined water volume.

vi. Proceed as indicated in Experiment 12, steps 2–4.

Experiment 14: Bioassay with volatiles

Materials required

Beakers, Petri dishes, filter paper, parafilm, pincette, incubator, seeds of test plant, scales.

Procedure

i. Line the bottom of wide beakers with filter paper and moisten it with distilled water.

ii. Put a small beaker in the middle of each beaker.

iii. Place the weighed sample of donor plant on the top of a filter paper.

iv. Prepare a wide beaker doing operations indicated in steps 1–3, but placing the portion of filter paper without addition of the sample.

v. Place Petri dishes with a determined amount of sterilized seeds on the top of small beakers. Cover the beakers with parafilm and put into an incubator.

vi. After a defined time, record the number of germinated seeds and estimate seedling lengths (root and shoot lengths) and dry weights.

Experiment 15: Sand bioassays

Materials required

Sterile Petri dishes, sterile sand, automatic pipette, pincette, incubator, seeds of test plant.

Procedure

i. Fill Petri dishes with a defined sand weight. Add a defined amount of extract or water (control) to each Petri dish.

ii. Using pincette, place the determined number of sterilised seeds of test plant on the top of the sand.

iii. Put Petri Plates with seeds into incubator with a temperature suitable for germination of test plant seeds.

iv. After the determined time, record the number of germinated seeds.

v. Elimínate the sand particles stuck on the roots by a washing to distilled water and a drainage, and estimate lengths and dry weights of seedlings (separately aerial parts and roots).

Suggested Readings

Blum, U. (1999). Designing laboratory plant debris–soil bioassays: Some reflections. *In*: *Principles and Practices in Plant Ecology: Allelochemicals Interactions.* (*Eds*., Inderjit, K.M.M. Dakshini and C. L. Foy), pp. 17–23. CRC Press, Boca Raton, FL.

Chon, S.U., Choi, S.K., Jung, S., Jang, H.G., Pyo, B.S. and Kim, S.M. (2002). Effects of alfalfa leaf extracts and phenolic allelochemicals on early seedling growth and root morphology of alfalfa and barnyard grass. *Crop Protection* **21:** 1077–1082.

Chon, S.U., Kim, Y.M. and Lee, J.C. (2003). Herbicidal potential quantification of causative allelochemicals from several compositae weeds. *Weed Research* **43:** 1–7.

Fujii, Y., Shibuya, T., Nakatani, K. Itani, T., Hiradane, S. and Parvez, M.M. (2004). Assessment method for allelopathic effect from leaf litter leachates. *Weed Biology and Management* **4**: 19–23.

Inderjit and Nilsen, Erik T. (2003). Bioassays and field studies for allelopathy in terrestrial plants: Progress and problems. *Plant Sciences* **22**: 221–238.

Jefferson, L.V. and Pennacchio, M. (2003). Allelopathic effects of foliage extracts from four Chenopodiaceae species on seed germination. *Journal of Arid Environments* **55**: 275–285.

John, J. and Nair, A.M. (1999). Allelopathic effect of leaf leachates of multipurpose trees. *Allelopathy Journal* **6:** 81–86.

Kobayashi, K. (2004). Factors affecting phytotoxic activity of allelochemicals in soil. *Weed Biology and Management* **4**: 1–7.

Macias, F.A., Castellano, D. and Molinillo, M.G. (2000). Search for a standard phytotoxic bioassay for allelochemicals. Selection of standard target species. *Journal of Agricultural and Food Chemistry* **48**: 2512–2521.

Matsuyama, M., Hiradane, S., Nakatani, K. and Fujii, Y. (2000). Developments of new bioassay and analysis method for volatile allelochemicals. *Weed Research, Japan* **45(S)**: 80–81.

Moradshahi, A., Ghadiri, H. and Ebrahimikia, F. (2003). Allelopathic effects of crude volatile oil and aqueous extracts of *Eucalyptus camaldulensis* Dehnh. leaves on crops and weeds. *Allelopathy Journal* **12:** 189–196.

Navaz, M., George, S. and Geethakumari, V.L. (2003). Influence of eupatorium [*Chromolaena odorata* (L.) King and Robinson] leachates on germination and seedling growth of rice and cowpea. *Allelopathy Journal* **11:** 235–240.

Nilsen, E.T., Walker, J.F., Miller, O.K., Semones, S.W., Lei, T.T. and Clinton, B.D. (1999). Inhibition of seedling survival under *Rhododendron maximum*

(*Ericaceae*): Could allelopathy be a cause? *American Journal of Botany* **86:** 1597–1605.

Ohno, T. and Doolan, K.L. (2001). Effects of red clover decomposition on phytotoxicity to wild mustard seedling growth. *Applied Soil Ecology* **16:** 187–192.

Ohno, T., Doolan, K., Zibilske, L.M., Liebman, M., Gallandt, E.R. and Berube, Ch. (2000). Phytotoxic effects of red clover amended soils on wild mustard seedling growth. *Agriculture, Ecosystems and Environment* **78:** 187–192.

Peterson, J.K., Harrison, H.F. Jr. and Snook, M.E. (1999). Comparison of three parameters for estimation of allelopathic potential in sweet potato [*Ipomea batatas* (L.) Lam.] germplasm. *Allelopathy Journal* **6:** 201–208.

Prasad, D., Pant, G. and Rawat, M.S.M. (1999). Phytotoxity of *Rhamnus virgatus* on some field crops. *Allelopathy Journal* **6:** 227–242.

Qasem, J.R. (2001). Allelopathic potential of white top and Syrian sage on vegetable crops. *Agronomy Journal* **93:** 64–71.

Qasem, J.R. (2002). Allelopathic effects of selected medicinal plants on *Amaranthus retroflexus* and *Chenopodium murale*. *Allelopathy Journal* **10:** 105–122.

Sharma, N.K., Samra, J.S. and Singh, H.P. (2000). Effect of aqueous extracts of *Populus deltoides* M. on germination and seedling growth. 1. Wheat. *Allelopathy Journal* **7:** 56–68.

Singh, D. and Narsing Rao, Y.B. (2003). Allelopathic evaluation of *Andrographis paniculata* aqueous leachates on rice (*Oryza sativa* L.). *Allelopathy Journal* **11:** 71–76.

Singh, N.B. and Singh, R. (2003). Effect of leaf leachate of *Eucalyptus* on germination, growth and metabolismm of greengram, blackgram and peanut. *Allelopathy Journal* **11:** 43–52.

Singh, N.B. and Thapar, R. (2003). Allelopathic influence of *Cannabis sativa* on growth and metabolism of *Parthenium hysterophorus*. *Allelopathy Journal* **12:** 61–70.

CHAPTER 3

Pollen Allelopathy

STEPHEN D. MURPHY[1]

1. INTRODUCTION

Apparently first described by I.N. Golubinskii in 1946 (Roshchina, 1999, 2001), allelopathic pollen is an interesting mechanism of competition that occurs in few species. It has been challenging to elucidate the mechanisms of pollen allelopathy (Ortega *et al.,* 1988; Roshchina and Melnikova, 1996, 1998, 1999; Roshchina, 1999, 2001, 2004, 2005), *e.g.* allelopathic pollen disrupts the membrane transport system, electron transport chain in mitochondria, ovule development and key enzymes in pollen germination (Thomson *et al.*, 1982; Ortega *et al.,* 1988; Anaya *et al.,* 1992; Roshchina, 2001; Murphy, 2005). Miotosis in sporophytes (as opposed to pollen) is disrupted but yet the specific reason for this is not known (Sukhada and Jayachandra, 1980a,b; Ortega *et al.*, 1988; Murphy, 1999). The relative ecological importance of impacts on sporophytes is questionable except in cropping systems (*e.g. Zea mays* L. var. *chalquiñocónico*), where high densities of pollen fall on crops and weeds below the canopy (Jimenez *et al.*, 1983; Anaya *et al.*, 1987). The pollen allelopathy is important ecological mechanism (*i.e.* competition), when heterospecific pollen transfers between the species that flower at the same time and have compatible pollination systems (Viswanathan and Lakshmanan, 1984; Murphy, 1992, 1999a,b, 2001a,b, 2005; Murphy and Aarssen, 1989, 1995a,b,c,d, 1996). While there are some small clusters of pollen allelopathic species (*e.g.,* within *Hieracium*), most species identified as pollen allelopathic are not closely related (Murphy, 1999a). Pollen allelopathy involves a diverse range of phenolics and terpenoids (Sukhada and Jayachandra, 1980a,b; Ortega *et al.*, 1988; Murphy, 1999). This diversity probably represents independent and convergent evolution in response to the general selection pressures of competition and the availability of secondary metabolites that evolved for other purposes that were co-opted as part of an allelopathic response (Murphy, 1999).

1. Department of Environment and Resource Studies, University of Waterloo, 200 University Avenue West,Waterloo, Ontario, N2L 3G1, Canada. E-mail: sd2murph@uwaterloo.ca

There are many unanswered questions that arise from the above knowledge:

i. What are the primary mechanisms of allelopathic pollen?

ii. What are the main chemicals and synthesis pathways involved?

iii. How conservative is pollen allelopathy?

iv. How did pollen allelopathy evolve (did it evolve independently as seems likely)?

v. How prevalent is pollen allelopathy?

vi. Is pollen allelopathy merely an interesting taxonomic and minor ecological phenomenon or does it have longer-term impacts on assembly of communities in "natural" and crop ecosystems?

To address these questions, following standard protocols are available in following sections.

Experiment 1: Screening of multiple species using solid *in vitro* media

Principle

A first step in bioassays for allelopathic pollen is to determine which species exhibit any toxicity *in vitro*. This method is rapid (24 h) for mass screening. *In vitro* toxicity can result from fungal or bacterial contamination, hence, sterile technique is essential. Additionally, apparent allelopathy can result if pollen is damaged by inappropriate humidity or if pollen does not have the proper concentrations of chemicals to germinate.

Materials required

Solid agar, autoclave, Erlenmeyer flasks, scientific grade filter paper, plastic boxes, Petri dishes, compound microscope with stage vernier.

Reagents

A total of 50 mg/L of magnesium chloride ($MgCl_2$) as a source of magnesium, 100 mg/L of boric acid (H_3BO_4) as a source of boron, 130 mg/L of potassium phosphate (KH_2PO_4) as a source of potassium, 270 mg/L of calcium chloride ($CaCl_2$) as a source of calcium, with pH adjusted to 5.7 with 0.1 N sodium hydroxide (NaOH), use 5 % raffinose for metabolism of pollen and as an osmoticum and 15 % polyethylene glycol (PEG) for osmoticum.

Procedure

i. *In vitro* screening requires immersing freshly harvested, rehydrated, measured mass of pollen from suspected allelopathic species in measured volumes of double-distilled water to produce a concentrated extract of 100 grains/μL.

ii. The extract is diluted, as needed, to produce a range of concentrations.

iii. All extracts are stored in sterilized Erlenmeyer flasks, double-covered with a stopper and sterile seal and refrigerated at 4 °C.

iv. When needed, the extract is homogenized in liquid agar media containing the above germination reagents. Preparation of the media begins 18–24 h before collecting the pollen (with practice, it takes about 6 h to prepare about 120 Petri dishes). Prepare agar with a growth base of 5 % raffinose for metabolism of pollen and as an osmoticum and 15 % polyethylene glycol (PEG) as an osmoticum. Autoclave for 3 h and add the above reagents. Then, add the appropriate concentrations of extract and leave some plates extract-free as controls. Replicate each extract treatment or control six times for appropriate statistical analysis. While liquid, thinly plate the agar (with or without extract) to about 0.1–0.5 mm thickness on sterile Petri dishes. Cover and seal each dish, thereafter let the agar solidify (2 h at room temperature).

v. Collect the fresh pollen of species likely to be susceptible to allelopathic pollen. It is critical that the pollen to be germinated is harvested from dehiscent anthers, *i.e.*, the pollen is being dispersed and has the maximum potential to germinate. In the field, collect pollen randomly from at least 50 shoots scattered across a field site. The pollen should be collected in small (50 × 50 cm) boxes made of plastic. Divide samples of each species into at least 10 boxes (per species) to reduce chances of mass fungal contamination. To provide sufficient humidity to prevent pollen dehydration, each box should have two pieces of filter paper soaked in double-distilled water and attached to the underside of the lid. This is necessary for *Poacae* and all families where there is tri-nucleate pollen.

vi. Take all samples to the lab and for each species, spread the pollen on the petri dishes with the agar media. Take care to ensure each species you wish to germinate is on a separate plate.

Observations

i. Examine each Petri dish under a stageless compound microscope (magnification 200×, using phase contrast).

ii. Choose five fields of view ("observations" or "germination counts") within the Petri dish are chosen arbitrarily and the pollen germination (%) is calculated. These fields of view are included to ensure that the results are not skewed by areas where pollen sinks into a too-thick agar medium. Pollen is considered to be germinated, if the grain successfully rehydrates (does not burst) and an intact pollen tube extends a minimum of 100 μm beyond the exine. If improper environmental controls cause most of the pollen grains and/or their tubes ruptured or if fungal hyphae were observed, the sample must be omitted from analyses.

iii. If there is pollen allelopathy, the resultant observations will resemble those in Fig 1.

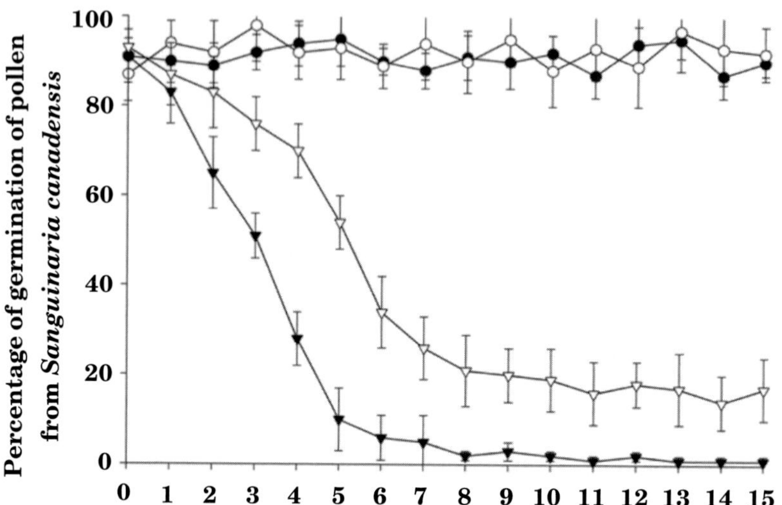

Fig 1. Effect of extract from the putative pollen allelopathic species *Chelidonium majus* on *in vitro* pollen germination in the confamilial *Sanguinaria canadensis*. There is a primary chemical that causes the allelopathic effect from the non-polar fraction but it is not yet identified

Statistical analysis

Analyze data with a simple analysis of variance with fixed and random effects on arcsine square-root transformed data (because the data are expressed as percentages and this will create heterogeneity in variances).

Precautions

Avoid using sulfates (*i.e.*, as in $MgSO_4$) and nitrates (*i.e.*, as in $CaNO_3$) in germination media, as they form more acids that make the pH hard to control. For some pollen, this makes little difference but other species are more sensitive to changes in pH. Once the medium is mixed (with or without the extracts), store it at 4 °C until needed. However, this still allows for a contamination risk and it also means the medium will have to warm to room temperature before it can be used. As noted, keeping pollen hydrated is critical.

Experiment 2: Quantifying pollen transfer rates

Principle

If the pollen allelopathic effect is to have any ecological importance, there must be sufficient pollens transferred from allelopathic to vulnerable species. The "sufficient amount" was indicated by *in vitro* tests in Experiment 1.

Materials required

Small bore tubes treated with silicone (so that pollen does not become electrostatically attracted to the tube), blocks of wood to hold tubes, compound microscope, fine dissecting kit to split stigmas.

Reagents

70 and 100 % ethanol, 0.1 % acidified aniline blue (dissolve 0.1g of aniline blue in an aqueous solution of 0.1 M K_3PO_4), 40 % aceto-carmine stain (boil 0.5g of carmine in 100 mL of 45 % acetic acid).

Procedure

i. Select 100–200 shoots from each plant species that are likely to be vulnerable to allelopathic pollen, *i.e.,* they flower concurrently with allelopathic species and sufficient pollen from the allelopathic species alights on the stigmas. Shoots are selected using stratified random sample designs that reduce spatial autocorrelation effects. Some of the plants are selected for observations of pollinator movements, where relevant. Other plants are harvested to allow the researcher to examine how many grains of pollen from allelopathic heterospecifics, non-allelopathic heterospecifics and conspecifics land on the stigma and how many conspecific grains produced viable tubes.

ii. Place harvested stigmas in small bore tubes and keep them upright in a block of wood with small holes drilled into it. Pure (100 %) ethanol removes pollen so it can be observed in depression well slides, with a lactose medium to reduce fungal contamination, under a compound microscope (magnification 200×).

iii. To observe pollen tubes, stigmas are not washed with 100 % ethanol. Instead, harvest stigmas, split them longitudinally, preserve them in 70 % ethanol and stain them in a 1:1 solution of 0.1 % acidified aniline blue (in distilled water) and 40 % aceto-carmine.

Observations

If the pollen allelopathic interactions are important, then the researcher should observe small numbers of heterospecific pollen on the stigmas causing a hyperbolic decline in germination and tube production of pollen conspecific to the stigmas.

Statistical analysis

A probity statistical analysis is performed to test whether there is a hyperbolic decline related to the treatments.

Precautions

Ensure harvest during pollination, avoid excessive jostling of stigmas and use fine dissection techniques.

Experiment 3: Screening multiple species using *in situ* pollen transfer

Principle

In this experiment, the researcher manipulates the pollen environment of entire fields. In all cases, the experiments are designed to be analyzed univariately with an ANOVA and Tukey's test for Honestly Significant Differences or they can be analyzed with a Mann-Whitney U test, if pairwise comparisons are being made.

Materials required

Battery powered clippers, mesh bags, compound microscope.

Reagents

70 and 100 % ethanol, 0.1 % acidified aniline blue (in distilled water), 40 % aceto-carmine.

Procedure

i. Choose fields containing the putative pollen – allelopathic species and potential targets. All fields must have similar edaphic conditions and species compositions; this is not a problem in agroecosystems.

ii. Experiments require several replicate fields that are isolated by forest tracts or some barrier that prevents almost all possibility of pollen and pollen vectors from traveling from one field to the next one. For example, in my own experiments, animal visitors, mainly insects in northern North America, are unlikely to fly over or through the forest. In wind-pollinated species, forest tracts create an overnight surface-based radiation inversion that essentially hermetically seals the field and no pollen gets in or out. This inversion exists for about 8 h, so this approach works only if the wind-pollinated species are flowering between approximately 11:00 p.m. and 07:00 a.m. If these conditions do not apply, then use and compare fields that are separated by several kilometers distance and there are no contamination sources in between.

iii. Use a stratified random sample to select at least 100 shoots of each species likely to be vulnerable to allelopathic pollen.

iv. Within a 2-m radius of each of these shoots, place mesh bags (*i.e.*, made of bridal veil) around each maturing inflorescence of the pollen allelopathic species. This reduces confounding effects from the next step – clipping – that might create changes in resource competition. The 2-m radius can vary, depending on the root and shoot architecture and physiology of the species in a community.

v. Cut the remaining maturing (pre-flowering) inflorescences of pollen allelopathic species in some fields and compare seed set in targets with those in fields where pollen allelopathic species flowered. This requires extensive labor if the field is relatively large, hence, you may wish to do this in a glasshouse experiment where the "fields" are isolated rooms.

vi. Measure heterospecific pollen transfer (see Experiment 2) and, later, seed set by harvesting a sub-set of stigmas and examining them under a microscope to count pollen or perform a germination test on the seeds (weigh them first – this gives a crude indication of whether the seed was matured or not). Germination tests are most accurate if seeds are placed in 8 cm pots and actually germinated in a growth room; with some species, germination will have to be induced by artificially breaking dormancy.

Observations

If the pollen allelopathic interactions are important, then the researcher should observe small numbers of heterospecific pollen on the stigmas causing a hyperbolic decline in germination and tube production of pollen conspecific to the stigmas.

Statistical analysis

As long as a stratified random design is used (with sufficient replications), the data are analyzed with a simple analysis of variance with fixed and random effects on arcsine square-root transformed data (because the data are expressed as percentages and this will create heterogeneity in variances).

Precautions

Ensure harvest during pollination, avoid excessive jostling of stigmas and use fine dissection techniques.

Experiment 4: Isolation of allelochemicals

Principle

Roshchina's research group (1996, 1998, 2000, 2001, 2004, 2005) indicates that many chemicals cause pollen allelopathy, *i.e.,* carotenoids (α- and β-carotene, crocetine and flavoxanthine), catecholamines (dopamine and noradrenaline), nitrogenous compounds (acetylcholine, histamine and serotonin), phenolics (benzoic acid, gallic acid, kaempferol, quercetin, rutin and vanillic acid), polyacetylines (capilline) and terpenoids (artemisinine, austricine, azulene, citral, cymol, damsin, desacacetylinulicine, gaillardine, grosshemine, inulicine, ledol, linalool and parthenin). This list includes stimulatory compounds (nitrogenous compounds, dopamine, and some terpenoids, carotenoids and polyacetylenes) as well as inhibitory compounds. The inhibitory compounds are ones more associated with "allelopathy" and these often are polar and acidic in nature. They may be amino acids or flavonoids that have phytotoxic activities and any derivatives may also be phytotoxic.

Materials required

Ground-glass pestle and mortar, silica gel columns, centrifuge, autoclave, high – performance liquid chromatograph, freeze-drier, pressure chamber, ion-exchange columns, desiccator.

Reagents

Polar solvent (95 % methanol, ethanol or double-distilled water), benzene:ethyl acetate eluent (90:10 and 90:30; v/v), sodium bicarbonate/ sodium carbonate, ethyl acetate (pure); $CHCl_3$:methanol (90:10; v/v), *n*-hexane:ethanol (70/30; v/v), petroleum ether:ethyl acetate (70:30; v/v), 2 N HCl in methanol.

Procedure

i. Macerate pollen using a ground glass mortar and pestle, then extract with methanol (95 %). Use of a polar solvent will promote the extraction of most pollen compounds as they have low polarity.

ii. To ensure that the sample will remain pure and relatively uniform in its properties, the sample may be freeze dried to –20 °C which will form a pellet. When required, part of the pellet may be resuspended in distilled water for use in bioassays for pollen – allelopathic interactions. The above method may be modified to allow for fractionation of distilled water extracts prior to storage.

iii. Concentrate the distilled water extracts under reduced pressure (50 °C) to 10 % of the original volume.

iv. All extracts are then acidified (pH = 2) with concentrated HCl and aliquots (25 mL) equi partitioned against ethyl acetate.

v. These extracts are concentrated under reduced pressure at 50 °C and the residue dissolved in a maximum of 15 mL of ethyl acetate.

vi. Use a solution of sodium bicarbonate/sodium carbonate (2 N) to produce acidic (carbonate soluble) and non acidic (carbonate insoluble, ethyl acetate soluble) fractions.

vii. Acidic fractions are acidified (pH = 2) with concentrated HCl and re extracted with ethyl acetate.

viii. Wash all fractions (distilled water), concentrate (vacuum pressure) and store (in a desiccator) for future characterization.

ix. Alumina columns, eluted with petroleum ether:ethyl acetate (70:30), are useful in separating non acidic allelochemicals, such as certain amino acids and alkaloids. Ion-exchange columns eluted with dilute HCl may also be used to separate basic compounds.

x. Neutral compounds are isolated by tandem stacking of AG 50 and AG 1 grade ion-exchange columns eluted with dilute HCl. Silica gel columns (70–120 µm of mesh) may be eluted with benzene:ethyl acetate (90:10 and 90:30) to separate acidic allelochemicals (Sukhada and Jayachandra, 1980b).

xi. Eluting anion exchange columns with 1 N HCl will also separate acidic allelochemicals, although there may be a loss of purity in the fractions.

xii. Silica gel columns eluted with $CHCl_3$:methanol and *n*-hexane: ethanol can separate a mixed fraction of weakly acidic phenolics.

xiii. To extract flavonoids for HPLC analysis, use 2 mL of methanolic HCl (2 N), followed by centrifugation (2 min, 15,600g), hydrolyzation of 150 µL of suspension in an autoclave (15 min at 120 °C). This isolates many compounds.

Observations

If the pollen allelopathic interactions are important, then the researcher should observe small numbers of heterospecific pollen on the stigmas causing a hyperbolic decline in germination and tube production of pollen conspecific to the stigmas.

Statistical analysis

As long as a stratified random design is used (with sufficient replications), the data can be analyzed with a simple analysis of variance with fixed and random effects on arcsine square-root transformed data (because the data are expressed as percentages and this will create heterogeneity in variances).

Precautions

Ensure harvest during pollination, avoid excessive jostling of stigmas and use fine dissection techniques.

Experiment 5: Determining mechanisms of allelochemical action

Principle

Pollen allelochemicals might interfere with the ATPase that promotes proper adhesion of pollen to stigmatic papillae, bind on specific plasmalemma receptors and stimulate pollen tube formation (Murphy and Aarssen, 1995), uncouple mitochondria oxidative phosphorylation (between coenzyme Q and cytochrome *c*, alter membrane permeability (theoretically, halting the vital rehydration and conservation of membrane integrity needed to germinate pollen of *Poaceae* and other *Monocotyledonae)* and stop mitosis (Anaya *et al.,* 1987; Murphy, 1992). Generally, the hypothesis with the most evidence has been in the effect on mitochondria. I have found useful to adopt the procedure described by Ortega *et al.* (1988), as it is still useful 20 years after it was first developed.

Materials required

Erlenmeyer flasks, high-speed blender, filter gauze, sonicator, ultracentrifuges, Teflon-tipped tissue grinder, Clark electrode for measures of oxygen consumption.

Reagents

Sucrose, ethylenediamine tetraacetic acid (EDTA), Tris base, 0.1 % bovine serum albumin (BSA), KC1, malate, glutamate.

Procedure

i. Work in a cold room (4 °C), immerse the target tissues (roots, shoots, pistils or pollen tubes) in 0.25 M sucrose – 1 mM ethylenediamine tetraacetic acid (EDTA) adjusted to pH 7.3 with a Tris base.

ii. Using a blender, macerate tissue (10–20 s).

iii. Filter tissue through a 100 μm gauze, collect filtrate in a sterilized Erlenmeyer flask and adjust filtrate to pH 7.3 with a Tris base.

iv. Centrifuge the filtrate and supernatant for 15 min at 4000 and 12,000 rpm, respectively.

v. Resuspend the pellet in the same medium with 0.1 % (w/v) bovine serum albumin (BSA) and combine to yield a final volume of 50 mL.

vi. Centrifuge two 25-mL aliquots at 12,000 rpm for 10 min. Resuspend the pellet in 0.75 mL of the medium using a Teflon-tipped tissue grinder.

vii. Measure the oxygen consumption polarographically using the Clark electrode at 25 °C. The reaction medium should contain 100 mM KC1, 10 mM phosphate – Tris buffer, pH 7.4, 0.1 % (w/v) BSA, and either 10 mM malate or 10 mM glutamate. These substrates are to be adjusted to pH 7.4 with Tris base.

viii. At start of the experiment, add 2 mg of mitochondrial protein to a final volume of 3 mL.

ix. For submitochondrial reactions, adjust the pH of mitochondria suspended within 0.25 M sucrose – 1 mM EDTA to pH 8.6 with a Tris base and sonicate for 1.5 min.

x. Remove non-sonicated mitochondria with centrifugation at 12,000 rpm for 10 min.

xi. The supernatant containing the submitochondrial particles should be centrifuged at 45,000 rpm in an air-driven ultracentrifuge.

xii. Resuspend the pellet in 1 mL of 0.25 M sucrose – 1 mM EDTA, pH 7.4.

xiii. Measure the rate of NADH oxidation with the Clark electrode.

xiv. For mitochondrial and submitochondrial trials, the experiment simply involves exposing the isolates of mitochondria to the extracts prepared using the methods described earlier in this chapter.

Observations

Ortega *et al.* (1988) report data as nanoatoms of oxygen consumed/min/mg of mitochondrial protein; the mitochondrial protein content should be estimated using their standard approaches.

Statistical analysis

Repeated measures of ANOVA are appropriate to determine if extracts significantly affect mitochondrial functions. This is used to compare the rates of oxygen consumption observed under different experimental conditions, *i.e.,* exposure to different extract types versus controls.

Precautions

The technique requires practice because of the delicate nature of preparation and the approach generally works best with polar allelopathic compounds because of the suspension methods used.

Suggested Readings

Anaya, A.L., Hernandez-Bautista, B.E., Jimenez-Estrada, M. and Velasco-Ibarra, L. (1992). Phenylacetic acid as a phytotoxic compound of corn pollen. *Journal of Chemical Ecology* **18**: 897–905.

Anaya, A.L., Ramos, L., Hernandez, J. and Ortega, R.C. (1987). Allelopathy in Mexico. *In*: Allelochemicals: Role in Agriculture and Forestry (*Ed*., G.R. Waller), ACS Symposium Series **330**: 89–101. American Chemical Society, Washington, DC.

Char, M.B.S. (1977). Pollen allelopathy. *Naturwissenschaften* **64**: 489–490.

Jimenez, J.J., Schultz, K., Anaya, A.L., Hernandez, J. and Espejo, O. (1983). Allelopathic potential of corn pollen. *Journal of Chemical Ecology* **9**: 1011–1025.

Murphy, S.D. (1992). The determination of the allelopathic potential of pollen and nectar. *In*: Modern Methods of Plant Analysis. Vol. 13. *Plant Toxin Analysis* (*Eds.,* H.F. Linskens and J.F. Jackson), pp. 333–357. Springer-Verlag, New York.

Murphy, S.D. (1999). Is there a role for pollen allelopathy in biological control of weeds? *In*: International Allelopathy Update – Vol. 2. *Basic and Applied Aspects* (*Ed.,* S.S. Narwal). pp. 204–215. Science Publishers, Enfield, NH, USA.

Murphy, S.D. (1999). Pollen allelopathy. *In*: Principles and Practices in Plant Ecology (*Eds.,* Inderjit, K.M. Dakshini and C.L. Foy), pp. 129–148. CRC Press, Boca Raton.

Murphy, S.D. (2001). Field testing for pollen allelopathy: A review. *Journal of Chemical Ecology* **26**: 2155–2172.

Murphy, S.D. (2001). The role of pollen allelopathy in weed ecology. *Weed Technology* **15**: 867–872.

Murphy, S.D. and Aarssen, L.W. (1989). Pollen allelopathy among sympatric grassland species: *In vitro* evidence in *Phleum pratense* L. *New Phytologist* **112**: 295–305.

Murphy, S.D. and Aarssen, L.W. (1995). *In vitro* allelopathic effects of pollen from three *Hieracium* species (Asteraceae) and pollen transfer to sympatric Fabaceae. *American Journal of Botany* **82**: 37–45.

Murphy, S.D. and Aarssen, L.W. (1995). Allelopathic pollen extract from *Phleum pratense* L. (Poaceae) reduces germination, *in vitro,* of pollen in sympatric species. *International Journal of Plant Science* **156**: 425–434.

Murphy, S.D. and Aarssen, L.W. (1995). Allelopathic pollen extract from *Phleum pratense* L. (Poaceae) reduces seed set in sympatric species. *International Journal of Plant Science* **156**: 435–444.

Murphy, S.D. and Aarssen, L.W. (1995). Allelopathic pollen of *Phleum pratense* reduces seed set in *Elytrigia repens* in the field. *Canadian Journal of Botany* **73**: 1417–1422.

Murphy, S.D. and Aarssen, L.W. (1996). Partial cleistogamy limits reduction in seed set in *Danthonia compressa* (Poaceae) by allelopathic pollen of *Phleum pratense* (Poaceae). *Écoscience* **3**: 205–210.

Ortega, R.C., Anaya, A.L. and Ramos, L. (1988). Effects of allelopathic compounds of corn pollen on respiration and cell division of watermelon. *Journal of Chemical Ecology* **14**: 71–86.

Roshchina, V.V. (1999). Mechanisms of cell-cell communication. *In*: *International Allelopathy Update - Volume 2. Basic and Applied Aspects* (*Ed.,* S.S. Narwal). pp. 3–26. Science Publishers, Enfield, USA.

Roshchina, V.V. (2001). Molecular-cellular mechanisms in pollen allelopathy. *Allelopathy Journal* **8**: 11–28.

Roshchina, V.V. (2004). Cellular models to study the allelopathic mechanisms. *Allelopathy Journal* **13**: 3–15.

Roshchina, V.V. (2005). Allelochemicals as fluorescent markers, dyes and probes. *Allelopathy Journal* **16**: 31–46.

Roshchina, V.V. and Melnikova, E.V. (1996). Microspectrofluorometry: A new technique to study pollen allelopathy. *Allelopathy Journal* **3**: 51–58.

Roshchina, V.V. and Melnikova, E.V. (1998). Allelopathy and plant generative cells: Participation of acetylcholine and histamine in signalling interactions of pollen and pistil. *Allelopathy Journal* **5**: 171–182.

Roshchina, V.V. and Melnikova, E.V. (1999). Microspectrofluorometry of intact secreting cells applied to allelopathy. *In*: Principles and Practices in Plant Ecology (*Eds.,* Inderjit, K.M.M. Dakshini, and C.L. Foy). pp. 99–126. CRC Press, Boca Raton.

Sukhada, K.D. and Jayachandra. (1980a). Pollen allelopathy – A new phenomenon. *New Phytologist* **54**: 739–746.

Sukhada, K.D. and Jayachandra. (1980b). Allelopathic effects of *Parthenium hysterophorus* L. Part IV. Identification of inhibitors. *Plant and Soil* **55**: 67–75.

Thomson, J.D., Andrews, B.J. and Plowright, R.C. (1982). The effect of a foreign pollen on ovule development in *Diervilla lonicera* (Caprifoliaceae). *New Phytologist* **90**: 777–783.

Viswanathan, K. and Lakshmanan, K.K. (1984). Phytoallelopathic effects on *in vitro* pollinnial germination of *Calotropis gigantea* R. Br. *Indian Journal of Experimental Botany* **22**: 544–547.

CHAPTER 4

Assays for Allelopathic Interactions Among Terrestrial Plants

D.A. Sampietro[1], José R. Soberón[1], Melina A. Sgariglia[1], Emma N. Quiroga[1] and Marta A. Vattuone[1]

1. INTRODUCTION

Plants affect growth of each other through several mechanisms and their combined effect is known as "interference" (Müller, 1969). Resource competition is major interference mechanism active in nature (Wu *et al.*, 2001). In competitive interactions, a plant takes up one or more resources that are also needed for growth of neighboring plants. Another interference mechanism is allelopathy which comprises the effect of one plant (a donor plant) on other plants (receptor plants) through the release of organic compounds to the environment. These compounds may have a direct action after uptake by receptor plants, acting alone or in mixture with other compounds but they also may have an indirect effect changing the environmental conditions, with negative or positive impact on growth of receptor plants (Sampietro and Vattuone, 2006a).

Allelopathy has been implicated in several interactions among terrestrial plants and has been observed in both natural and managed terrestrial ecosystems, although conclusive proofs are often lacking. Obstacles to prove allelopathy arise from difficulties to simulate natural environmental conditions in laboratory and greenhouse experiments and problems to distinguish other interference mechanisms occurring in natural settings (Inderjit and Weiner, 2001). Although no protocol is available to prove allelopathy, the following criteria have broad acceptance (Patterson, 1986; Romeo and Weidenhamer, 1998; Romeo, 2000):

i. Evidences suggesting an allelopathic phenomenon must be recognized in the field and alternative explanations to allelopathy must be empirically discarded. Field studies indicate which factors should be considered for further design of assays. Laboratory and greenhouse

1. Instituto de Estudios Vegetales "Dr. Antonio R. Sampietro", Facultad de Bioquímica, Química y Farmacia, Universidad Nacional de Tucumán, España 2903, 4000, San Miguel de Tucumán, Tucumán, Argentina. E-mail: sampietro@tucbbs.com.ar

assays allow analyzing how different components (*i.e.,* microbial activity, soil properties and climatic parameters) may affect a whole allelopathic process.

ii. Mode of release of allelochemicals from donor plants must be detected (*i.e.,* leachates from aerial parts, root exudates, volatilization from foliage or litter decomposition). Then, allelochemicals involved should be isolated and identified. Assays should consider soil or air, where allelochemcals are released.

iii. Crop plants such as radish (*Raphanus sativus* L.) and lettuce (*Lactuca sativa* L.) have been broadly used as receptor plants in laboratory and greenhouse assays. However, plant species assayed should be those suspected to be receptor plants in the field. Assays also should consider densities of receptor and donor plants usually found in the field. Different densities of donor plants can determine accumulation of allelochemicals into the soil. Receptor plants exposed to allelochemicals may be inhibited in their growth at low densities, while they can be stimulated or unaffected at high densities (Weidenhamer, 1996).

iv. Since microorganisms often participate in allelopathic interactions, assays to elucidate microbial involvement should be included (Sampietro and Vattuone, 2006a).

v. Appropriate growth parameters and/or physiological processes must be selected for assay. Germination and radicle and/or shoot elongation of receptor plants are commonly measured in both greenhouse and laboratory experiments, radicle being the most sensitive to allelochemicals. Assays involving radicle elongation should be performed with pre-germinated seeds. When germination and radicle elongation are measured in the same assay, response of radicle elongation may be confounded with delays in seed germination. Assessment of the effect of allelochemicals/plant leachates or root exudates on physiological processes of receptor plants may provide insight about how allelopathic substances act on plant growth (Sampietro *et al.*, 2007). Osmolality and pH of plant leachates/root exudates sometimes affect growth of receptor plant in both laboratory and greenhouse experiments. The effect of these parameters should be evaluated on plant growth through the development of proper assays.

In this chapter, we propose several assays and criteria to study allelopathic interactions among terrestrial plants. They were selected to provide answers about the biological – ecological aspects of allelopathic interactions and to hold the basis for further studies on isolation and identification of allelochemicals.

2. PRELIMINARY OBSERVATIONS

The study of an allelopathic phenomenon always should begin with preliminary observations consisting of detailed description of the natural settings where chemical interference would occur. Descriptions should include:

i. Observation of disappearance or growth reduction of a plant species related to the presence of other one, persistence of vegetation succession stage, or poor yield in a crop infested with a particular invasive weed. Quantitative analysis of these observations often is achieved through counting the number of individuals of each plant species in the field and determining their dry weight.

ii. Densities of suspected donor and receptor plants in the field. History of the field (*i.e.,* past uses, cultural practices applied, knowledge about distribution of plant species observed previously to the current situation).

iii. Geographical location of the field under study (latitude and longitude), climate (*i.e.,* temperature and precipitation regimes) indicating seasonal changes should also be registered.

iv. Soil characteristics, including texture, chemistry and classification.

v. An experimental frame analyzing alternative explanations to chemical interference should be performed (*i.e.,* traps and baiting experiments will quickly indicate whether or not herbivory is likely; analysis of competition for water, light and nutrients; search for phytopathogenic organisms related to growth response of receptor plants).

Preliminary description should allow hypothesizing which are the affected receptor plants, sensitive growth stage and inhibitory mechanism on receptor plant, possible mode of allelochemical release from donor plant, possible microbial involvement, role of environmental factors and the part of the environment in which the allelochemicals would be released.

3. SOIL ASSAYS

Soil is the first fate of several allelochemicals released after leaching from aerial plant parts, decomposition of dead plant residues and root exudation. The behavior of these allelochemicals in soil is governed by its physico-chemical properties (*i.e.,* adsorption and desorption from soil particles) and microbial activity (*i.e.,* transforming/degrading plant allelochemicals). Allelochemicals found in the soil environment are often different from those released from living and decomposing tissues of donor plants. Hence, evaluation of allelopathic activity of soils is critical to understand if an allelopathic interaction is possible (Inderjit and Weston, 2000). The following guidelines are useful to design soil assays:

i. Substrates used should be rhizosphere and root-zone soils from donor plant. Rhizosphere soil is expected to contain the highest contents of plant allelochemicals (Cheng, 1995). Soils from sites not exposed to the donor plant but with similar properties as those of the mentioned substrates should also be assayed. Some authors evaluate the activity of soils arising from beneath and outside donor plant canopy (Lesica and DeLuca, 2004).

ii. Soil sampling from natural settings should also consider that accumulation of allelochemicals may be related to a specific growth stage of the donor plant or a specific place or season of the year (Politycka, 2005). According to field situation, soil sampling design may be at random, systematic or stratified.

   Random sampling: It is the most common sampling design. It assumes that the bulk soil material is made of a large number of equally sized and discrete portions. A random sampling is only possible if each portion has an equal chance of being selected.

   Systematic sampling: It consists of collecting samples at locations and times according to spatial or temporal patterns. The most common systematic sampling approach is to overlay a grid on an area and then to collect a sample at each grid node.

   Stratified sampling: It is achieved by separating the sample population into similar, non-overlapping groups, called strata, and then selecting either a simple random or systematic sample from each stratum. In many types of ecological sampling, stratification increases the reliability of the data relative to unstratified studies with a similar cost. The success of a stratified design is only possible when each stratum is homogeneous and different from the others.

iii. Before sampling, quadrants of the total sampling area and number of samples per quadrant are established. The best tools for soil sampling are the bucket auger and the pit sampler that are punched into the soil at the proper depth to extract a small core of soil (Fig 1). When these devices are not available, soil sampling may be done with shovels or trowels. In some cases (*i.e.,* rhizosphere soil and root-zone soil), the use of shovels or trowels may be a better choice than using bucket auger or pit sampler.

iv. Assays are often conducted in pots with limited soil volume in controlled environments. Response of receptor plants may be influenced by pot size (John *et al.*, 2006). These effects can be minimized increasing soil volume available for receptor plant. Soil volume available is a problem when biological activity of rhizosphere or root-zone soil samples is assayed. This problem may be overcome by mixing the samples and obtaining sufficient soil volume for assay.

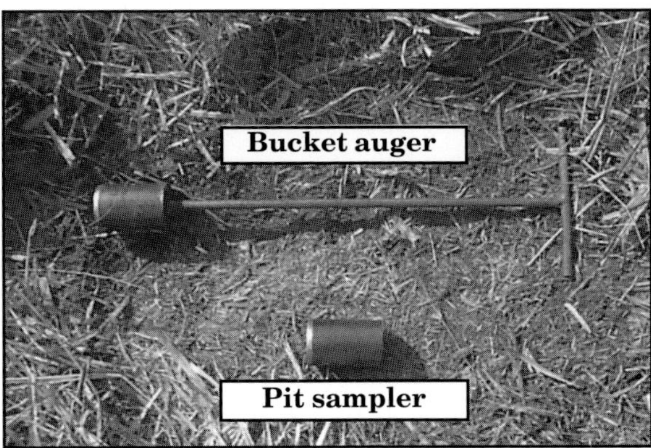

Fig 1. Soil sampling devices

v. Analysis of pH, organic matter, electrical conductivity and mineral components from soil samples before (and sometimes after) beginning of bioassays can provide information about interference mechanisms that may influence growth of receptor plants together or instead of a direct action of allelochemicals (Sampietro and Vattuone, 2006a).

Experiment 1: Use of a bucket auger

Principle

The bucket auger is a narrow tool for boring holes in the ground. This experiment describes the use of the bucket auger for collection of soil samples in allelopathy research.

Materials required

Bucket auger (7 cm in diameter), 9 cm diameter heavy walled PVC sleeve cut into 10 cm length (one or more), rubber mallet (one or more), 5 × 10 cm wood cut into 2.3 cm length (one or more), 40 × 50 cm plastic bags (1 per sample), 5 gallon buckets with handles filled with clean water, 5 gallon bucket filled with soapy water, water sprayer, wash brushes, double distilled water, isopropyl alcohol, disposable gloves, plastic bags (label containers or plastic bags before soil sampling; record date, location, number and depth of each soil sample), sieve (1 mm mesh size).

Procedure

i. Clean the top area where the auger will be punched. Place the PVC sleeve over the site to be sampled and drive it into the ground approximately 10 cm using the wood and the rubber mallet.

ii. Place a clean auger in the sleeve. Turn the handle to advance the auger until it fills with soil. Pull the auger up and remove all the soil into a plastic bag. Return the auger to the sleeve and continue turning until the top of the auger is levelled with the soil surface. Again pull the auger up and load all the soil into the same bag. Surface soils tend to be compacted and it may be necessary to fill the auger two or more times to go from 0 to 10 cm (see *Observation* 1).

iii. If deep soil samples are needed, drive the auger as indicated in step (ii) but discard the soil extracted until the desired depth is reached. Collect samples in bags as indicated in step (ii) at the desired depth.

iv. While wearing gloves, hold the top of the plastic bag closed, while shaking the soil to mix it thoroughly.

v. Clean the auger by scrubbing it with a brush in a 5 gallon bucket filled with soapy water. Then, rinse in a 5 gallon bucket filled with clear water. Then, rinse again spraying with double distilled water. Give a final rinse with isopropyl alcohol dispensed from a clean bottle.

vi. Prevent contamination from 0 to 10 cm soil, cleaning the PVC sleeve with a gloved hand and scooping it out of the bottom of the hole before taking the second sample.

vii. Label and transport the samples to the laboratory on dry ice. If samples are not immediately processed they can be stored at –20 °C. Samples should be thawed overnight in a refrigerator (about 10 °C) before processing.

viii. Sieve soil samples and air dry composite soil samples, spreading them on paper or other suitable material in the laboratory. After 5 or 6 weeks, store the dry samples in labeled paper bags at room temperature or assay/analyze them (see *Observation* 2).

Observations

1. A pit sampler may be used for collection of samples from top soil layers (0–10 cm). To do it, place the pit sampler inside the sleeve and punch into the ground using the wood and the rubber mallet.
2. Several compounds are irreversibly adsorbed to soil particles during drying process. Hence, these substances will be not available when soil is re-hydrated for assay/analysis. To avoid this problem, fresh samples can be frozen at –20 °C until use. In this case, variations in water content can affect responses of receptor plants or soil analysis. Then, water content should be the same among samples or should be adjusted to be the same.

3.1. Examples of soil collection

Collection of soil rhizosphere from a plant producing allelochemicals (Fujii *et al.*, 2005): Grow the plants from a selected species in plastic pots in controlled conditions for a suitable time. Take out the plants from the plastic pot without disturbance. Then, shake softly plant root and collect root-zone soil in plastic bags (Fig 2A). Then, remove and collect the soil adhered to the surroundings of roots (rhizosphere soil) in plastic bags. Sieve soil samples through 1 mm mesh. Discard root residues of rhizosphere soils. Place in plastic bags and freeze (–20 °C) before processing. These samples should be processed during the following 3 months after collection.

Collection of soil samples in a walnut – corn cropping system (Jose and Gillespie, 1998): Take soil cores (0–10 cm depth) at four different distances from the tree rows: 0, 0.9, 2.5 and 4 m from the row (Fig 2B). Collect and combine 10 soil cores at each plot and collection distance. Place each soil sample in a plastic bag and immediately transport to the laboratory in dry ice for processing. Collect on each sampling date, a composite sample of 10 soil cores from a nearby corn field which served as a control for recovery experiments. Take samples as mentioned at different times during the year to evaluate fluctuations in soil biological activity.

Collection of soil samples from crop plots in conventional and no-till cropping systems (Blum *et al.*, 1992): Assuming plots of 30 × 8 m with eight rows subjected to conventional and no-till systems, divide each plot into four sections (15 × 4 m) for sampling. Take two soil cores (5.5 cm in diameter; 0–10 cm depth) per section and combine them. Take also additional soil cores (0–2.5 cm depth) adjacent to the previous sampling locations. The 0–2.5 cm core samples are taken because allelochemicals release from plant residues often are more concentrated in top soil, being possible to detect vertical gradients in the soil. Take samples as mentioned at several times after crop harvest to evaluate changes in soil biological activity.

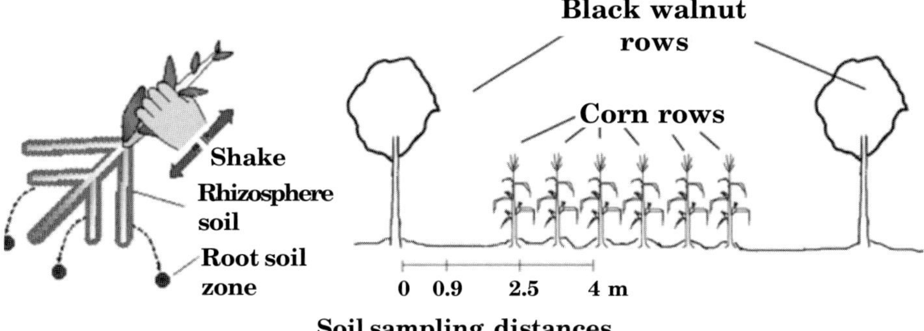

Fig 2. (A) Separation between root-zone soil and rhizosphere soil from a donor plant (*Source*: Fujii *et al.*, 2005). (B) Walnut–corn cropping system (*Source*: Jose and Gillespie, 1998)

Experiment 2: Soil sandwich method

Principle

This method evaluates the allelopathic potential of soil samples from a donor plant on growth of receptor plants. Soil is incorporated into an agar layer. Then, a second agar layer is poured covering the soil–agar mixture. Seeds of the receptor plant are placed onto the second agar layer (Furubayashi *et al.*, 2002). Soil sandwich method is appropriate to assay small-seeded receptor plants, small soil samples with restricted availability or for preliminary screening of allelopathic activity from a broad range of donor plants.

Materials and equipments

Autoclave, greenhouse, ice bath, plastic pots, seeds of receptor plant, sieve (1 mm mesh), soil samples (rhizosphere soil or other), six well-dishes, 0.5 % agar (dissolve 5g of agar powder in 1 L of water, sterilize at 1 atm for 30 min, leave to cool until 40 °C before use).

Procedure

i Please follow one of the following alternatives:

  a. Fill plastic pots with soil, and cultivate the donor plant in a greenhouse at appropriate conditions for several weeks. Take out the donor plants from the plastic pots without soil disturbance. Shake softly the plant roots to remove root-zone soils (Fig 2). Collect the rhizosphere soil which is adhered to the roots surroundings. Plastic pots containing soil without the donor plant will be used as controls.

  b. Collect soil samples from the field. Carefully extract roots of the donor plants with their related soil. Proceed as indicated in step (a) for soil collection.

ii. Sieve soil through a 1-mm mesh to remove root hair and other particles (see *Observation* 1).

iii. Perform the assay in six-well dishes. Add 5 mL of 0.5 % agar cooled at 40 °C on fresh soil [root–zone soil or rhizosphere soil) previously placed in each well (Fig 3)]. Amount of fresh soil per well may be 3g (expressed in dry weight; see *Observation* 2).

iv. After agar solidification, add 3.2 mL of 0.5 % agar cooled at 40 °C (Fig 3). Placed five seeds of a receptor plant in each well (see *Observation* 3). Each treatment should have at least five replications and experiment should be twice repeated.

v. Incubate dishes in the dark at controlled conditions for 3 days. Then, measure radicle elongation (see *Observation* 4).

Calculations

Calculate means of radicle elongation for each treatment. Then, calculate root elongation expressed as percentage of control. Data are expressed as mean ± S.D.

Statistical analysis

The difference between means of radicle length in soil rhizosphere or root-zone soil and control is tested by Student's *t*-test.

Observations

1. Soil processing may have a strong influence on the receptor plant response. Most authors suggest that soil samples should be air-dried and stored in the dark in paper bags under non-humid conditions before use in bioassays (Cheng, 1995). However, other authors argue that soil drying leads to irreversible adsorption of allelochemicals, as occur with commercial herbicides, being more representative to assay soils with their original water content.
2. Measure gravimetric water content of soil samples. Then express soil weight on dry basis.
3. The method should be adjusted for Petri dishes (3 cm or 6 cm diameter) when large-seed species are assayed.
4. Sandwich method can also be used to assay allelopathic potential of dry leaf litters (Fujii *et al.*, 2004).

4. PLANT LEACHATES

Plant leachates arise from aqueous washing of donor plant parts. In nature, above-ground plant parts and residues are washed by water from rain, snow, dew and fog drip.

Fig 3. Soil-sandwich assay. Steps to prepare a well from the 6-well-dish.

4.1. Collection of natural leachates

4.1.1. Stemflow

Stemflow is the rainwater running down the bole of trees and shrubs. Significant amounts of plant allelochemicals are washed by stemflow, which is finally incorporated into the soil. Collars made of polyurethane foam applied around the bole of trees or shrubs allow stemflow collection (Fig 4A). The foam is tightly sealed against the bole creating an effective water barrier. After placing the foam, a trough is dug in the top of the collar and a funnel and a bottle collection apparatus is positioned under a low point in the trough where the water will drip out. During precipitation, stemflow is intercepted by the collar and drains into the funnel and bottle that is used for bioassay.

4.1.2. Throughfall

Throughfall is the rainwater that falls through the plant foliage. Throughfall is collected using common pluviometers or funnel collectors made of non-contaminating polyethylene plastic (Fig 4B). These collectors are placed in the canopy of the tree, shrub or herbaceous donor plant under study. Control consists in precipitations collected with the same device in open fields close to the donor plants under study. In funnel collectors, an inert screen on the funnel prevents contamination with outside elements. The funnel and the collector bottle are connected with a tube which has a loop to reduce evaporation. A layer of glass wool is inserted in the neck of the funnel to prevent splashing and to retain organic debris.

4.1.3. Dew and fog drips

Dew and fog drips are cloud of vapors that clings on plants and then dips to the ground. Devices similar to those described for stemflow and throughfall collection may be appropriate for fog drip collection.

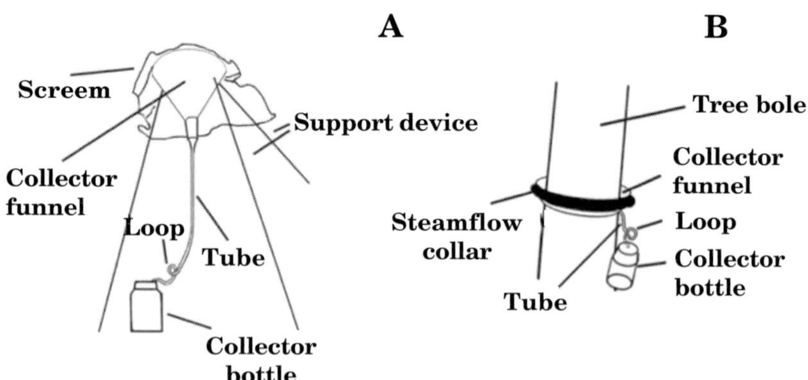

Fig 4. (A) Polyurethane stemflow (*Source*: Likens and Eaton, 1970) and (B) throughfall collectors (*Source*: Hart and Parent, 1974)

4.1.4. Leachates from plant residues

Experiment 3: Litter bag assay

Principle

Leachates from plant residues are made of dead leaves, bark, twigs and other plant residues produced by natural abscission. In litter bag assay, dead plant materials are left to degrade under natural conditions and leachates from plant residue are collected at different decomposition degrees (Souto *et al.*, 1994).

Materials and equipments

Nylon bags of 2 mm mesh, distilled water, tools for collection of plant residues, refrigerator and balance, tape measure.

Procedure

i. Divide the field sampling area in plots of 1 m^2. Define the number and location of each plot according to the selected sampling design and statistical analysis.

ii. Collect 200g litter from each plot over many days. Litter collected from the same plot should be thoroughly mixed to obtain homogeneous samples.

iii. Prepare a composite sample mixing the litter samples.

iv. Put composite sample in the nylon bags. In each bag add 30g of litter. Prepare several bags (*i.e.,* 30) and place them in groups (*i.e.,* 5 per group) at random points on the sampling site. Remove a bag from each group after 1, 15, 30 and 60 days.

v. Prepare litter leachates saturating the collected material with a known volume of distilled water, allowing it to soak for 24 h in the dark at 15 °C. Filter and use for bioassays.

Experiment 4: Influence of plant debris on soil properties

Principle

Components released from plant debris left-onto or incorporated-into the ground can modify growth of receptor plants through changes in chemical and/or physical conditions of top soil layers (Inderjit and Dakshini, 1999). Chemical changes often involve organic compounds, pH and availability of macro- and micronutrients while physical changes frequently comprise light, water and temperature levels (Oliva *et al.*, 2002; Sampietro and Vattuone, 2006a). The assay described below determines the effects of plant debris on growth of receptor plants.

Materials and equipments

Plastic containers, sieve (1 mm mesh), plant debris from suspected donor plant, seeds from receptor plant, greenhouse, balance, pH meter, conductivity meter.

Procedure

A. Soil treatments

i. Collect the plant debris of the suspected donor plant species.

ii. Select two sites for soil sampling: Site 1 must be free of suspected plant species for several years. After mixing, the samples will constitute the control soil. In Site 2, the suspected donor plant has been grown for several years. After mixing, these samples will constitute the donor-plant soil.

iii. Take top soil samples (up to 20 cm) depth from the two sites. Air dry, sieve and store soil samples. Mix samples as indicated in step (ii).

iv. Fill containers with a known amount of control soil and donor-plant soil. The following treatments must be prepared:

   a. Containers filled with control soil: without debris (control),
   covered with debris,
   amended with debris.

   b. Containers filled with donor-plant soil: without debris (control),
   covered with debris,
   amended with debris.

   Amount of assayed debris should be selected according to field observations. Also consider, if plant material is uniform or unevenly distributed onto the soil (see *Observations* 1 and 2). Each treatment should have at least four replicates and should be once repeated (see *Observations* 3 and 4).

v. Irrigate the containers once with a known volume of distilled water using a sprayer. Irrigate slowly to prevent water run out at the bottom of the pots.

vi. Air dry and sieve soils from each container after 14 days. Store in paper bags. Soil of each container should be analyzed for pH and electrical conductivity (EC). Concentration of water soluble phenolic compounds is often determined in these assays (see *Observations* 5 and 6).

vii. Place containers in greenhouse at controlled conditions (*i.e.,* for several plant species 25 ± 2 °C; 60 % of relative humidity, with a 16-h photoperiod of 90 mmol m^{-2} s^{-1} photosyn tetically active radiation, fluorescent light).

B. Growth experiment:

i. Pre-germinate seeds of a receptor plant species (see *Observation* 7).

ii. Transfer remaining soil, obtained from the above–mentioned treatments, to new containers.

iii. Put equal numbers of pre-germinated seeds in each container.

iv. Irrigate the containers spraying a proper volume of distilled water. Irrigate slowly.

v. Incubate the containers by spraying within the greenhouse as previously indicated.

vi. Extract the plants from pots after a week, without damage to roots and shoots. Measure the root and shoot length.

Statistical analysis

Combine the two repetitions of each experiment and subject data to analysis of variance. If data are not normally distributed, transform them. Calculate Pearson correlation coefficients for soil chemical variables and biological parameters (Snedecor and Cochran, 1978) to establish if changes in chemical parameters are associated with registered biological responses. Calculate significant differences among means using Tukey's test.

Observations

1. Uneven distribution is simulated assaying different amounts of plant debris, instead of only one.

2. Microbial immobilization of soil nutrients may inhibit the growth of receptor plants. Microbial population can be evaluated after incorporating the organic matter and using the non-amended soils with plant residues and amended with inert organic matter (*i.e.*, cellulose and pit moss) equal to those assayed for donor plant residues.

3. Plant debris can be burnt and incorporated into soil to determine the influence of inorganic compounds on soil properties and growth of receptor plants (Sampietro and Vattuone, 2006a).

4. Influence of soil microflora on activity of leachates through soil sterilization. Soil is often autoclaved 3 times at 120 °C and 103 kPa pressure for 30 min. Each sterilization is done at 24 h intervals (Sampietro and Vattuone, 2006a,b). Comparison of growth response of receptor plant in sterile vs no sterile soil will indicate if soil microorganisms increase or reduce the bioactivity of plant leachates.

5. To determine phenolics concentration, soak 5g of soil in 15 mL of water and shake, followed by filtration. Then, proceed as indicated in

Experiment 3, Chapter 7, using Folin & Ciocalteu's reagent, but do not add ethanol to the mixture.

6. Soil electrical conductivity provides estimate of changes in organic and inorganic ions. Changes in pH, allow estimation of changes in solubility of soil micro and macronutrients. A more detailed analysis of chemical soil properties is highly recommended to understand how debris modifies soil properties. Soil chemical analysis often includes cation exchange capacity, organic matter, available phosphate, total nitrogen and mineral nutrients (*i.e.,* Mg, K, Ca, Na, Cu, Fe and Mn).

7. Seeds are pre-germinated to obtain constant densities of receptor plants (Weidenhamer, 1996). Assay can be adapted to evaluate seed germination, using no pre-germinated seeds.

4.2. Collection of artificial stemflow or throughfall leachates

Collection of artificial leachates is often achieved by spraying a known volume of distilled water on aerial parts (stems and leaves) of receptor plants for a proper time. Water leachates are collected in containers placed under the plant parts. Artificial as well as natural plant leachates can be collected for several days, freezing aqueous leachates immediately after each collection. They are then filtered and assayed in Petri dishes using filter paper or a layer of filter paper over soil as substrate (Gonzalez *et al.*, 1997).

Experiment 5: Amendment of activated carbon

Principle

Activated carbon is a wide range adsorbent with little affinity to inorganic electrolytes (Cheremisinoff and Ellerbusch, 1978). Water-soluble organic constituents (including allelochemicals) leached from plants can be scavenged adding a layer of activated charcoal onto soil or incorporating this substance into the bulk soil. These operations allow separating allelopathic effect from other interference mechanisms (Nilsson, 1994; Sampietro and Vattuone, 2006b).

Materials required

Plastic containers, PVC pipes (5 cm diameter × 10 cm long, 2 mm thickness) open at both ends, seeds of receptor plant, fine powdered activated charcoal, balance.

Procedure

i. Collect pieces of intact upper soil profile (up to 20 cm depth) from areas mainly covered by the suspected donor species.

ii. Transfer soil to plastic containers. Prepare an equal number of tube and non-tube plots into the plastic containers as follows:

a. **Tube plots:** Insert the PVC tubes into the soil under foliage of the donor plant until the upper surface is at soil surface level. PVC tubes should be inserted vertically to the bottom of the container and do not pond water inside them. Lower part of tubes should be in contact with the bottom of the container. Put a proper amount of seeds from the receptor plant in each PVC tube (see *Observation* 1). Monitor the early seedling growth and left only one individual of the same stage in each PVC tube.

b. **Non-tube plots:** No PVC tubes are inserted in these plots. Put a proper amount of seeds from the receptor plant in selected positions (see *Observation* 1). Monitor the early seedling growth and left only one individual of the same stage in each plot.

Prepare about 20 plastic containers.

iii. Spread a proper amount of activated carbon onto half of the tube and non-tube plots in each container. Hence, half of the tube and no-tube plots in each container will be free of activated carbon (see *Observation* 2). The obtained treatments are (Fig 5) given as follows:

a. **Non–tube plots without activated carbon:** Foliage/litter allelopathy and root resource competition are evaluated on receptor plants.

b. **Tube plots without activated carbon**: Foliage/litter allelopathic effects are evaluated, excluding root resource competition.

c. **Non–tube plots with activated carbon**: Root resource competition is evaluated, excluding allelopathy.

d. **Tube plots with activated carbon:** Both foliage/litter allelopathy and root resource competition are evaluated on receptor plants.

iv. Containers are maintained in controlled conditions of temperature, light and relative humidity, simulating those observed in natural conditions. Spray known volumes of water onto containers simulating natural rain when needed. Carefully extract the receptor plants from soil and wash them, after an appropriate vegetative growth period.

Calculations

Measure elongation and dry weight of roots and shoots (see *Observation* 3).

Statistical analysis

Subject data to two-way analysis of variance. Significant differences among means are determined by Tukey's multiple range test.

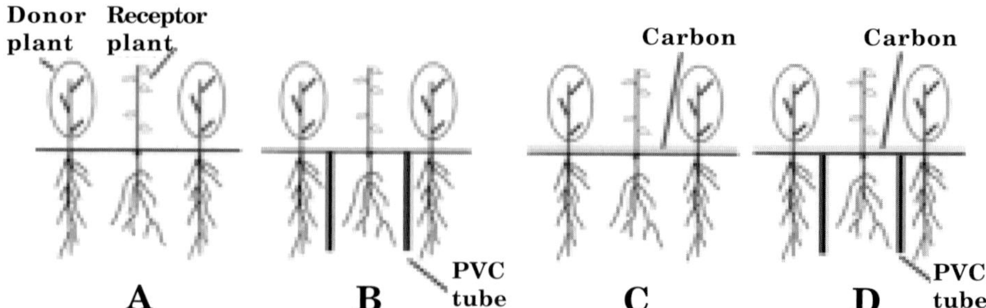

Fig 5. Experimental treatments proposed for relieving allelopathy and distinguishing allelopathy from resource competition (adapted from Nilsson, 1994): (A) Non-tube plot without activated charcoal, (B) tube plots with activated charcoal, (C) non-tube plot with activated charcoal and (D) tube plot with activated charcoal.

Observations

1. Consider natural density and seed viability of receptor plants. Adjust distribution of receptor plants according to these criteria, repeating the same distribution pattern of receptor plants in both PVC tube and non-PVC tube plots.

2. Allelochemicals leached from the foliage and litter of donor plants will be adsorbed by the activated carbon layer. Some researchers prefer incorporation of activated carbon to bulk soil, because allelochemicals adsorbed may not only be from plant leachates but also from the root exudates. However, note that incorporation of activated carbon implies undesirable soil disturbance.

3. If growth of receptor plants is improved by addition of activated carbon (layered on top of the mineral soil below the litter later), it means that inhibition mediated by soluble organic compounds has been eliminated and allelopathy is a possible interference mechanism.

5. ASSAY INVOLVING PLANT VOLATILE ALLELOCHEMICALS

Several plants release volatile allelochemicals which would move at vapor state in natural conditions, affecting growth of neighboring plants. Most of the plant species releasing these compounds are rich in essential oils and most of these allelochemicals are identified as mono and sesquiterpenes (Weidenhamer *et al.*, 1993).

Experiment 6: Assays in closed chambers

Principle

The effect of volatile compounds released by donor plants can be evaluated on receptor ones in closed chambers. Seeds of receptor plant are exposed to different amounts of aerial parts (*i.e.*, stems and leaves) of the donor plant in a closed glass beaker. Contact between seeds of receptor plants and donor plant materials is only possible through the air (Fig 6).

Materials and equipments

500-mL glass beakers, 10-mL glass beakers, seeds of receptor plant, donor plant material (aerial plant parts), balance, cotton, vinyl wraps.

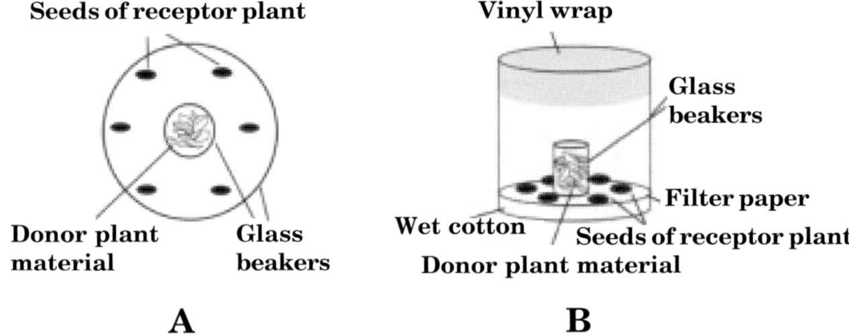

Fig 6. Assay of volatiles in a closed chamber: (A) top and (B) lateral view of the system described in the text

Procedure

i. Prepare 500-mL glass beakers as follow: Add a known weight of cotton (moistened with a known volume of distilled water) at the bottom of the beaker. Then, place a proper amount of seeds from receptor plant on a layer of filter paper onto the cotton layer. Before seeding, place one or more 10-mL beakers containing a known amount of donor plant material, in the centre of the 500-mL beaker. Glass beakers prepared as previously mentioned but without addition of donor plant material in the 10-mL beakers will serve as controls (see *Observations* 1 and 2).

ii. Prepare treatments considering to assay more than one amount of donor plant material on the receptor plant species. Each treatment should have at least three replications.

iii. Close the glass beakers with vinyl wraps and place them in a growth chamber at controlled conditions for a proper time (*i.e.*, for several plant species is suitable 25 °C, 16-h photoperiod at 400 μmol m^{-2} s^{-1} photosynthetically active radiation for 4 days).

Calculations

Total germination (G_T) can be calculated using the relation:

$$G_T = \frac{N_T \times 100}{N};$$

where N_T is the number of germinated seeds for each treatment at the end of the assay and N is the total number of seeds used in the assay.

Observations

1. The assay described above can be transformed in to a post-germination assay, using uniform pregerminated seeds of the receptor plant and measuring root and shoot elongation after a proper time.

2. Sometimes controls are prepared as indicated above. However, humidity of receptor plant can also influence the growth of receptor plant. To avoid it, replace donor plant material in controls by an equal weight of wet cotton.

Statistical analysis

Analyze data using one-way Analysis of the Variance and Mann-Whitney U test.

Experiment 7: Assay in twin-chamber cages

Principle

The effects of plant volatiles released by a donor plant on growth of a receptor one can be evaluated using the twin-chamber cage system. Each unit of the system consists of a chamber cage (inducing chamber, IC) which is sequentially connected through a hole to the next chamber cage (responding chamber, RC). Donor and receptor plants are placed in IC and RC chambers, respectively. Air circulates from IC to RC chambers allowing exposure of receptor plant to volatiles released by donor plants. Then, air is drawn out through a vacuum tank equipped with a fan (Fig 7). This assay was applied to study changes in biomass allocation induced in receptor plants by volatiles emitted from intact donor plants (Ninkovic, 2003; Ninkovic *et al.*, 2006).

Materials and equipments

Acrylic chamber cages, seedlings of donor and receptor plants, greenhouse, sand, plastic bags, draining-tubes, oven, foam plastic plugs, fan-equipped vacuum tank, balance, black plastic curtains, planimeter.

Procedure

i. Set up twin-chamber cages (Fig 7). Holes of IC chamber should be done at a level to ensure transfer of volatile compounds from donor plant to receptor plant located into RC chamber.

ii. The following treatments should be performed:

   a. Exposure of receptor plant to air from donor plant (Fig 7A).

   b. Exposure of receptor plant to air from other individual of the same species (Fig 7B).

   c. Exposure of receptor plant to clean air (Fig 7C).

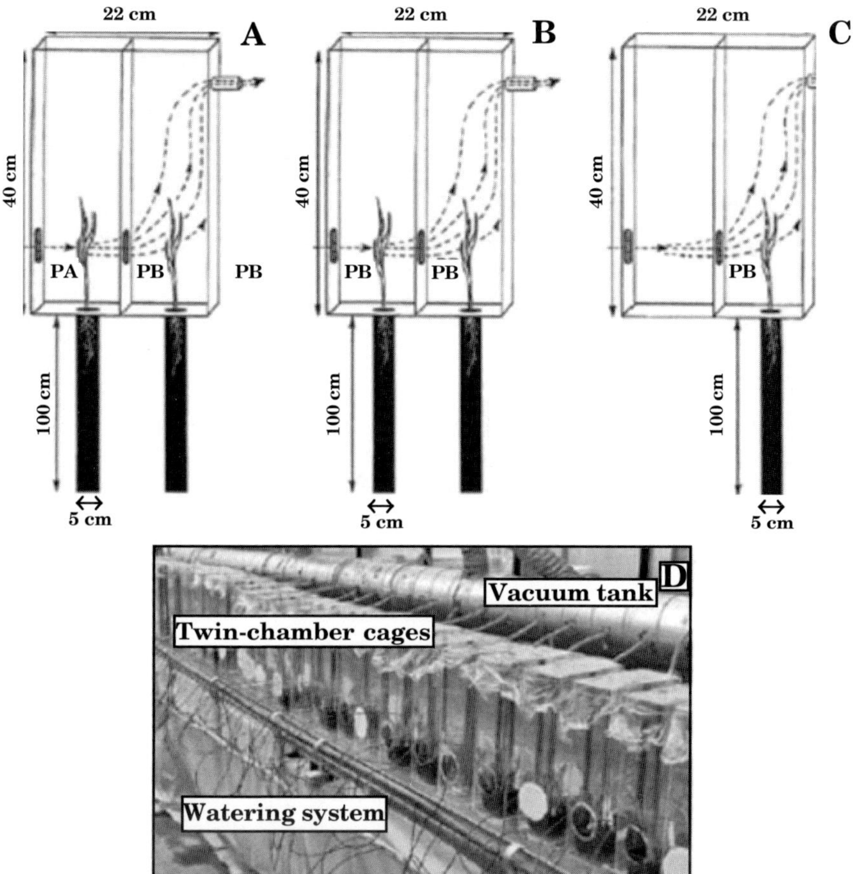

Fig 7. System of twin-chamber cages for exposure a receptor plant (PB) to volatiles from a donor one (PA). (A) PB plant is exposed to volatiles emitted from PA plant. (B) PB plant is exposed to volatiles from other PB plant. (C) PB plant is not exposed to volatiles of PA plant. (D) A view of the whole device (*Source*: Ninkovic, 2003; Ninkovic *et al.*, 2006)

iii. Germinate seeds of receptor and donor plants on moistened filter paper in Petri dishes at controlled conditions. After 24 h, uniform germinated seeds (about 1 mm root elongation) are transferred to the cage system. One pregerminated seed is placed at 1 cm depth in the sand of each plastic bag.

iv. Fill 100 × 5 cm perforated plastic bags with clean sand. Pull the upper, open end of the bag through a hole made at the top of a table, to which the bag is fixed by a staple. The mentioned holes are continued in the middle of the bottom of each cage. The lower end of the bags should have two drainage holes (3 mm diameter) connected to draining-pipes. Place a foam plastic plug between the plastic tube and the wall of the hole to prevent light from above the table affecting the roots below and to minimize the influx of air into the cage from beneath the table. Fix a black plastic curtain around the tables to provide darkness for roots.

v. Place the twin-cages in the greenhouse at random on the table, at the beginning of the experiment to compensate for any spatial variation in irradiance (see *Observation* 1). Adjust temperature, light and relative humidity regimes according to the needs of both receptor and donor plants. A standard nutrient solution should be provided during the experiment to receptor and donor plants (*i.e.*, a proper dilution of Hoagland´s solution manually provided or through an automatic watering system) to maintain a constant nutrient regime.

vi. Proper growth vegetative intervals should be fixed according to known growth rates of donor and receptor plants. Define at least five intervals. At the end of each interval, choose five plants per treatment from RC (responding chamber) and separate into roots, stems and leaves. Measure leaf area immediately after separation. Carefully wash roots with water. Then, dry plant parts for 48 h at 60 °C until constant dry weight.

Calculations

The morphological indices are calculate using the relation:

$$TDW\ (g) = DLW\ (g) + DRW\ (g) + DSW\ (g);$$

where TDW is total dry weight, DLW is dry leaf weight, DRW is dry root weight and DSW is dry stem weight.

$$LMF = \frac{DLW}{TDW};$$

where LMF is leaf mass fraction.

$$LMF = \frac{DLW}{TDW};$$

where RMF is root mass fraction.

$$SMF = \frac{DSW}{TDW};$$

where SMF is stem mass fraction.

Graph LMF, RMF and SMF as function of TDW for each treatment at different sampling times. These graphs will indicate if volatile allelochemicals from donor plant affect biomass allocation on receptor plant.

Statistical analysis

Analyse linearity for each of the parameters previously calculated in the three treatments using an ANOVA goodness-of-fit test (Draper and Smith, 1966). Analyse differences among treatments' means using Tukey's test.

Observations

Prepare appropriate number of replicates per treatment (*i.e.,* 30 twin-chamber cages per treatment).

6. ASSAYS OF ROOT EXUDATES

Root exudates are a common source of allelochemicals with potent biological activity (Inderjit and Weston, 2000). The following methodological difficulties are often associated to root exudates:

i. Root exudates must be collected considering the natural mode of allelochemical release. Exudates should be collected from undisturbed root systems of donor plants. Extraction of root allelochemicals by shaking or after damage of root tissues should be avoided.

ii. In most systems for assay/trapping root exudates, roots of donor plants are grown or maintained in substrates other than soil. Hence, the influence of several soil factors on bioactivity of exuded allelochemicals (*i.e.,* soil microbial activity, chemistry and texture) is not determined.

iii. Trapping techniques for root exudates often do not provide real assessment of allelopathic potential. These techniques allow us to know if bioactive compounds are found in root exudates but do not provide real information about rates of allelochemical release or bioactive concentrations in natural environments.

iv. In general, allelopathic potential of root exudates cannot be explained through the presence of one particular compound. In fact, exuded components with no apparent allelopathic effect can modify the biological activity of known allelochemicals. For example, non-inhibitory concentrations of glucose and phenylalanine increase the inhibitory activity of *p*-coumaric acid on morning glory seedling biomass, because glucose and phenylalanine reduced microbial utilization and/or sorption of *p*-coumaric acid onto soil particles (Pue *et al.*, 1995). This indicates that some root exudates' components (*i.e.,* glucose and amino acids) not considered as bioactive compounds (Whittaker and Feeny, 1971) can modify the activities of allelochemicals and should be considered during the analysis of root exudates bioactivity. Mode of action of several individual allelochemicals is often known but mode of action of compounds in mixtures is yet needed.

The following experiments allow the collection and testing of root exudates. They do not provide solutions for some of the methodological problems mentioned above but they constitute approaches currently used to study allelopathic activity of root exudates.

Experiment 8: Stairstep technique

Principle

Stairstep technique involves the use of the stairstep apparatus, where a nutrient solution flows through pots containing donor plants alternated with pots of receptor plants (Fig 8A). The nutrient solution recirculates in the system. This assay allows evaluation of the effect of root exudates from donor plants on receptor ones. It also allows separating the allelopathic mechanism from competition for mineral nutrients. A modification of the classic stairstep apparatus is proposed in the experiment below, where nutrient solution circulates from trays containing donor plants to several consecutive rows of pots containing receptor plants (Liu and Lovett, 1993).

Materials and equipments

Plastic trays and pots, reservoir tanks, collector tanks, seeds of receptor plants, seedlings of donor plants (grown in trays), gravel (5–10 mm particle size), clean sand, opaque tubes, electric pump, 10 mm black pipes, 3 mm black tubes, bent 3 mm diameter black tubes, nylon meshes (0.8 mm), Hoagland´s solution (Table 1), pH meter, conductimeter.

Procedure

i. Build the stairstep apparatus (Fig 8B) which consists of four units. Each unit has six steps: a top reservoir is filled with nutrient solution, which flows by gravity first to trays containing the donor plants or without them (control) and then through three rows of receptor plant pots to a collector tank at the bottom of the apparatus. Each unit has a tray and four replications per row of receptor plant pots. The liquid is recirculated from the collector tank to the reservoir by an electric pump. Opaque tubes connect trays with receptor pots and receptor pots from different rows each other as shown in Fig 8B. Topmost opaque tube is joined to the tray by 10 mm-diameter black pipes. Connections from this tube to the next level and between lower levels are with 3 mm diameter black tubes. A bent 3-mm-diameter black tube with a 150° angle is inserted into the opaque tube under each tray. This bent tube is designed for a fine adjustment of the flow rate of solution by turning the tube clockwise or counter clockwise. Three centimeters of gravel layer is sandwiched by two nylon mesh (0.8 mm) liners on the base of each tray. Twelve centimeters sand (0.7–1.5 mm particle size) layer cover the nylon mesh. Individual pots are filled in similar fashion and contain 500g gravel and 2500g sand.

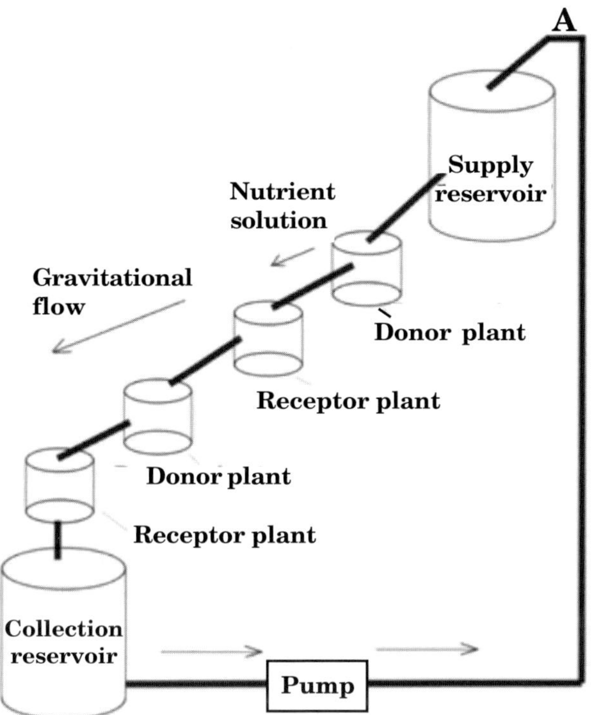

Fig 8. (A) Diagram showing one line of a classical stairstep apparatus. (B) Modified stairstep apparatus described in the text (*Source*: Liu and Lovett, 1993)

ii. Fill the collection tank with 50 L of half-strength Hoagland's solution on day 1. Flow rates are gradually increased from 1 to 6 L/h during the experiment and the solution is cycled for 2 h twice a day. Before each recycling, the electrical conductivity (EC) and pH of the solution are monitored. A 10 % Hoagland's solution is added to each collecting tank at day 12 and 12 % Hoagland's solution is further added at day 17 to maintain the EC range of 1.0–1.5 mS/cm^2 (see *Observation* 1).

iii. The treatments should be:

 i. Tray without donor plant (control).

 ii. Tray with donor plants, and pots seeded with receptor plant on day 1.

 iii. Tray with donor plants, and pots seeded with receptor plants on day 28.

 iv. An additional tray containing an amount of oven dried donor plant materials, corresponding to a density expectable in the field can also be added to evaluate the response of receptor plants to degrading donor plant residues.

Density of receptor and donor plants should be representative of field conditions. The same number of receptor plant are sown directly into each test pot and thinned to few seedlings in uniform positions at day 4 after sowing. The receptor plant is harvested after growing for 21 days in controlled conditions. The plants are separated into leaves, stems, and roots. The leaf area can be determined using a planimeter. The fresh and oven dry weights of each component are measured.

Observations

Electrical conductivity and pH should be maintained as constant as possible to avoid influence of these factors on growth of receptor plants.

Experiment 9: Capillary mat system

Principle

The capillary mat system allows production of large quantities of living seedling roots and their respective exudates. Seeds from a donor plant are germinated between two layers of double cheesecloth placed on an aluminium window screen which is on a matting system. The matting system ensures water absorption and keeps humidity in the overall layers. The system is able to produce root exudates within 10–14 days without contamination of fungal or bacterial pathogens (Czarnota *et al.*, 2001).

Materials and equipments

Double cheesecloth 60 cm long, aluminium window (28 × 43 cm), ridge insulation mat (61 × 248.8 cm), weed mat (53.3 × 304.8 cm), capillary mat, water trough, plastic (4 mm thickness), egg crate light baffles (60.3 × 121.3 cm), redi-Heat propagation mat (53.3 × 304.8 cm), single-edge razor blade, rotary evaporator, growth chamber, PTFE 0.2-µm filter, solenoid valve.

Reagents

Methylene chloride and 0.25 % glacial acetic acid.

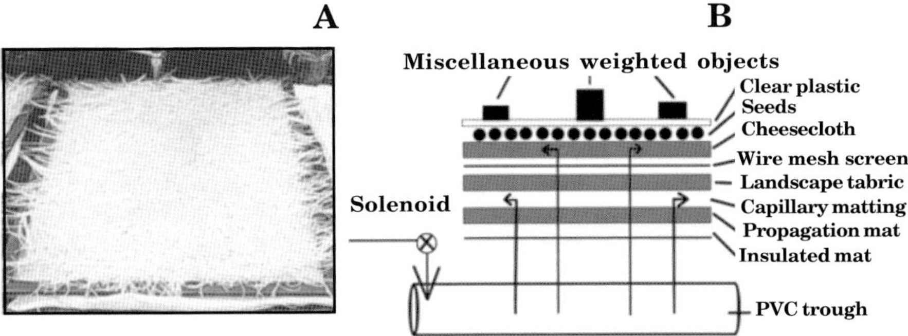

Fig 9. Capillary system for the production of large quantities of living seedling roots and their exudates (*Source*: Czarnota, 2001)

Procedure

i. Place a proper amount of seeds (*e.g.,* 300g for sorghum) between two double layers of cheese cloth.

ii. Put cheese clothes on the aluminium window screen.

iii. Place seed mat onto the capillary mat, which is on the redi-Heat propagation mat. Ridge insulation mat is under propagation mat.

iv. Drop the cheese cloth and the mats, in the order previously indicated, into a water trough.

v. Cover the entire system with a 4 mm clear plastic and top it with two sheets of egg crate light baffles.

vi. Press the entire system with about 10 kg weight of miscellaneous objects. This ensures the contact of screen in contact with the capillary mat and the system components as the seeds germinated.

vii. Maintain humidity levels so that the capillary mat is uniformly but not over moist. A solenoid valve is used to maintain water levels.

viii. Leave roots to develop for 3–4 day in appropriate conditions (*i.e.,* for sorghum 20–25 °C temperature are appropriate) and a proper time to ensure coverage of the bottom of the aluminium screen with roots of the donor plant.

ix. Cut roots of donor plants with the single-edge razor blade. Immerse immediately roots in methylene chloride, acidified with 0.25 % glacial acetic acid, for approximately 30s.

x. Filter the solvent extract with a PTFE 0.2-µm filter and evaporate to dryness in a rotary evaporator at 40 °C. The dry material can be dissolved at different concentrations and used for laboratory bioassays.

xi. Repeat the experiment twice. Each treatment should be replicated twice.

Experiment 10: Root exudate recirculating system

Principle

This system allows direct irrigation of acceptor plants with the exudates of donor plants. A nutrient solution is recirculated carrying the exudates in a closed system (Stevens and Tang, 1985).

Materials required

Brown glass solvent bottles (1 L) with bottoms removed, 12g of Amberlite XAD-4 polymeric adsorbent resin, basaltic rock (2 cm size), glass wool, perforated teflon disks, coarse grade silica sand, plastic pots, vermiculite, glass tubings, bored rubber stoppers, Teflon sealant tapes, air pump, aluminium foil, Hoagland´s solution (Table 1), receptor and donor plants.

Procedure

i. Add a thin layer of glass wool and a perforated Teflon disk in the bottom of pots. Then add an 8-cm layer of crushed basaltic rock

ii. Add sand onto the rock, in the bottom of the pots to provide the rooting medium.

iii. Top the silica sand in acceptor pots with 1 cm of vermiculite to maintain moisture for germinating seeds.

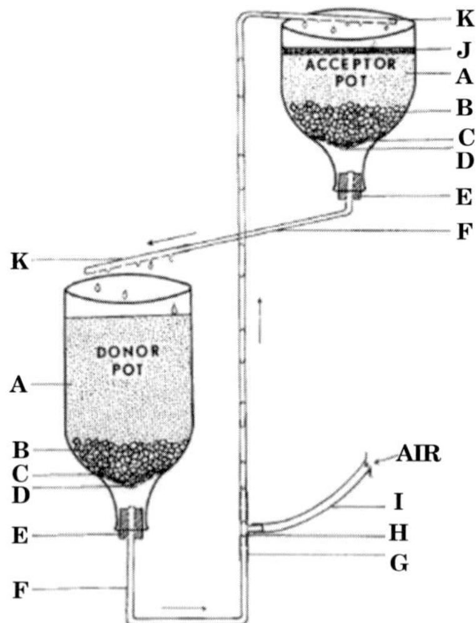

Fig 10. The root exudate recirculating system. A, silica sand; B, crushed basaltic rock; C, glass wool; D, perforated Teflon disk; E, rubber stopper wrapped with Teflon sealant tape; F, glass tubing; G, Teflon sleeve connector; H, glass T; I, Teflon tube to air pump; J, vermiculite; K, perforated Teflon tube. Arrows indicate direction of flow. A glass column containing XAD-4 resin was attached to the bottom of the donor pot for resin controls (*Source*: Stevens and Tang, 1985).

Table 1. Full strength Hoagland's nutrient solution. Stock solutions and volumes of each one needed to prepare 1 L of nutrient solution (Hoagland and Arnon, 1950)

	Stock solutions		**Volumes of stock solution to prepare 1L of nutrient solution[c] (mL)**
	Weight (g)	**Distilled water (mL)**	
Macronutrients[a]			
$NH_4H_2PO_4$	2.3	20	1
KNO_3	10.1	100	6
$Ca(NO_3)_2$	11.5	70	4
$MgSO_4$	4.8	40	2
Micronutrients[b]	**Weight (mg)**		
H_3BO_3	57.2	20 mL all together	1
$MnCl_2{\cdot}4\,H_2O$	36.2		
$ZnSO_4{\cdot}7\,H_2O$	4.4		
$CuSO_4{\cdot}5\,H_2O$	1.6		
$H_2MoO_4{\cdot}H_2O$	0.4		

[a] Weigh chemicals and keep stock solutions in separate bottles.

[b] Weigh chemicals and dissolve them all together in 20 mL distilled water.

[c] The amount of chemicals required for the nutrient solution is taken from the stock solutions and diluted with distilled water to 200 mL. Before use, the 200 mL nutrient solution are further diluted with 800 mL of distilled water to a total volume of 1 L.

iv. Elevate receptor pots relative to donor to allow gravitational flow and to prevent shading of receptor seedlings by donor plants.

v. Connect glass tubing at the bottom of the pots with bored rubber stoppers wrapped with Teflon sealant tapes.

vi. Connect an air pump. Use Teflon tubings for air inlet lines and glass tubing connections.

vii. Wrap glass tubing with aluminium foil to exclude light.

viii. Recirculate one-half strength Hoagland's nutrient solution through both containers (see *Observations*).

ix. Nutrient solution from the donor pot is air-lifted through 6-mm-diameter glass tubing to the receptor pot using the air pump, at a rate of about 1 L/h.

x. Three recirculating systems should be set up considering the following treatments:

   a. **Receptor plants exposed to donor plants:** Establish a proper constant number of donor seedlings in each donor pot. These plants are previously grown in controlled conditions. Place a proper constant number of pregerminated seeds of receptor plants in the receptor pot, below the vermiculite layer. Seedlings arising of this seeds are thinned when needed.

   b. **Receptor plants exposed to donor plants connected to a XAD-4 trapping system:** Proceed as indicated in **c)** but connect in the recirculating system a glass column filled with 12g of XAD-4 prior to the air inlet line, between the donor and the receptor pot. XAD-4 resin will adsorb hydrophobic secondary compounds. Hence, the use of XAD-4 allows the comparison of effects of donor on receptor plants. Resin control columns should be changed every day to ensure efficient trapping of secondary compounds.

   c. **Receptor plants without exposition to donor plants:** Donor pot must be free of donor seedlings, while receptor pot are prepared as indicated in a).

   Treatments should be repeated twice, with at least two replications.

xi. Receptor plants may be collected 14 days after seeding.

Calculations

Measure root and shoot lengths and dry weight of receptor seedlings.

Statistical analysis

Comparison between the treatments (a) and (b) allows analyzing how root exudates affect growth of receptor plants. Differences between the treatments (a) and (c) allows to analyze how hydrophobic allelochemicals exuded from roots of donor plants affect growth of receptor ones.

Observations

Replenish with water and Hoagland's solution considering the electrical conductivity and pH measured each day, which should be maintained at constant level. Nutrient levels should be maintained at 0.5-strength Hoagland's solution at pH 7.0 with H_2SO_4.

Experiment 11: Equal compartment agar method

Principle

In equal compartment agar method, a donor plant species is co-grown with a receptor one in a water–agar medium. The effect of allelochemicals, diffused from intact roots of donor plants, is evaluated on growth of a receptor plant species. The method allows screening of allelopathic potential of root exudates of a large number of donor species on one or more receptor plant species (Wu *et al.*, 2000).

Materials required

500 mL glass beaker, agar powder, seeds of wheat and ryegrass (proposed as donor and receptor plants, respectively), white paper board, autoclave, 500 mL glass beakers, aluminum foil, Parafilm, Erlenmeyer's flasks; sodium hypoclorite, laminar flow, growth chamber.

Procedure

i. Suspend 3g of agar powder in 1 L of distilled water and heat to boiling until the medium is completely dissolved.

ii. Transfer warm agar media to 500 mL glass beakers, adding 30 mL to each beaker. Seal the mouth of each beaker with paper. Then, cover the paper seal with aluminum foil. Adjust the cover (aluminum foil + paper) to the glass walls of the beakers with cotton cord. Add 500 mL of distilled water to an Erlenmeyer's flask and close it with a cotton plug. Wrap white paper boards with aluminum foil.

iii. Sterilize the beakers, Erlenmeyer's flask and paperboards in an autoclave at 1 atm and 121 °C for 20 min.

iv. Pregerminate seeds of donor and receptor species: sterilize wheat and ryegrass seeds with 1 % sodium hypoclorite for 15 min considering that you need 100 mL of hypoclorite solution to sterilize 0.5g wheat or ryegrass seeds. Rinse with sterilized distilled water and place seeds in sterile Petri dishes. Work in a laminar flow. Use seeds germinated after 24 h.

v. Go to the laminar flow. Place in each sterilized beaker 12 sterile pre-germinated wheat seeds (see *Observation* 1) on one-half of the agar surface, in 3 rows. Prepare controls consisting in beakers without wheat seeds.

vi. Wrap the beakers with a piece of Parafilm. Place the beakers in a growth chamber in controlled conditions (*i.e.,* proper conditions for wheat may be a daily light/dark 13 h/11 h, a temperature of 25 °C/13 °C under fluorescent light). Prepare three replications per treatment.

vii. After a proper time (*i.e.,* 7 days for wheat), place 12 sterile pre-germinated seeds of ryegrass on the other half of the agar surface in 3 rows.

viii. Put a piece of a pre-autoclaved white paperboard 1 cm above the agar surface, dividing the entire beaker into two equal compartments occupied separately by seedlings of the donor and the receptor species.

ix. Wrap the beakers with Parafilm and place back in the growth chamber for 10 days.

x. Experiments should be repeated twice (see *Observation* 2).

xi. Measure root length and plant height of both donor and receptor plants.

Calculations

Calculate means of the above – mentioned growth parameters.

Statistical analysis

If data do not fix with normal distribution, subject them to logarithmic transformation prior to analysis. Data should be subjected to analysis of variance and means may be tested for least significant difference (LSD). Correlations between the mentioned parameters of receptor and donor plants may also be done to analyze the influence of plant species each other.

Observations

1. Adjust seed densities when the selected species were changed by other spp.
2. Seeds of the donor plant species arising from different locations/years should be tested to obtain reliable results of biological activity.

Experiment 12: Plant box method

Principle

The plant box method is a sort of mixed planting which use agar (without nutrients) as a medium (Fujii, 1999). It allows visualizing allelopathic potential of root exudates from a donor plant on receptor plant species. This method has been used to screen allelopathic potential among root exudates of crop varieties.

Materials and equipments

Plastic boxes (70 × 70 cm wide × 100 cm deep), seedlings of donor plant, clean sand, deionised water, nylon-mesh cylinder (3.2 cm diameter, 6.6 cm tall, 0.224 mm mesh), 0.75 % agar (dissolve 7.5g of agar powder in 1 L of distilled water; autoclave at 1 atm for 30 min; leave cool until 40 °C before use), autoclave, ice bath, seeds of receptor plants (*i.e.,* lettuce), growth chamber, plastic wrap.

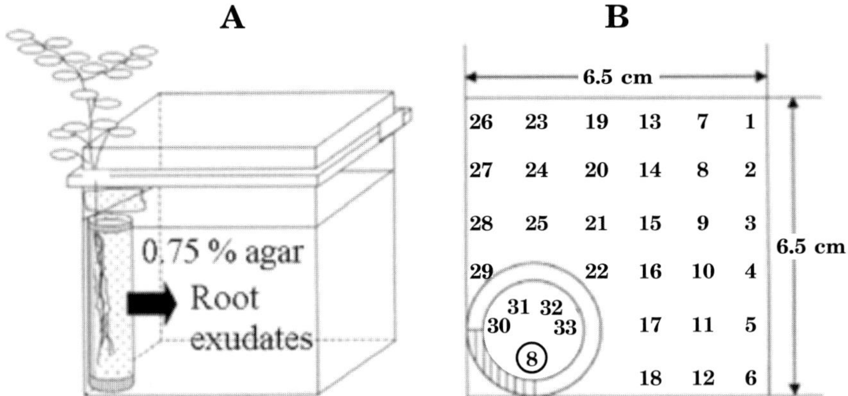

Fig 11. Diagram showing the device designed for plant-box method. In the plant-box: (A) a tamarind seedling, and (B) sowing position (numbers 1–33) of seeds of different weed and edible crop species, which were tested at different distances from the tamarind seedling (S) (*Source*: Parvez *et al.*, 2003)

Procedure

i. Seedlings of a donor species are grown in pots filled with clean sand, with the appropriate water irrigation.

ii. After a proper vegetative growth period, seedlings are extracted from sand washing roots with deionised water.

iii. Place each seedling into a nylon-mesh bounded cylinder. Place each cylinder upright in the corner of a clear plastic box. Tape the stem of the seedling to the corner of the box to stabilize the plant in the box. Pour 250 mL of agar at 40 °C in each box. Boxes are previously kept on ice. Agar should cover all root material without covering stems.

iv. After agar solidification, insert 33 seeds of the receptor plant into the agar surface of each box using porcelain forceps. Seeds should be inserted partially with the germ (point) end down and distributed in an equidistant grid pattern (Fig 11). Cover boxes at the level of seedling roots. Control boxes are prepared as previously mentioned but without putting a seedling receptor plant.

v. Seal the boxes with the plastic wrap to minimize water loss.

vi. Place in controlled conditions for 5 days (*i.e.,* for lettuce, 25 °C, 14-h days; 20 °C, 10-h nights).

vii. Remove seedlings of receptor plant and measure root and shoot length.

Calculations

Measure root length of receptor plants.

Statistical analysis

Regression analysis is used to estimate root length of receptor plants at zero distance from roots of donor plant (y-intercept), and slope of the inhibition curve.

Suggested Readings

Blum, U., Gerig, T.M., Worsham, A.D., Holappa, L.D. and King, L.D. (1992). Allelopathic activity in wheat conventional and wheat-no-till soils: Development of soil extract bioassays. *Journal of Chemical Ecology* **18**: 2191–2221.

Cheng, H.H. (1995). Characterization of the mechanisms of allelopathy: Modelling and experimental approaches. *In*: Allelopathy: Organisms, Processes and Applications (*Eds*., Inderjit, K.M. Dakshini and F.A. Einhellig). ACS Symposium Series **582**. pp. 132–141. Washington D.C., USA. American Chemical Society.

Czarnota, M.A., Paul, R.N., Dayan, F.E., Nimbal, C.I. and Weston, L.A. (2001). Mode of action, localization of production, chemical nature, and activity of sorgoleone: A potent PSII inhibitor in *Sorghum* spp. root exudates. *Weed Technology* **15**: 813–825.

Fujii, Y. (1999). Discrimination and proof methods for allelopathy by bioassay, greenhouse and field tests. *In*: Recent Advances in Allelopathy (*Eds*., F.A. Macias, J.C.G. Galindo, J.M.G. Molinillo and H.G. Cutler). International Allelopathy Society, Universidad de Cadiz.

Fujii, Y., Furubayashi, A. and Hiradate, S. (2005). Rhizosphere soil method: A new bioassay to evaluate allelopathy in the field. *In*: Proceedings, Fourth World Congress on Allelopathy: Establishing the Scientific Base (*Eds*., J.D. Harper, M. An and J. Kent). pp. 517–520. Wagga Wagga, Australia.

Fujii, Y., Shibuya, T., Nakatani, K., Itani, T., Hiradate, S. and Parvez, M.M. (2004). Assessment method for allelopathic effect from leaf litter leachates. *Weed Biology and Management* **4**: 19–23.

Furubayashi, A., Hiradate, S., Araya, H., Horimoto, S. and Fujii, Y. (2002). Soil sandwich method: A new method for bioassay to evaluate the allelopathic activity in rhizosphere soils. *Third World Congress on Allelopathy*. Japan. p. 93.

Gonzalez, L., Souto, X.C. and Reigosa, J.C. (1997). Weed control by *Capsicum annuum*. *Allelopathy Journal* **4**: 101–110.

Hart, G.E. and Parent, D.R. (1974). Chemistry of throughfall under douglas fir and rocky mountain juniper. *American Midland Naturalist* **92**: 191–201.

Hoagland, D.R. and Arnon, D.I. (1950). *The water-culture method for growing plants without soil*. University of California, Agriculture Experimental Station, Berkeley, CA.

Inderjit and Dakshini, K.M.M. (1999). Bioassays for allelopathy: Interactions of soil organic and inorganic constituents. *In*: Principles and Practices in Plant

Ecology: Allelochemical Interactions (*Eds*., Inderjit, K.M.M. Dakshini and C.L. Foy). pp. 35–44. CRC Press Boca Raton, FL.

Inderjit and Nilsen, E.T. (2003). Bioassays and field studies for allelopathy in terrestrial plants: Progress and problems. *Critical Reviews in Plant Sciences* **22**: 221–238.

Inderjit and Weiner, J. (2001). Plant allelochemical interference or soil chemical ecology? *Perspectives in Plant Ecology, Evolution and Systematics* **4**: 3–12.

Inderjit and Weston, L.A. (2000). Are laboratory bioassays for allelopathy suitable for prediction of field responses? *Journal of Chemical Ecology* **26**: 2111–2118.

John, J., Patil, R.H., Joy, M. and Nair, A.M. (2006). Methodology of allelopathy research: 1. Agroforestry systems. *Allelopathy Journal* **18**: 173–214.

Jose, S. and Gillespie, A.R. (1998). Allelopathy in black walnut (*Juglans nigra* L.) alley cropping. I. Spatio-temporal variation in soil juglone in a black walnut–corn (*Zea mays* L.) alley cropping system in the midwestern USA. *Plant and Soil* **203**: 191–197.

Likens, G.E. and Eaton, J.E. (1970). A polyurethane stemflow collector for trees and shrubs. *Ecology* **51**: 938–939.

Liu, D.L. and Lovett, J.V. (1993). Biologically active secondary metabolites of barley. I. Developing techniques and assessing allelopathy in barley. *Journal of Chemical Ecology* **19**: 2217–2230.

Muller, C.H. (1969). Allelopathy as a factor in ecological process. *Vegetatio* **18**: 348–357.

Nilsson, M.C. (1994). Separation of allelopathy and resource competition by the boreal dwarf shrub *Empetrum hermaphroditum* hagerup. *Oecologia* **98**: 1–7.

Ninkovic, V. (2003). Volatile communication between barley plants affects biomass allocation. *Journal of Experimental Botany* **54**: 1931–1939.

Ninkovic, V., Glinwood, R. and Petterson, J. (2006). Communication between undamaged plants by volatiles: The role of allelobiosis. *In*: Communication in Plants (*Eds*., F. Baluška, S. Mancuso and D. Volkmann). pp. 421–434. Springer Verlag, Berlin.

Oliva, A., Lahoz, E., Contillo, R. and Aliotta, G. (2002). Effects of *Ruta graveolens* leaves on soil characteristics and on seed germination and early seedling growth of four crop species. *Annals of Applied Biology* **141**: 87–91.

Parvez, S.S., Parvez, M.M., Fujii, Y. and Gemma, H. (2003). Allelopathic competence of *Tamarindus indica* L. root involved in plant growth regulation. *Plant Growth Regulation* **41**: 139–148.

Patterson, D.T. (1986). Allelopathy. *In*: Research Methods in Weed Science (*Ed*., N.D. Camper). pp. 111–134. Southern Weed Science Society, USA.

Peter Lesica, P. and DeLuca, T.H. (2004). Is tamarisk allelopathic? *Plant and Soil* **267**: 357–365.

Politycka, B. (2005). Pot cultures: Simple tool and complex problem. *Allelopathy Journal* **16**: 47–62.

Pue, K.J., Blum, U., Gerig, T.M. and Shafer, S.R. (1995). Mechanism by which non-inhibitory concentrations of glucose increase inhibitory activity of *p*-coumaric acid on morning-glory seedling biomass accumulation. *Journal of Chemical Ecology* **21**: 833–847.

Romeo, J.T. (2000). Raising the beam: Moving beyond phytotoxicity. *Journal of Chemical Ecology* **26**: 2011–2014.

Romeo, J.T. and Weidenhamer, J.D. (1998). Bioassays for allelopathy in terrestrial plants. *In*: Methods in Chemical Ecology: Bioassay methods (*Eds*., H.F. Haynes and J.G. Millar). pp. 179–211. Springer Verlag, New York.

Sampietro, D.A. and Vattuone, M.A. (2006a). Nature of the interference mechanism of sugarcane (*Saccharum officinarum* L.) straw. *Plant and Soil* **280**: 157–169.

Sampietro, D.A. and Vattuone, M.A. (2006b). Sugarcane straw and its phytochemicals as growth regulators of weed and crop plants. *Plant Growth Regulation* **48**: 21–27.

Sampietro, D.A., Sgariglia, M.A., Soberón, J.R., Quiroga, E.N. and Vattuone, M.A. (2007). Role of sugarcane straw allelochemicals in the growth suppression of arrowleaf sida. *Environmental and Experimental Botany,* (in press).

Souto, X.C., Gonzalez, L. and Reigosa, M.J. (1994). Comparative analysis of allelopathic effects produced by four forestry species during decomposition process in their soils in Galicia (N.W. Spain). *Journal of Chemical Ecology* **20**: 3005–3015.

Stevens, G.A. and Tang, C.S. (1985). Inhibition of seedling growth of crop species by recirculating root exudates of *Bidens pilosa* L. *Journal of Chemical Ecology* **11**: 1411–1425.

Weidenhamer, J. (1996). Distinguishing resource competition and chemical interference: Overcoming the methodological impasse. *Agronomy Journal* **88**: 866–875.

Weidenhamer, J., Macías, F., Fisher, N. and Williamson, B. (1993). Just how insoluble are monoterpenes? *Journal of Chemical Ecology* **19**:1799–1807.

Whittaker, R.H. and Feeny, P.P. (1971). Allelochemics: Chemical interactions between species. *Science* **171**: 757–770.

Wu, H., Pratley, J., Lemerle, D. and Haig, T. (2000). Evaluation of seedling allelopathy in 453 wheat (*Triticum aestivum*) accessions against annual ryegrass (*Lolium rigidum*) by the equal-compartment agar method. *Australian Journal of Agricultural Research* **51:** 937–944.

Wu, H., Pratley, J., Lemerle D., Haig, T. and An, M. (2001). Screening methods for the evaluation of crop allelopathic potential. *Botanical Review* **67**: 403–415.

CHAPTER 5

Genetic Toxicology and Environmental Mutagenesis in Allelopathic Interactions

José Marcello Salabert de Campos[1], Lyderson Facio Viccini[2], Larissa Fonseca Andrade[1], Lisete Chamma Davide[1] and Geraldo Stachetti Rodrigues[3]

1. INTRODUCTION

Assays on seed germination and seedling growth are widely performed in allelopathy research. Seeds of the selected plant species usually are treated with plant leachates or solutions of allelochemicals at varying concentrations. Then, germination rate and root and shoot lengths are monitored in reference to control samples (Vyvyan, 2002). Although these assays provide general information on the biological effects of plant leachates or allelochemicals, they do not offer knowledge about their action mechanisms. Cytogenetic and mutagenic tests allow one to know if leachates or allelochemicals have effects on division of plant cells. Many plant species can be used as indicators of cytogenetic and mutagenic effects (Juchimiuk and Maluszynska, 2005). *Allium cepa* and *Vicia faba* are often assayed for chromosome damage because they have large chromosomes amenable to the study of chromosome aberrations in somatic cells. Some plants such as *Zea mays* and *Tradescantia* are used for the study of both gene mutations and chromosome aberrations (Grant, 1994). Furthermore, genetic assays involving higher plants have the following advantages: (i) relatively low cost and easy handling; (ii) size of plant chromosomes, make higher plants suitable to cytological analysis since they have shown good correlation with other bio-testing systems; (iii) short generation times; (iv) wide range of environmental conditions, pH, and temperature under which assays can be conducted; (v) evaluation of the genotoxicity of single chemicals to complex mixtures; (vi) wide range of genetic endpoints like cytological alterations and gene mutations in whole plants, leaves, embryos and pollen; (vii) detection of intermediate mutagenic metabolites (promutagens) when combined with

1. Departamento de Biologia, Universidade Federal de Lavras, Lavras-MG, Brazil. E-mail: jmscampos@yahoo.com.br
2. Departamento de Biologia, Universidade Federal de Juiz de Fora, Juiz de Fora-MG, Brazil.
3. Embrapa Meio Ambiente (Empresa Brasileira de Pesquisa Agropecuária), Jaguariúna-SP, Brazil.

microbial assay; (viii) high sensitivity (few false negatives) in predicting carcinogenicity of test agents; (ix) many hundreds of genetic loci can be monitored; and (x) technicians can be trained to conduct plant assays relatively quickly (Grant, 1994). This chapter provides techniques used in plant cytogenetic/mutagenic assays for evaluation of chromosome alterations or gene mutation that can be applied together to traditional tests in allelopathy research.

2. BASIC PRINCIPLES OF MUTAGENESIS: MUTATION AND CHROMOSOME ALTERATIONS

Mutational changes in the genome of an organism include modifications in chromosome number and structure, as well as in the structures of individual genes. Mutations that involve changes at specific sites in a gene are referred to as point mutations. They include the substitution of one base pair for another or the insertion or deletion of one or a few nucleotide pairs at a specific site in the genome (Fig 1; Snustad and Simmons, 2000).

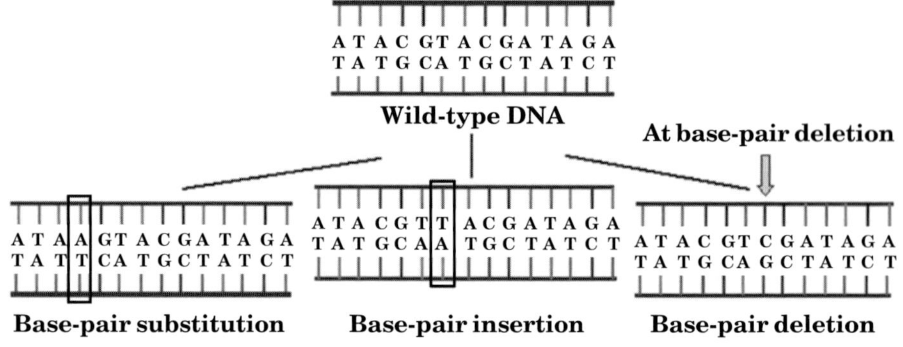

Fig 1. Types of point mutations that occur in DNA

Many chemicals are mutagenic and produce modification in the genetic material. To be recognized, mutations must result in some detectable phenotypic change (Snustad and Simmons, 2000). The effects of mutations on phenotype can be detected by mutagenic tests such as Trad-SHM in *Tradescantia* assay in which expression of the heterozygous dominant blue character of the stamen hair cells is prevented, resulting in the appearance of the recessive pink color (Rodrigues *et al.*, 1997).

Variations in chromosome number and structure are also induced by mutagenic agents. The numerical changes are usually described as variations in ploidy of the organism. Cells or organisms that carry extra sets of chromosomes are said to be polyploid. Another type of numerical changes is aneuploidy. The distinction between aneuploidy and polyploidy is that aneuploidy refers to a numerical change in part of the genome, usually just a single chromosome, whereas polyploidy refers to a numerical change in a whole set of chromosomes (Snustad and Simmons, 2000).

Chemicals with action on microtubule assembly generally induce the two types of numerical changes mentioned above. The microtubules carry out important cell functions during growth and mitotic cycle, such as chromosome migration in anaphase (Jordan and Wilson, 1999). Plant leachates or chemicals affecting the normal operation of the microtubules can increase the number of chromosomes at metaphase. In general, those alterations are caused by disturbances in polymerization and depolymerization dynamics of microtubules. As a consequence, the cell cycle is interrupted at metaphase. This fact, largely known as c-metaphase, was observed for the first time after colchicine treatment (Sharma and Sen, 2002). As a consequence of a lingering effect on cycle inhibition at metaphase, some polyploidy cells with duplicated chromosome number can be observed. Additionally, the observation of some cells with micronuclei and multipolar anaphases reinforces the hypothesis that a plant leachate or allelochemical

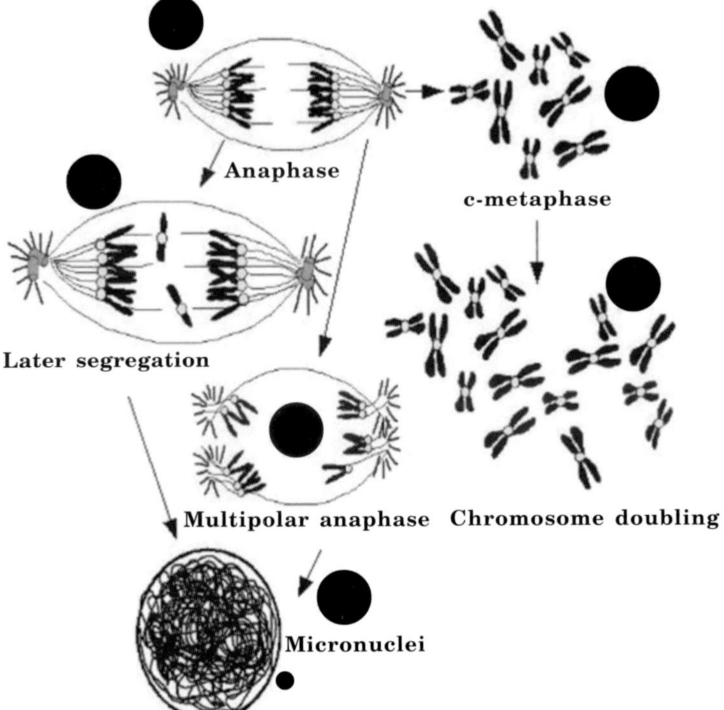

Fig 2. (A) In anaphase the sister chromatids of each chromosome disengages and moves toward opposite poles of the cell. Movement is the result of microtubule shortening. (B) Disturbances in the dynamics of the microtubules by substances that interfere with the polymerization and depolymerization of the microtubules cause interruption in the cell cycle at metaphase. This alteration is known as c-metaphase. (C) Some polyploidy cells with chromosome duplicated number can be observed as a consequence at cell cycle arrest; Additionally, the disturbances in microtubule assembly can induce later segregation. (D) Later segregation or (E) Multipolar anaphase, alterations related with (F) Formation of micronuclei.

can act on microtubules. Some studies demonstrated that micronuclei can appear starting from the condensed metaphase chromosomes spread into a c-metaphase. When the microtubules are not formed, the micronucleation occurs as a consequence of mitosis modification, in which micronuclei directly form from metaphase chromosomes or fragments (Falconer and Seagull, 1987; Morejohn *et al.*, 1987; Verhoeven *et al.*, 1990). The relations between these alterations are given in Fig 2.

Regarding the chromosome structure alterations, a common cause and indicator of chromosomal mutagenesis is the formation of anaphase bridges – chromosomes pulled simultaneously to both poles of the microtubule spindle during the chromosome segregation (Hoffelder *et al.*, 2004). The bridges resulted from fusion of broken chromosome ends leading to the formation of dicentric chromosomes that attach to both spindle poles (Hoffelder *et al.*, 2004). The chromosome breaks lead to the formation of chromosome fragments, which can be observed as other indicator of chromosomal mutagenesis. Micronuclei also might arise from acentric fragments generated by chromosome breaks. Thus, micronuclei have been strongly correlated with the presence of anaphase bridges (Thomas *et al.*, 2003). The mechanism involved in the formation of chromosome bridges are given in Fig 3.

The presence of sticky chromosomes is another alteration related with exposure to genotoxic agents. Metaphases with sticky chromosomes lose

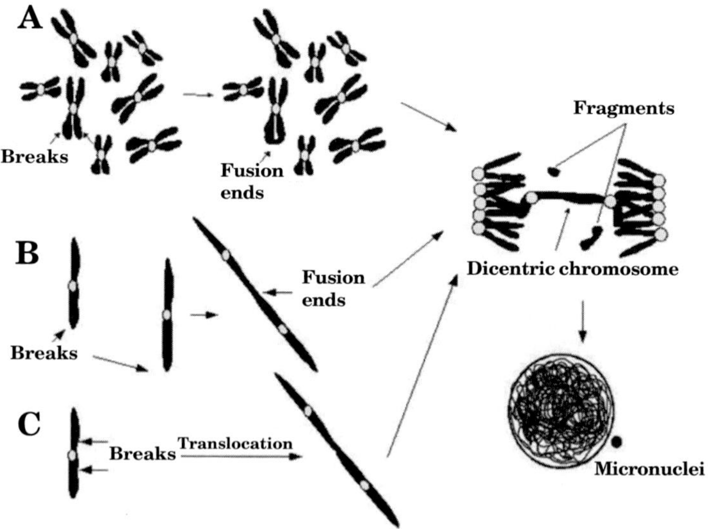

Fig 3. The bridges resulted from fusion of broken chromatids ends (A: after chromosome duplication), from fusion of broken chromosomes ends (B: before chromosome duplication) or centric fragment translocated to another chromosome (C) leading to the formation of dicentric chromosomes that attach to both spindle poles. Fragments generated by breaks leading to the formation of micronuclei.

their normal appearance, and they are seen with a sticky "surface", causing chromosome agglomeration (Babich *et al.*, 1997). This response can be attributed to a toxic effect on chromatin and chromosome organization.

In addition to mutation effects and chromosome aberrations, the cell death constitutes an important variable to be analyzed in toxicity studies.

3. PLANT CITOGENOTOXIC ASSAYS FOR MUTAGENESIS EVALUATION

3.1. Cytogenetic assays

Experiment 1: Allium test

Principle

In the *Allium* test, roots of onion (*Allium cepa*) are exposed to test solutions (treatments and controls) for a proper time. Mitotic cells are observed after fixation, hydrolysis and staining of meristematic root tips. According with the sample preparation and the staining procedures, chromosomes can be examined under optical microscopy. Alterations in cell cycle, chromosome number and/or structure are scored. The *Allium* test is a very useful tool for evaluating and ranking environmental chemicals with reference to their toxicity (Fiskesjö, 1988).

Materials and equipments

Allium cepa onion bulbs with uniform size, assay tubes, slides, cover glass, oil immersion, razor blade, needle, filter paper, Parafilm, aluminum foil, freezer, refrigerator, microscope, stereomicroscope, thermostatic water bath set at 60 °C.

Reagents

Fixative solution: 3 volumes of methanol or ethanol (95 to 100 %) are dissolved in 1 volume of glacial acetic acid (see *Observation* 1); 1 N HCl: add 24.84 mL HCl to 275.16 mL of distilled water. Schiff reagent (Feulgen staining): gradually dissolve 1 g of basic Fuchsin in 200 mL of boiling distilled water and shake throughly. Cool to 50 °C and filter. Add 30 mL of 1 N HCl to the filtrate and then 3g of potassium metabisulfite. Close the mouth of the container with a stopper, seal with Parafilm, wrap the container with aluminum foil, and store in a dark chamber at room temperature for 24 h. If solution shows faint straw color, add 0.5g of activated charcoal powder, shake throughly, and keep overnight in a refrigerator at 4 °C. Filter and store the stain in a colored bottle, in a refrigerator. 45 % acetic acid: add 45 mL of acetic acid to 55 mL of distilled water.

Procedure

i. **Treatment concentrations:** The choice of treatment concentrations is a critical point of the test. High concentrations imply elevated toxicity and difficult identification of the effects on chromosomes or cell cycle. On the other hand, lower concentrations are not effective in inducing cell alterations. For choice of concentrations, a preliminar test of germination and root growth can be carried out with the purpose of finding the 50 % effect concentration (EC_{50}) where the roots reach half length of the control (Fig 4).

Fig 4. Preliminary *Allium* germination test

For evaluation of plant leachates, concentrations of 0, 1, 2, 3, 4, 5, 10, 15, 20, 25, 30, 35, 40, 45 and 50 % can be used in the preliminar test. Onions are placed in plastic containers filled with a known volume of test solutions at different concentrations. Six onions are assayed per concentration. Only bulbs of good condition and approximately the same weight are used in experiments and tap water or distilled water can be used as negative controls. Every day the onions are transferred to fresh test solutions. On day 2, the onions with the poorest growth are removed from each concentration and the experiment is finished on day 5 by measuring root length (Fiskesjö, 1985). Ten roots are measured from each onion (60 roots in each concentration). Mean is calculated for each concentration and expressed as percent of the control. The EC_{30} and EC_{50} concentrations can be used as reference in the choice of the concentrations. For example, with EC_{30} and EC_{50} of 3 and 4 % respectively, concentrations of 0, 1, 2, 3, 4, 5 and 10 % can be chosen. On the other hand, concentrations of 0, 5, 10, 15, 20, 25 and 30 % can be chosen for EC_{30} and EC_{50} of 15 and 20 %, respectively.

ii. **Pre-treatment:** Before the test is started, the outer scales of the bulbs and the brownish bottom plate are removed, leaving intact the ring of root primordia. Bulbs may be exposed to water prior to assay test solution to stimulate root emergence.

iii. **Environmental conditions:** The experiment should be performed at relatively constant room temperature at about 20 ± 2 °C and protected against direct sunlight.

iv. **Exposure and recovery time:** The exposure time of bulbs to test liquids is variable and dependent of the chemical toxicity. Generally, 12,

24 or 36 h of exposure time has been used. After exposure to test liquids, the roots may be submitted to a recovery time of 24, 42 or 48 h. Recovery time after exposure to active compounds improves detection of micronucleus in the *Allium* root micronucleus (MCN) test. The damage on chromosomes by mutagens in interphase nuclei (breaks) or alterations in microtubules assembly leads to the MCN formation. The MCN are revealed in the subsequent generation, in the interphase or prophase cells (F1 cells). Thus, in the root MCN system, a longer recovery time after treatment is required. This is designed to capture the maximum chromosome damage which was inflicted in the interphases (G1, S, G2) of meristematic cells, that later becomes micronuclei in the F1 cell population (Ma and Xu, 1986).

v. **Cytogenetic analysis:** Root tips (about 10 mm) are excised with a razor blade and fixed immersing in a tube containing ethanol or methanol/ glacial acetic acid (3:1). Then, they are kept in the freezer at –20 °C for at least 24 h. For slides preparation by the squash method, the fixed root tips are hydrolyzed in 1 N HCl at 60 °C for 5–10 min, rinsed in distilled water and stained in Schiff's reagent. The end of root tips at 1 mm (mitotic zone) and/or at 2 mm (F1 cells, for micronucleus analysis) are immersed in a drop of 45 % acetic acid on a clean slide and squashed under a cover glass. In order to spread the cells evenly on the surface of the slide, squashing is accomplished with a bouncing action by striking the cover glass with a pencil eraser.

vi. For each slide, 100–200 dividing cells are analyzed and mitotic index is scored from 400–600 cells (see *Observation* 2).

Observations

1. The fixative solution should be prepared fresh each time before use.

2. In other cytogenetic models such as *Zea mays, Crepis capillaris* and *Vicia faba,* the procedures used are the same as that in *Allium cepa*, with few modifications. Good descriptions are found for *Zea mays* (Grant and Owens, 2006), *Crepis capillaris* (Grant and Owens, 1998) and *Vicia faba* (Sang and Li, 2004).

Calculations

The following cytogenetic parameters can be calculated after cytogenetic analysis: number of micronuclei, mitotic index (number of dividing cells/ total number of cells observed), metaphase index (number of dividing cells at metaphase/total number of cells observed), early anaphases (number of dividing cells at early anaphase/total number of cells observed). Other parameters that can also be evaluated are stickiness, later segregation, c-metaphase, bridges, fragments, multipolar anaphase, polyploidy cells and cell death (Fig 5).

Statistical analysis

For statistical analysis, generally the data of abnormalities are expressed as percentage mean ± S.E. The micronuclei are expressed in terms of MCN/ 1000 cells. The tests are arranged in a completely random design with four replications. Analysis of variance (ANOVA), Student's *t* test, Dunnett's *t*

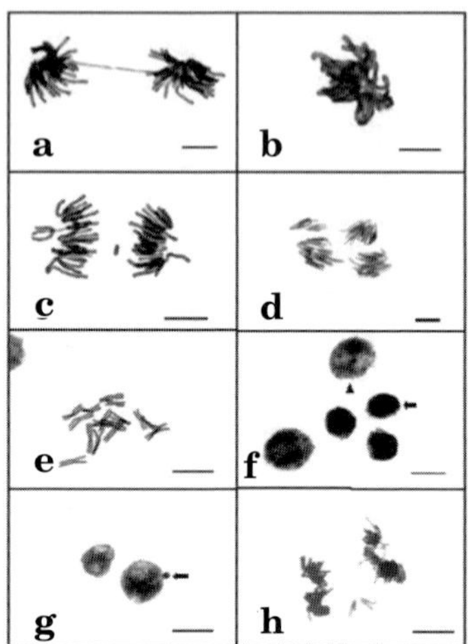

Fig 5. Chromosome alterations in meristematic cells of plant bioassays: (a) chromosome bridge, (b) stickiness chromosomes, (c) chromosome fragment, (d) multipolar anaphase, (e) c-metaphase, (f) cell death (arrow), note the normal interphase nuclei (head arrow), (g) micronuclei (arrow), (h) later segregation. Scale bar = 10 µm.

test, a 2 × 2 contingency χ^2 test or Tukey test are generally employed to determine the significant differences between treatments. Linear regression can be used to establish dose–response curves.

3.2. *Tradescantia* assays: Trad-MCN (micronuclei); Trad-SHM (Stamen hair mutation)

Experiment 2: *Tradescantia* micronucleus (Trad-MCN) methodology

Principle

The *Tradescantia* micronucleus (Trad–MCN) test is one of the bioassays most commonly used with higher plants for detection of genotoxic effects. It is based on the formation of micronuclei resulting from chromosome breakage in the meiotic pollen mother cells of *Tradescantia* spp. inflorescences (Rodrigues *et al.*, 1997) (Fig 6). *Tradescantia* clone 4430 is a suitable lineage for this genotoxic experiment (Rodrigues *et al.*, 1997). It is a diploid hybrid between *T. hirsutiflora* Bush (2461C) that possesses blue flowers and *T. subacaulis* Bush (2441) with pink flowers (Schairer and

Sautkulis, 1982). This clone is especially adapted for laboratory experiments and has been described in detail (Emmerling-Thompson and Nawrocky, 1980). Due to their sterility, there is no risk of losing the genetic identity of 4430 clone by recombination.

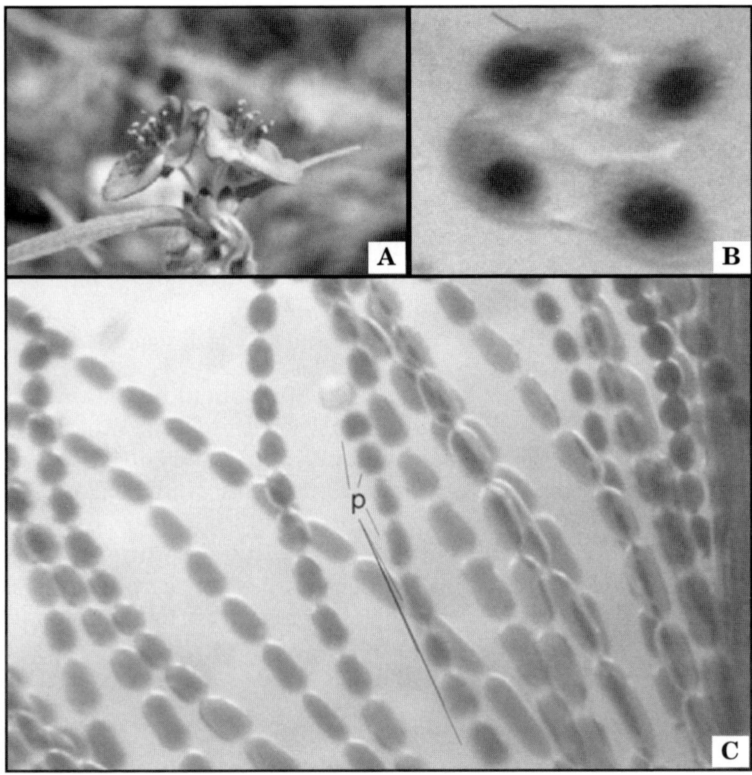

Fig 6. (A) Flower of *Tradescantia* clone 4430, (B) micronuclei in tetrad detected in *Tradescantia* micronucleus assay (Trad-MCN), (C) stamen hair analyzed in Trad-SHM (Stamen Hair Mutation).

Materials and equipments

Tradescantia clone 4430 (nearly 500 mature plants), plastic pots, beaker flasks, aeration bomb, slides, cover glass, oil immersion, razor blade, needle, filter paper. Freezer, microscope, stereomicroscope.

Reagents

Solution A: add 280 mg of H_3BO_3, 340 mg of $MnSO_4{\cdot}H_2O$, 10 mg of $CuSO_4{\cdot}5H_2O$, 22 mg of $ZnSO_4{\cdot}7H_2O$ and 10 mg of $(NH_4)_6Mo_7O_{24}{\cdot}4H_2O$ to 50 mL of deionized water; adjust volume to 100 mL and store at 4 °C until use. Solution B: add 0.5 mL H_2SO_4 to 95.5 mL of deionized water and store at 4°C. Solution C: add 3.36g of Na_2EDTA and 2.79g $FeSO_4$ to 200 mL of distilled water. Adjust volume to 400 mL. Heat the solution to 70 °C while stirring until the color turns yellow-brown. Cool down, adjust the volume to 500 mL and store at 4 °C until use. Hoagland's Stock Solution (Gamborg and Wetter, 1975): dissolve 4.7g of $Ca(NO_3)_2 \cdot 4H_2O$, 2.6g of $MgSO_4 \cdot 7H_2O$, 3.3g of KNO_3, 0.6g of $NH_4H_2PO_4$, 5 mL of solution A and 0.5 mL of solution B in 250 mL of deionized water; Adjust volume to 500 mL with deionized H_2O and store at 4 °C. Hoagland's nutrient solution (3X): mix 300 mL of 10X stock solution with 15 mL of solution C; Adjust volume to 1000 mL with deionized H_2O. Fixative solution: add 3 volumes of methanol or ethanol to 1 volume of glacial acetic acid (see *Observation* 1). 70 % ethanol. Aceto carmine stain: boil 0.5g of carmine in 100 mL of 45 % acetic acid.

Procedure

i. **Population size for test performance:** Nearly 500 mature plants containing an average of 4–6 branches in different development phases should be maintained during the experimental period. Such a population can produce at least 200 inflorescences every two weeks.

ii. **Plant maintained conditions:** The plants should be maintained in a greenhouse with 27/21 ± 2 °C (day/night), with supplemental illumination to ensure a 17 h photoperiod. The plants should be cultivated in plastic pots from 15–20 cm (diameter) containing soil/sand mixture, ensuring good drainage. In general, the plants should be watered on alternate days and weekly they should receive a diluted solution of nutrients. Inflorescence are collected every 2 weeks. At each collection, all pots should be inspected carefully and the plants must be trimmed, in order to maintain a clean environment free of pests. No pesticides can be applied during the experimental period.

iii. **Trad-MCN procedure:** In each experiment, 15–20 branches (for each treatment and replication) of *Tradescantia* containing young inflorescences should be collected 24 h before the beginning of the treatments and transferred to an environment with controlled light (fluorescent light should be about 378 $Em^{-2}s^{-1}$ and the incandescent light should be around 38 $Em^{-2}s^{-1}$; to induce flowering 16/8 h a light/dark photoperiod is necessary) and temperature (21–25 °C). Before exposure to plant leachates or allelochemicals, these inflorescences should be maintained in beaker flasks containing sterile water under constant aeration during 24 h. Wither plants should be eliminated at this time and only turgid branches with inflorescences in expansion should be treated. The genotoxic agents to be tested (plant leachates or allelochemicals) should be diluted in the 3X Hoagland´s nutrient solution (Downs and Hellmers, 1975). Negative control consists in this solution without addition of the genotoxic agents. The positive control is usually represented by 30 h of treatment (without recovery period), with a solution containing 100 ppm of ethyl methanesulfonate (EMS). When lower concentrations of active substances are tested, samples are recommended to be fixated immediately after exposure of 30 h.

iv. **Inflorescence fixation and cell staining:** After treatment, the inflorescences should be removed from the branches and fixed in acetic acid:ethanol (1:3; v/v) during 24 h and then preserved in ethanol 70 % (v/v). For micronucleus counts, the anthers extracted from the young floral buds should be softened on a microscope slide in aceto-carmine stain solution. After positioning the cover glass, the slide should be quickly heated (nearly 80 °C) and gently pressed between leaves of filter paper to remove the excess of stain reagent.

iv. **Micronucleus counting:** Micronucleus counting should be performed at a magnification of 400× (see *Observation* 2).

Observations

1. The fixative solution should be prepared fresh each time before use.
2. The slides should be always codified and analyzed by the same observer.

Calculations

The number of micronucleus should be sampled at random in group of 300 young tetrads, in five slides (Ma, 1979; Ma, 1981), resulting 1500 tetrads analyzed (see *Observation* 2).

Statistical analysis

In this method, each button containing the appropriate young tetrads for counting (one valid slide) is considered a sample population, and buds of the other branches are replications of a treatment. The data can be subjected to analysis of variance (ANOVA) and differences among means can be established using Student's *t* test, considering two-tail distribution (no tendency assumed *a priori* in the data) and $\alpha = 0.01$.

Experiment 3: *Tradescantia* stamen hair mutation (Trad-SHM)

Principle

The Trad-SHM (Stamen Hair Mutation) is a somatic (mitotic) mutation bioassay in which expression of the heterozygous dominant blue character of the stamen hair cell is prevented, resulting in the appearance of the recessive pink color (Underbrink *et al.*, 1973) (Fig 6). *Tradescanthia* 4430 clone is a suitable lineage for this genotoxic experiment because the genetic base for expression of pink cells in stamen hair of *Tradescantia* 4430 clone was established from reciprocal crossings with parental lineage pink and white *T. subacaulis* Bush (Emmerling-Thompson and Nawrocky, 1980). It was determined that pink pigmentation was dependent of an allele pair from a unique locus with the allele B (blue) dominant on allele b (pink).

Materials and equipments

See *Materials and equipments*, Experiment 2.

Reagents

Hoagland solution: see *Reagents*, Experiment 2. Ethanol:glycerin solution (1:1; v/v).

Procedure

i. **Population size and plant maintained conditions:** see *Procedure*, Experiment 2.

ii. **Trad-SHM procedure:** The Trad-SHM bioassay is accomplished with 25–30 floral branches per treatment containing young inflorescences picked up 24 h before the exposure and transferred to a controlled environment (light and temperature, see Trad-MCN procedure). The inflorescences should be maintained in aerated sterile water during 24 h for adaptation. The inflorescences should be transferred to beaker flasks wrapped in aluminum paper containing Hoagland solution (dilution of 1/3) as a negative control, solution containing 100 ppm of ethyl methanesulfonate (EMS) as a positive control, and samples containing dilutions of the test solutions. After exposure, the samples should be maintained in the nutrient solution under aeration in an illuminated environment during the recovery period. The Trad-SHM assay requires a recovery period of at least 14 days. During the recovery time the inflorescences should be maintained under continuous aeration, with daily change of nutrient solutions and remove of withered leaves. From the day 8 until day 14, the flowers recently opened should be collected and the stamens removed, placed in a microscopy slide containing a drop of ethanol:glycerin solution (1:1; v/v) and evaluated at 40× with a stereomicroscope.

Calculations

Number of mutagenic events are counted in each hair stamen. A mutagenic event consists of one or a sequence of adjacent pink cells, among the dominant blue cells, in any position of the stamen hair (Underbrink *et al.*, 1973). At least five flowers should be evaluated per day and treatment, totaling 1800 hairs/treatment/day. The number of stamen hair can be estimated each day by counting hairs from two anti-sepal and two anti-petal stamens from flowers obtained at random (Ichikawa and Ishii, 1991). At least five flowers should be evaluated per day and treated, totaling 1800 hairs/treatment/day.

Statistical analysis

The data can analyzed by analysis of variance (ANOVA) of the mutations percentage and averages comparison accomplished by Student's *t* test, Dunnett's *t* test, a 2 × 2 contingency χ^2 test or Tukey test.

3.3. Mutagenesis assays

Experiment 4: Soybean (*Glycine max*) assay

Principle

The leaf mosaicism assay in soybean (*Glycine max*) is based on genes that affect the chlorophyll synthesis. The gene Y_{11}, that exhibit incomplete dominance, is suitable for mutagenic analysis. The heterozygous plants ($Y_{11}y_{11}$) exhibit light green leaves, the homozygous plants, $Y_{11}Y_{11}$ and $y_{11}y_{11}$ genotypes, exhibit dark green and greenish yellow leaves, respectively. Soybean seeds (T219H clone) treated with genotoxic agents can produce plants that exhibit different colors (mosaicism) in their leaves.

Materials and equipments

Soybean seeds (T219H clone), plastic pots, stereomicroscope.

Procedure

i. For the performance of assays, seeds are washed in distilled water as negative control or submitted to treatment with the mutagenic agents.

ii. After seedling emergence, only seedlings with light green leaves are selected for mutagenic assays. These seedlings are kept until expansion of the first leaf and evaluated for mosaicism expression.

iii. *Mosaicism evaluation*: The evaluation is based on counting the number and the kind (dark green or greenish–yellow or both) of leaf spots. The seedlings are analyzed during the development until fourth leaf expansion.

Statistical analysis

The data analysis can be accomplished by analysis of variance (ANOVA) of the total number of sectors with mosaicism, as well as of each type (dark green, greenish–yellow or both).

Experiment 5: Waxy locus assay in *Zea mays*

Principle

At least 13 specific loci have been used for mutagenesis evaluation in maize (*Zea mays* L.), involving modifications in color or morphology observed on leaf, endosperm or pollen grains (Plewa, 1982). The alterations in pollen grains are recognized as excellent indicators of mutagenic analysis. First, the haploidy condition makes easy the expression and detection of mutations. Second, millions of pollen grains can be analyzed, increasing the statistical power of the test. The waxy locus (wx) controls amylose production and provides a valuable genetic assay for mutagenesis analysis. The waxy pollen (wx) grains have reddish color when stained with iodine in contrast with starchy Wx pollen grains that have black color. This difference is due to the presence of amylose in Wx pollen grains that react with iodine.

Materials and equipments

Maize seeds for waxy locus test, plastic pots, slides, cover glass, oil imersion, razor blade, needle, filter paper, microscope, stereomicroscope.

Reagents

70 % ethanol, iodine solution (0.18 %; w/v).

Procedure

i. The maize seeds (WxWx or wxwx) are soaked in distilled water (negative control) for 24 h. Other seeds are exposed to the treatments with the tests solutions. The seeds are washed in distilled water and sown individually in pots.

ii. Portions of the inflorescences with unopened florets (pre-anthesis) are collected and fixed in acetic acid:ethanol (1:3; v/v) during 24 h and then preserved in ethanol 70 % (v/v). Scoring consists in five florets randomly selected. They are washed in 70 % ethanol to remove any strange pollen grains, and dissecting them to extract the anthers. The anthers are then minced in a Petri dish containing iodine, under a stereomicroscope.

iii. The stained pollen grains are mounted on a slide and counted (Rodrigues *et al.*, 1998).

Calculations

Pollen grains (20,000) are analyzed per plant and 10 plants can be used for each treatment, totalling an average of 200,000 pollen grains per treatment. The frequency of mutated pollen grains (reddish color) is calculated for each plant, and the mean mutation frequency for the 10 plants compared among treatments.

Statistical analysis

The mutation frequency in treatments and control are compared statistically by analysis of variance (ANOVA) and the treatments significance is tested by Student's t test, Dunnett's t test, a 2 × 2 contingency χ^2 test or Tukey test.

Suggested Readings

Babich, H., Segall, M.A. and Fox, K.D. (1997). The *Allium* test – A simple, eukaryote genotoxicity assay. *The American Biology Teacher* **59**: 580–583.

Downs, R.J. and Hellmers, H. (1975). Environment and the experimental control of plant growth. Academic Press Inc., New York.

Emmerling-Thompson, M. and Nawrocky, M.M. (1980). Genetic basis for using Tradescantia clone 4430 as an environmental monitor of mutagens. *The Journal of Heredity* **71**: 261–265.

Falconer, M.M. and Seagull, R.W. (1987). Amiprophos-methyl (APM): A rapid reversible, anti-microtubule agent for plant cell cultures. *Protoplasma* **136**: 124–145.

Fiskesjö, G. (1988). The *Allium* test – An alternative in environmental studies: The relative toxicity of metal ions. *Mutation Research* **197**: 243–260.

Fiskesjö, G. (1985). The *Allium* test as a standard in environmental monitoring. *Hereditas* **102**: 99–112.

Gamborg, O.L. and Wetter, L.R. (1975). *Plant tissue culture methods*. National Research Council of Canada, Saskatoon.

Grant, W.F. (1994). The present status of higher plant bioassays for the detection of enviromental mutagens. *Mutation Research* **310**: 175–185.

Grant, W.F. and Owens, E.T. (1998). Chromosome aberration assays in Crepis for the study of environmental mutagens. *Mutation Research* **410**: 291–307.

Grant, W.F. and Owens, E.T. (2006). *Zea mays* assays of chemical/radiation genotoxicity for the study of enviromental mutagens. *Mutation Research* **613**: 17–64.

Hoffelder, D.R., Luo, L., Burke, N.A., Watkins, S.C., Gollin, S.M. and Saunders, W.S. (2004). Resolution of anaphase bridges in cancer cells. *Chromosoma* **112**: 389–397.

Ichikawa, S. and Ishii, C. (1991). Validity of simplified scoring methods of somatic mutations in Tradescantia stamen hairs. *Environmental and Experimental Botany* **31**: 247–252.

Jordan, M.A. and Wilson, L. (1999). The use and action of drugs in analyzing mitosis. *Methods in Cell Biology* **61**: 267–295.

Juchimiuk, J. and Maluszynska, J. (2005). Transformed roots of Crepis capillaries – A sensitive system for the evaluation of the clastogenicity of abiotic agents. *Mutation Research* **565**: 129–138.

Ma, T.H. (1979). Micronuclei induced by X-rays and chemical mutagens in meiotic pollen mother cells of Tradescantia – A promising mutagen test system. *Mutation Research* **64**: 307–313.

Ma, T.H. (1981). Tradescantia micronucleus bioassay and pollen tube chromatid aberration test for *in situ* monitoring and mutagen screening. *Environmental Health Perspectives* **37**: 85–90.

Ma, T.H. and Xu, Z. (1986). Validation of a new protocol of the *Allium* micronucleus test for clastogens. *Environmental Mutagen* **8**: 65–66.

Ma, T.H., Xu, Z., Xu, C., MacConnell, H., Rabago, E.V., Arreola, G.A. and Zhang, H. (1995). The improved *Allium/Vicia* root tip micronucleus assay for clastogenicity of environmental pollutants. *Mutation Research* **334**: 185–195.

Mericle, L.W. and Mericle, R.P. (1967). Mechanism of somatic mutation for flowers of hybrid Tradescantia (clone 02). *Genetics* **56**: 576–577.

Mericle, L.W. and Mericle, R.P. (1971). Somatic mutations in clone 02 Tradescantia: A search for genetic identity. *Journal of Heredity* **62**: 323–328.

McClintock, B. (1941). Simultaneous alterations in chromosome size and form in Zea mays. *Cold Spring Harbor Symposia on Quantitative Biology* **9**: 72–81.

Morejohn, L.C., Bureau, T.E., Molébajer, J., Baje, A. and Fosket, D.E. (1987). Oryzalin, a dinitroaniline herbicide, binds to plant tubulin and inhibits microtubule polymerization *in vitro*. *Planta* **172**: 41–147.

Nayar, G.G. and Sparrow, A.H. (1967). Radiation-induced somatic mutations and the loss of reproductive integrity in Tradescantia stamen hairs. *Radiation Botany* **7**: 257–267.

Plewa, M.J. (1982). Specific-locus mutation assays in *Zea mays*: A report of the U.S. Environmental Protection Agency Gene-Tox Program. *Mutation Research* **99**: 317–337.

Reigosa, M.J., Sánches-Moreiras, A. and Gonzáles, L. (1999). Ecophysiological Approach in Allelopathy. *Critical Reviews in Plant Sciences* **18**: 577–608.

Rodrigues, G.S., Ma, T.H., Pimentel, D. and Weinstein, L.H. (1997). Tradescantia bioassays as monitoring systems for environmental mutagenesis: A review. *Critical Reviews in Plant Sciences* **16**: 325–359.

Rodrigues, G.S., Pimentel, D. and Weinstein, L.H. (1998). *In situ* assessment of pesticide genotoxicity in an integrated pest management program: II. Maize waxy mutation assay. *Mutation Research* **412**: 245–250.

Sang, N. and Li, G. (2004). Genotoxicity of municipal landfill leachate on root tips of Vicia faba. *Mutation Research* **560**: 159–165.

Schairer, L.A. and Sautkulis, R.C. (1982). Detection of ambient levels of mutagenic atmospheric pollutants with the higher plant Tradescantia. *In*: Environmental Mutagenesis, Carcinogenesis, and Plant Biology (*Ed*., E.J. Klekowski), pp. 155–194. Praeger, New York.

Sharma, A. and Sen, S. (2002). *Chromosome botany*. Science Publishers, Enfield.

Singh, R.J. (2002). *Plant Cytogenetics*. CRC Press, Boca Raton.

Snustad, D.P.and Simmons, M.J. (2000). Principles of Genetics. John Wiley & Sons Inc., New York.

Thomas, P., Umegaki, K. and Fenech, M. (2003). Nucleoplasmic bridges are a sensitive measure of chromosome rearrangement in the cytokinesis-block micronucleus assay. *Mutagenesis* **18**: 187–194.

Underbrink, A.G., Schairer, L.A. and Sparrow, A.H. (1973). Tradescantia stamen hairs: A radiobiological test system applicable to chemical mutagenenis. *In*: Chemical Mutagens – Principles and Methods for their Detection (*Ed*., Hollaender, A.), pp. 171–207. Plenum Press, New York.

Verhoeven, H.A., Ramulu, K.S. and Dijkhuis, P.A. (1990). Comparison of the effects of various spindle toxins on metaphase arrest and formation of micronuclei in cell-suspension cultures of *Nicotiana plumbaginifolia*. *Planta* **182**: 408–411.

Vyvyan, J.R. (2002). Allelochemicals as leads for new herbicides and agrochemicals. *Tetrahedron* **58**: 1631–1646.

CHAPTER 6

Plants Bioassays: Comet Assay on Higher Plants

ANITA MUKHERJEE[1] AND TOMÁŠ GICHNER[2]

1. INTRODUCTION

Terrestrial and aquatic plants are exposed to various types of environmental xenobiotics, either deliberately (agricultural pesticides) or accidentally as compounds present in air, soil or water. Some of these compounds induce genetic damage, which can be detected through *in vitro* and *in vivo* tests known as genotoxicity assays. Such tests include: (i) assays that directly assess the key types of genetic alterations (gene mutations and chromosomal aberrations) and (ii) indirect genotoxicity tests that respond to types of DNA damage leading to these alterations. The latter category of tests may assess either DNA damage (*e.g.*, DNA adducts or DNA strand breakage) or cellular responses to DNA damage (*e.g.*, unscheduled DNA synthesis, single cell gel electrophoresis or commonly known as Comet assay). It is clear that no single test is capable of detecting all relevant genotoxic agents. Therefore, a battery of *in vitro* and *in vivo* assays are done for genotoxicity. Such plant bioassays can be an integral part of genotoxic assays particularly in areas of the world, where financial support for environmental evaluation is limited.

For most plant species, detection of these xenobiotics is not currently available. This limitation prevents the detection of genotoxicity of environmental xenobiotics in plants growing on polluted soil. To overcome this, a plant-based molecular assay – the Comet or Single Cell Gel Electrophoresis (SCGE) assay, can be applied to detect induced DNA damage (Collins, 2002; Tice *et al.*, 2000). To make the assay more sensitive and reliable, unique procedures and specialized measures have been developed for DNA migration (Raisuddin and Jha, 2004). Although this technique was primarily applied to animal cells, the use of plant tissues in the comet assay (Gichner and Plewa, 1998; Koppen and Verschaeve, 1996) has lead to basic and applied studies in environmental mutagenesis. In theory, the comet assay can be used to in all plant species. Currently, the assay is used for genotoxicity testing by regulatory authorities [Organization for Economic Cooperation and Development (OECD), United States Environmental Protection Agency (US EPA), and other Federal Government agencies (Hartman *et al.*, 2003; Kumaravel and Jha, 2006; Tice *et al.*, 2000)].

1. Department of Botany, Center of Advance Study in Cell and Chromosome Research Center, University of Calcutta, 35 Ballygunge Circular Road, Kolkata 700 019, India, E-mail: anitamukherjee28@gmail.com
2. Institute of Experimental Botany, Academy of Sciences of Czech Republic, Na Karlovce 1a, 160 00, Prague 6, Czech Republic, E-mail: gichner@ueb.cas.cz

In this chapter, we describe the methods to evaluate the induced DNA damage by the comet assay in tobacco (*Nicotiana tabacum*) and onion (*Allium cepa*).

2. PRINCIPLES OF THE COMET ASSAY

The comet assay is based on the ability of negatively charged loops/fragments of DNA to be drawn through an agarose gel in response to an electric field. The extent of DNA migration depends directly on the DNA damage present in the nuclei. In this assay, a suspension of cells or nuclei is mixed with low melting point agarose and spread on a microscope glass slide. The nuclear DNA is organized in nucleosomes as loops with supercoils and the nucleus is covered with a nuclear membrane. After lysing and unwinding with a detergent at high salt concentration, the nuclear membrane is digested and most of the proteins are denatured. The supercoils are relaxed and loops are spilled out and a "halo" is formed around the nuclei. This occurs both in nuclei of treated and untreated cells. When subjected to an electric field, the negatively charged DNA migrates out of the nuclei, in the direction of the anode. When such nuclei are stained with a specific DNA – binding dye such as ethidium dye, under a fluorescent microscope, an image like a comet appears (Fig 1). The size and shape of the comet and the distribution of DNA within the comet have been correlated with the extent of DNA damage (Fairbairn *et al.*, 1995).

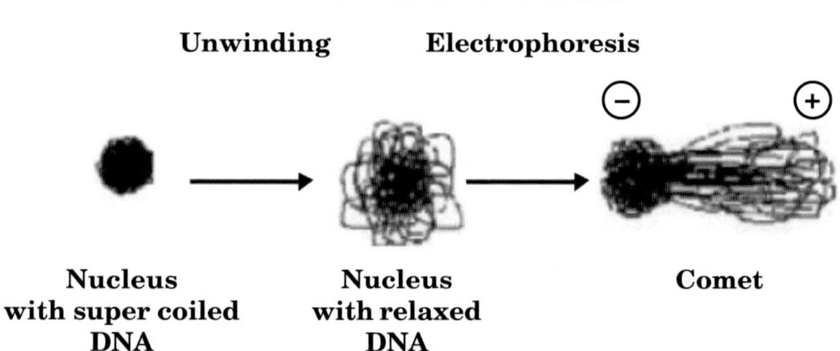

Fig 1. Schematic representation of the DNA migration pattern produced in the Comet assay

DNA unwinding and electrophoresis can be done at neutral pH (7–8), to facilitate the detection of double-strand breaks and cross-links. Unwinding and electrophoresis at pH 12.1–12.4 facilitate the detection of single and double-strand breaks, incomplete excision-repair sites and cross-links, while, unwinding and electrophoresis at a pH higher than 12.6 express alkali-labile sites in addition to all types of lesions listed above (Miyamae *et al.*, 1997). For monitoring purposes, the alkaline version of the comet protocol with electrophoresis at pH > 13 is performed.

Experiment 1: Comet assay using tobacco (*Nicotiana tabacum*) as test system

Materials and equipments

Tobacco seeds, test solutions (unknown compound dissolved in a proper solvent, a positive consisting in a recognized genotoxic compound like ethyl methanesulphonate (EMS) and negative controls consisting in the solvent used to dissolve the test compound), analytical balance, beakers, measuring cylinder, conical flasks, pipettes, Magenta boxes, filter paper, Petri dishes, pots, refrigerator, plant growth chambers, microwave oven/hot plate, pH meter, horizontal gel electrophoresis box with power supply unit, fluorescence microscope equipped with an excitation filter of 515–560 nm and a barrier filter of 590 nm, ocular micrometer, computerized image analysis system (Kinetic Imaging Ltd., Liverpool, UK), linked to a CCD camera (optional).

Reagents

Tobacco seed sterilizing solution: Add 0.5 mL of 5 % sodium hypochlorite and 5 µL 10 % Triton-X to 4.5 mL distilled water (dH_2O).

Tobacco cultivation medium: Dissolve 4.6g of Murashige and Skoog salts (Sigma Chemical Co., USA), 15g sucrose, 10 mL vitamin stock solution, 500 mg 2-(*N*-morpholino)ethanesulphonic acid and 2.2g Phytagel in 1000 mL of dH_2O; then set to pH 5.6. Autoclave medium 20 min at 120 °C before use.

Vitamin stock solution: 5g myo-inositol and 500 mg thiamine hydrochloride in 500 mL dH_2O.

0.4 M Tris buffer: Dissolve 9.7g Tris in 180 mL dH_2O, set pH to 7.5 with concentrated hydrochloric acid, add dH_2O to make up the volume to 200 mL, store at room temperature.

1 % Normal melting point agarose (NMPA): Dissolve 1g NMPA in 100 mL dH_2O, microwave or heat until near boiling.

1 % low melting point agarose (LPMA): Dissolve 250 mg (LMPA) in 25 mL PBS, microwave or heat until near boiling. Prepare 2–5 mL aliquots in small vials and store in the refrigerator.

0.5 % low melting point agarose (LMPA): Dissolve 125 mg (LMPA) in 25 mL PBS, microwave or heat until near boiling. Prepare 2–5 mL aliquots in small vials and store in the refrigerator;

Phosphate buffered saline (PBS): Add 990 mL dH_2O to 1 L packet of PBS adjust pH to pH 7.4, make up the volume to 1000 mL, store in a refrigerator.

200 mM disodium EDTA: Dissolve 14.89g in 175 mL dH_2O, slightly warm, stir, set to pH 10 with 10 N NaOH and add dH_2O to 200 mL. Store at room temperature.

10 N NaOH: Dissolve carefully 200g NaOH in 500 mL dH_2O. Store at room temperature.

Electrophoresis buffer: Add 60 mL 10 N NaOH and 10 mL of 200 mM EDTA to 1930 mL dH_2O.

Ethidium bromide: Dissolve 10 mg ethidium bromide in 50 mL dH_2O. For staining mix 1 mL stock solution with 9 mL distilled H_2O. Store in dark bottle.

Procedure

Preparation of slides: For slide preparation, microscope slides with one-fourth frosted ends are dipped into solution of 1 % normal melting point agarose (NMPA) prepared with water at 50 °C. The backside of the slides are wiped to remove the agarose and placed horizontally on a level surface for drying at room temperature. These slides serve as the base coated ones. After the slides are prepared, they are kept dry in slide boxes until further use.

Plant cultivation:

i. **Cultivation of tobacco seedlings *in vitro*:** Seeds are sterilized in 70 % ethanol for 2 min followed by a 20 min immersion in a sterilizing solution. The sterilizing solution is aspirated and the seeds are washed 5× in sterile distilled water. Each seed is placed in a sterile container filled with 50 mL of sterile solid growth medium. The seedlings are grown in sterile vented containers, in a plant growth chamber at 26 °C with 16 h photoperiod until the 5–6 leaf stage (Fig 2a).

ii. **For outdoor culture:** Seeds are soaked in tap water for 2 h and allowed to germinate in Petri dishes on moist filter paper at room temperature (18–25 °C) in dark. After a few days, when the first cotyledon leaves emerge, the seedlings are then transferred to pots containing mixture of sand and soil in equal portions and allowed to grow under sunlight (Fig 2b).

iii. The tobacco seedlings are removed from the cultivation containers or from pots with a mixture of soil and sand and the roots are washed in water. The roots of the seedlings are immersed in plastic/glass vials containing the test solutions assayed for DNA damaging effects. The tested agent may be dissolved in water, buffer or in an organic solvent. A minimum of three concentrations of the unknown compound and a single concentration of 8 mM EMS should be assayed together with the corresponding negative controls. Seedlings are treated in dark at 26 °C for 18–72 h. After treatment, the treated roots are thoroughly rinsed in water.

iv. Plants have cell wall that cannot be removed by lysing with high salt concentrations and detergents, unlike the cell membrane in animals. The nuclei for the comet assay have to be thus isolated mechanically. For the isolation of nuclei from leaves and roots of tobacco, a small part of the leaf (about 2 × 1 cm) or roots from a single seedling are placed in a 60 mm plastic Petri dishes. About 200–300 µL of cold 0.4 M Tris buffer is spread onto each leaf or root. Using a fresh razor blade, the leaf or root is sliced into fine pieces (Fig 3). The Petri dish is kept tilted in the ice so that the released nuclei would be collected in the buffer.

v. On each slide with the dried NMPA layer, 50 µL of the nuclear suspension and 50 µL of 1% molten low melting point agarose (LMPA) are added at 40 °C. Repeated pipetting using a cut tip gently mixes the nuclei and the LMPA and then a coverslip (24 × 50 mm) is placed on the mixture. The slide is placed on ice for a minimum of 5 min after which the coverslip is removed and a final layer of 100 µL of 0.5 % LMPA is added and covered with a coverslip (left side of Fig 4).

Fig 2. (A) Tobacco grown *in vitro* and (B) tobacco grown in pots with a mixture of soil and sand

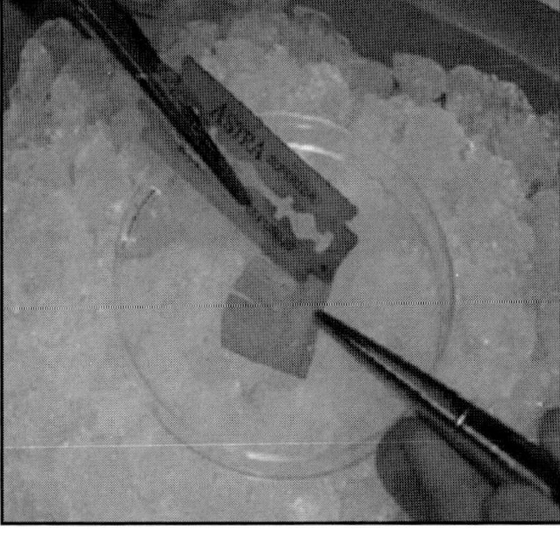

Fig 3. Tobacco leaves are gently sliced with a razor blade and the isolated nuclei are collected in the isolation buffer contained into the Petri dish

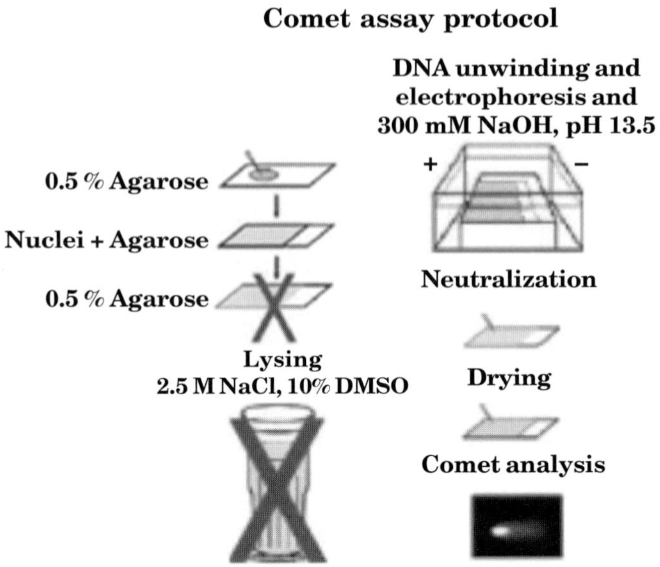

Comet assay protocol

0.5 % Agarose
Nuclei + Agarose
0.5 % Agarose
Lysing
2.5 M NaCl, 10% DMSO
DNA unwinding and electrophoresis and 300 mM NaOH, pH 13.5
Neutralization
Drying
Comet analysis

Fig 4. The scheme of the plant Comet assay protocol

DNA unwinding, electrophoresis and staining:

i. The coverslips are removed and the slides are immersed in an electrophoresis tank with cold (4 °C) electrophoresis buffer. The slides are placed end to end in contact with another for 15 min and incubated to allow DNA unwinding. After unwinding, electrophoresis is conducted using the same buffer at 0.74 V/cm (26 V, 300 mA) for 25 min at 4 °C.

ii. Following electrophoresis, the slides are neutralized 3 times with 0.4 M Tris buffer and stained or left overnight to dry, and stored in slide boxes (up to several months). The slides are stained with 100 µL of ethidium bromide (20 µg/mL) for 5 min dipped in ice-cold water to remove the excess stain and covered with a coverslip (right side of Fig 4).

Evaluation

For each sample, 2 × 25 nuclei per slide are analyzed using a fluorescence microscope with an excitation filter of BP 546/10 nm and a barrier filter of 590 nm (in the case of staining DNA with ethidium bromide). Each treatment is repeated at least twice.

Observations

1. Plant materials, methods, chemicals and equipment described in the experiment are used in Laboratory of Mutation Genetics, Institute of Experimental Botany, Prague, Czech Republic; and in Laboratory of Genetic Toxicology, Department of Botany, Center of Advance Study in

Cell & Chromosome Research, University of Calcutta, India. Details can be found in publications (Gichner and Plewa, 1998; Gichner *et al.*, 1999, 2000, 2001, 2006). For more information please contact gichner@ueb.cas.cz or anitamukherjee28@gmail.com.

2. It is important to remark that plant cell has a wall that cannot be removed by lysis in high concentrations of detergents and salts, and so the nuclei have to be isolated mechanically. It is exceedingly important to gently isolate the nuclei from the leaves or roots, otherwise high levels of DNA damage is also induced in control nuclei. All operations are conducted under dim or yellow light.

3. Depending on the purpose of the study, the used plant material and the methods can be modified. Other plant species can also be used, *e.g., Vicia faba* (Koppen and Verschaeve, 1996), *Arabidopsis thaliana* (Menke *et al.*, 2001), barley (Jovtchev *et al.*, 2001), *Calamagrostis epigejos* (Ptáèek *et al.*, 2002). Leaves of various weeds (Gichner and Mühlfeldová, 2002) and agronomic crops (Gichner *et al.*, 2003, 2005,) or tobacco-cell cultures (Stavreva *et al.*, 1998) can also be used in the comet assay. For each plant species, the comet assay protocol has to be modified, especially duration of unwinding and the duration, voltage and amperage of electrophoresis. The main task is to generate increased DNA migration values in treated cells without affecting the migration in the controls. Other types of power supplies, gel electrophoresis boxes, fluorescence microscopes and special software are commercially available to analyze the induced DNA damage.

Calculations

i. **Computer image analysis:** Several companies supply software that, linked to a closed–circuit digital camera mounted on the microscope, automatically analyses individual comet images. The programmes differentiate comet head from tail, and can measure a variety of parameters including tail length; Percentage of total fluorescence in head and tail; and 'tail moment', calculated in different ways but essentially representing the product of tail length and relative tail intensity. The percentage of DNA in tail is linearly related to DNA break frequency up to 80 % in tail and this defines then usual range of the assay. Tail length tends to increase rapidly with dose at low levels of damage, but soon reaches its maximum. Tail moment (TM) can be calculated as follows:

$$TM\ (\mu m) = \frac{[TL\ (\mu m) \times TI\ (\%)]}{100};$$

where TL is tail length and TI is tail intensity (Figs 5 and 6).

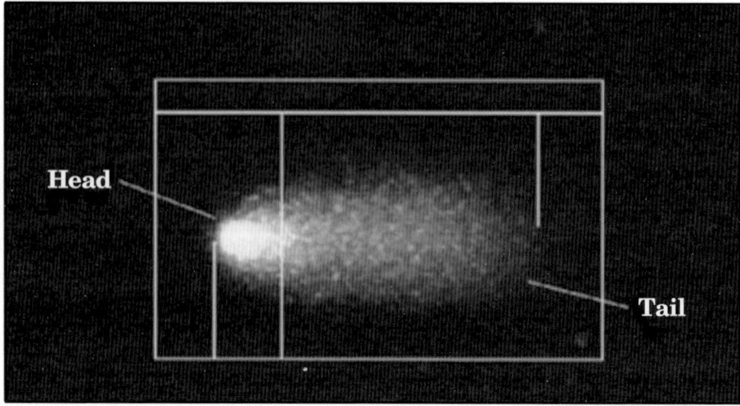

Head DNA (%)	Tail DNA (%)	Tail Length (µm)	Tail Moment (µm)
34	**66**	**140**	**92**

Fig 5. The computerized image analysis system measures the amount of DNA in the head (34 %), in the tail (66 %), and the length of the tail (140 µm). For expressing the DNA damage, the parameters percentage tail DNA or tail moment (TM) can be used. The TM value represents the percentage of tail DNA multiplied by the Tail Length divided by 100.

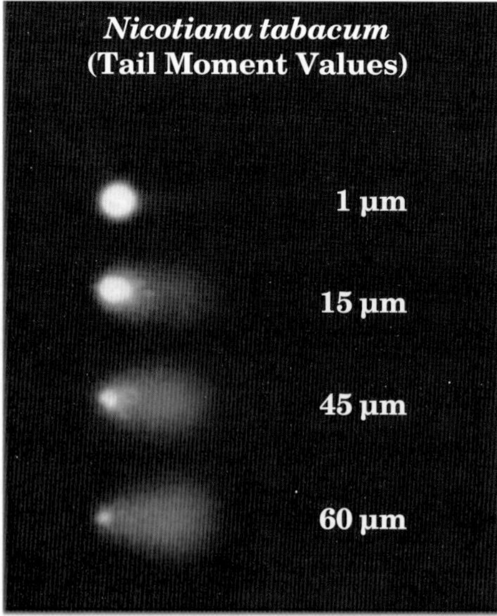

Fig 6. Illustration showing nuclei of tobacco with different levels of DNA damage expressed as Tail Moment values in µm

ii. **Visual analysis:** It is possible to analyze the comets quantitatively without image analysis software. The human eye can readily discriminate comets representing different levels of damage. Dušinská (2000) has developed a scheme for visual scoring based on five recognizable classes of comet, from class 0 (undamaged; no recognizable tail) to class 4 (almost all DNA in tail; insignificant head). Fifty comets are selected at random from each slide. Each comet is given a value according to the class it is assigned into, so that an overall score can be derived for each slide, ranging from 0 to 400 arbitrary units. With practice visual scoring can be quick (Fig 7).

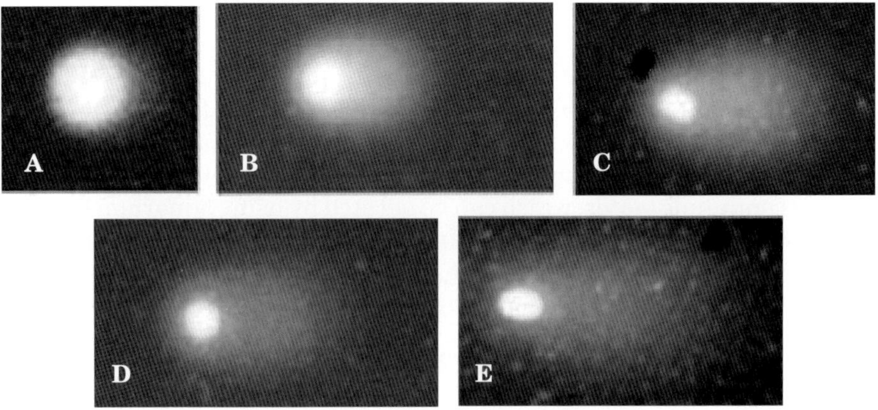

Fig 7. Images illustrating nuclei of onion roots with different levels of DNA damage for visual scoring (A) No tail, (B) Class I, (C) Class II, (D) Class III, and (E) Class IV.

Statistical analysis

Two or three slides are evaluated per treatment and each treatment is repeated at least twice. From the repeated experiments, the averaged median tail moment value or the averaged median percentage of tail DNA, and tail moment are calculated from each slide values. Data are analyzed using the statistical and graphical functions of Sigma Plot 8.0 and SigmaStat 3.0 (SPSS Inc., Chicago, IL, USA). With the slide as the unit of measure, the median tail moment values or the percentage of tail DNA, are used in a one-way analysis of variance test. If a significant F-value ($P = 0.05$) is obtained, a Dunnett's multiple Comparison versus the control group analysis is conducted. Differences between two groups are statistically evaluated by the Paired t-test.

One-way ANOVA and Dunnett's multiple comparison tests can be done without the help of statistical programme. One can consult Sokal and Rohlf (1981) for details.

Precautions

The unwinding buffer is highly alkaline, therefore, avoid exposure with bare hands. Ethidium bromide is a carcinogenic and mutagenic agent.

Experiment 2: Comet assay using onion (*Allium cepa* L.) as test system

Materials, equipments and reagents

Onion bulbs; aquarium bubblers and those indicated for Experiment 1.

Procedure

Preparation of slides: See *Procedure*, Experiment 1.

Plant cultivation:

i. *Sprouting of onion roots*: The outer scales of onion bulbs are removed and the root primordia at the base of each bulb are scrapped off to remove the dead, old roots. The onion bulbs are placed on the mouth of glass vials containing water, which can be aerated by using aquarium bubblers. The water is changed everyday and the entire growth set is maintained in dark. When the roots grow to suitable size, they are transferred to the test solution used in the experiment.

ii. The onion bulbs are removed from the cultivation containers and the roots are washed in water. The roots are immersed in plastic/glass vials containing the agent tested for DNA damaging effects. The tested agent may be dissolved in water, buffer or in a solvent. Corresponding controls have to be also applied. Bulbs are treated in dark at 26 °C for 18–72 h. After treatment, the treated roots are thoroughly rinsed in water.

iii. For isolation of nuclei from onion roots, the bulb roots are held in tufts of 4–6 in a 60-mm plastic Petri dish. The Petri dish is placed on ice pack. 200–300 µL of cold 0.4 M Tris buffer is added in each Petri dish and roots are chopped into fine pieces using a fresh razor blade. The Petri dish is kept tilted in the ice so that the nuclei would collect in the buffer.

iv. Prepare slides with isolated nuclei as indicated in step (v), *Procedures*, Experiment 1.

DNA unwinding, electrophoresis and staining

i. DNA unwinding and electrophoresis procedure is the same as for the Experiment 1, with the exception that the unwinding time is 20 min.

ii. Following electrophoresis, the slides are neutralized 3× with 0.4 M Tris buffer, rinsed in water and left overnight to dry. The dried slides can be stored in slide boxes for several months before staining. The slides are stained with 100 µL ethidium bromide (20 µg mL^{-1}) for 5 min dipped in ice-cold water to remove the excess of stain and covered with a coverslip.

Observations

1. See *Observations* of Experiment 1.

2. The advantage of onion roots for the Comet assay is that in the same test system we can analyze induced DNA damage (as measured by the comet assay), and the chromosomal aberration and micronuclei (MN) frequency in the same test system. The chromosomal aberration and MN frequency studies are not a topic of the present protocol.

Evaluations, calculations, statistical analysis:

See Experiment 1.

Suggested Readings

Collins, A.R. (2002). The comet assay: Principles, applications, and limitations. *Methods in Molecular Biology* **203**: 163–177.

Dušinská, M. (2000). The comet assay modified for detection of oxidized bases with the use of bacterial repair endonucleases. Comet assay interest group website, http://cometassay.com/.

Fairbairn, D.W., Olive, P.L. and O'Neill, K.L. (1995). The Comet assay: A comprehensive review. *Mutation Research* **339**: 37–59.

Gichner, T. and Plewa, M.J. (1998). Induction of somatic DNA damage as measured by single cell gel electrophoresis and point mutation in leaves of tobacco plants. *Mutation Research* **401**: 143–152.

Gichner, T. and Mühlfeldová, Z. (2002). Induced DNA damage measured by the Comet assay in 10 weed species. *Biologia Plantarum* **43**: 509–516.

Gichner, T., Stavreva, D.A. and Van Breusegem, F. (2001). *o*-Phenylenediamine-induced DNA damage and mutagenicity in tobacco seedlings is light dependent. *Mutation Research* **495**: 117–125.

Gichner, T., Patková, Z. and Kim, J.K. (2003). DNA damage measured by the comet assay in 8 agronomic plants. *Biologia Plantarum* **47**: 185–188.

Gichner, T., Mukherjee, A. and Velemínský, J. (2006). DNA staining with the fluorochrome EtBr, DAPI and YOYO-1 in the Comet assay with tobacco plants after treatment with ethyl methanesulphonate, hyperthermia and DNase-I. *Mutation Research* **605**: 17–21.

Gichner, T., Ptáèek, O., Stavreva, D.A. and Plewa, M.J. (1999). Comparison of DNA damage in plants as measured by single cell gel electrophoresis and somatic leaf mutations induced by monofunctional alkylating agents. *Environmental and Molecular Mutagenesis* **33**: 279–286.

Gichner, T., Menke, M., Stavreva, D.A. and Schubert, I. (2000). Maleic hydrazide induces genotoxic effects but no DNA damage detectable by the Comet assay in tobacco and field beans. *Mutagenesis* **15**: 385–389.

Gichner, T., Patková, Z., Száková, J. and Demnerová, K. (2006). Toxicity and DNA damage in tobacco and potato plants growing on soil polluted with heavy metals. *Ecotoxicology and Environmental Safety* **65**: 420–426.

Hartmann, A., Agurell, E., Beevers, C., Brendler-Schwaab, S., Burlinson, B., Clay, P., Collins, A., Smith, A., Speit, G., Thybaud, V. and Tice, R.R. (2003). *In*: Proceedings of the Fourth International Comet Assay Workshop.

Recommendations for conducting the *in vivo* alkaline Comet assay. *Mutagenesis* **37**: 45–51.

Jovtchev, G., Menke, M. and Schubert, I. (2001). The Comet assay detects adaptation to MNU-induced DNA damage in barley. *Mutation Research* **493**: 95–100.

Koppen, G. and Verschaeve, L. (1996). The alkaline comet test on plant cells: A new geno-toxicity test for DNA strand breaks in *Vicia faba* root cells. *Mutation Research* **360**: 193–200.

Kumaravel, T.S. and Jha, A.N. (2006). Reliable Comet assay measurements for detecting DNA damage induced by ionising radiation and chemicals. *Mutation Research* **605**: 7–16.

Menke, M., Chen, I-Peng, Angelis, K.J. and Schubert, I. (2001). DNA damage and repair in *Arabidopsis thaliana* as measured by the comet assay after treatment with different classes of genotoxins. *Mutation Research* **493**: 87–93.

Miyamae, Y., Iwasaki, K., Kinae, N., Tsuda, S., Murakami, M., Tanaka, M. and Sasaki, Y.F. (1997). Detection of DNA lesions induced by chemical mutagens using the single-cell gel electrophoresis (Comet) assay. 2. Relationship between DNA migration and alkaline condition. *Mutation Research* **393**: 107–113.

Ptáèek, O., Mühlfeldová, Z., Dostálek, J. and Gichner, T. (2002). Monitoring DNA damage in wood small-reed (*Calamagrostis epigejos*) plants growing in a sediment reservoir with substrates from uranium mining. *Journal of Environmental Monitoring* **4**: 592–595.

Raisuddin, S. and Jha, A.N. (2004). Relative sensitivity of fish and mammalian cells to sodium arsenate and arsenite as determined by alkaline single cell gel electrophoresis and cytokinesis-block micronucleus assay. *Environmental and Molecular Mutagenesis* **44**: 83–89.

Sokal, R.R. and Rohlf, F.J. (1981). *Biometry.* 3rd ed. Freeman, San Francisco, CA.

Stavreva, D.A., Ptáèek, O., Plewa, M.J. and Gichner, T. (1998). Single cell gel electrophoresis analysis of genomic damage induced by ethyl methane sulfonate in cultured tobacco cells. *Mutation Research* **422**: 323–330.

Tice, R.R., Agurell, E., Anderson, D., Burlinson, B., Hartmann, A., Kobayashi, H., Miyamae, Y., Rojas, E., Ryu, J.C. and Sasaki, Y.F. (2000). Single cell gel/comet assay: Guidelines for *in vitro* and *in vivo* genetic toxicology testing. *Environmental and Molecular Mutagenesis* **35**: 206–221.

Plant – Microorganisms Interactions

CHAPTER 7

Mycorrhiza – Plant Bioassays

Rukhsana Bajwa[1], Ghazala Nasim[1] and Arshaid Javaid[1]

1. INTRODUCTION

The phenomenon of allelopathy where one plant exerts a detrimental influence on another through the production of allelochemicals has been widely reported in plants (Renne *et al.*, 2004; Sampietro *et al.*, 2007; Amoo *et al.*, 2008). Allelochemicals are secondary metabolites produced via the acetate and shikimic acid pathways that encompass a wide range of chemical types including phenolic acids, coumarins, flavonoids, terpenoids, alkaloids and sulfides (Chou, 1995; Narwal, 1994). They affect all physiological plant processes such as photosynthesis, chlorophyll production, respiration, hormonal balance, protein synthesis and plant water relations (Yamane *et al.*, 1992). These compounds enter into the soil by rain leaching from foliage, root exudation and decomposition of plant residues or as microbial by-products of residue decomposition, and selectively inhibit the growth of other plants, soil microorganisms or both (Einhellig, 1996; Popa *et al.*, 2008).

Arbuscular mycorrhiza (AM) are probably the most abundant fungi in agricultural soils. They account for 5–50 % of the biomass of soil microbes (Olsson *et al.*, 1999). Colonization of roots by mycorrhizal fungi has been shown to improve growth and productivity of several field crops (Javaid *et al.*, 1993; Cavagnaro *et al.*, 2006) by increasing nutrient element uptake (Al-Karaki, 2002). They also impart other benefits to plants including production of certain secondary metabolites (Schliemann *et al.*, 2008), increased rate of photosynthesis (Wu and Xia, 2006), enhancement of nitrogen fixation by symbiotic or associative N_2-fixing bacteria (Javaid *et al.*, 1994), osmotic adjustment under drought stress (Ruiz-Lozano, 2003), increased resistance to pests (Khaosaad *et al.*, 2007), tolerance to various abiotic stress factors (Takeda *et al.*, 2007), and improving soil aggregation (Wu *et al.*, 2008) and thus improving soil physical properties and stability. Studies have shown that AM fungi alleviate allelopathic stress and improve crop growth and yield (Bajwa *et al.*, 1999, 2003; Javaid, 2007, 2008).

1. Institute of Mycology and Plant Pathology, University of the Punjab, Quaid-e-Azam Campus, Lahore–54590, Pakistan, E-mail: rukhsanabajwa_mppl@yahoo.com

Ectomycorrhizae (ECM) are another type of mycorrhizal association commonly found in nature. Most ECM are established with trees of north temperate countries as for example pine, spruce, larch, fir, poplar, willow, lime and oak. In the southern hemisphere, many important genera of trees including *Eucalyptus* and *Nothofagus* have ectomycorrhizae, as do certain tropical genera (Jackson and Simson, 1984). On a global scale, several thousand fungi are thought to form ECM with tree hosts in natural and managed forest habitats (Cairney, 1999). Different ECM fungal species appear to play differential activities in the soil. Some fungi can produce enzymes involved in the uptake of organic nutrients from the soil (Smith and Read, 1997), while some seem to be important in weathering of primary minerals in the soil (Landeweert *et al.*, 2001). Wallander (2000) showed that some ECM fungi stimulated P uptake from apatite by exuding organic acids into the soil solution. Ectomycorrhizal fungi are able to alleviate toxic effects of allelochemicals in host plants (Zeng and Mallik, 2006). Some ECM fungal species and isolates are tolerant to phenolics and other allelochemicals. Inoculation of such well adapted tolerant ECM species generally alleviate the allelopathic stress and enhance the host plant growth (Mallik and Zhu, 1995). Potential of ECM fungi to alleviate allelopathic stress can be best exploited by selecting mycorrhizal species well adapted to the habitat where allelopathy occurs (Cote and Thibault, 1988; Zeng and Mallik, 2006).

Experiment 1: Extraction and identification of AM spores from rhizospheric soil of allelopathic plants

AM fungi produce zygospores, azygospores, chlamydospores or sporangia, either ectocarpically within the soil or in roots, or in hypogeous or epigeous sporocarps. The taxonomy of the AM fungi has traditionally been based on the morphology of these spores. Families and genera were mainly distinguished by the hyphal attachment and mode of formation of the spore, whereas the structure of the spore walls played an important role in species identification (Trappe, 1982). About 150–200 obligatory biotrophic fungal species capable of forming mycorrhiza have been described so far (Karandashov and Bucher, 2005). These spores can be isolated from the soil by various methods. A modification of standard procedure for isolation of AM fungal spores from soil samples is presented here (Tommerup, 1992). This method has been used successfully in a wide range of soil types.

Principle

Spores of AM fungi being lighter in weight, float while soil particles being heavier settle down when soil is thorough mixed in water. Since spores of different species are generally different in diameters, they are separated according to their sizes when supernatant are passed through a series of sieves having different pore sizes.

Materials and equipments required

Stacking sieves with a range of pore size between 40 and 1000 mm, wash bottle, centrifuge (2000 rpm), vacuum filter apparatus, filter papers, Petri dishes, powdered kaolin clay, dissecting microscope, slides.

Reagents

50 % sucrose solution (w/v), polyvinyl alcohol-lacto-glycerol (PVLG) mountant (Koske and Tessier, 1983): 18.33g of polyvinyl alcohol, 50 mL of distilled water, 50 mL of lactic acid and 5 mL of glycerine.

Procedure

i. Collect soil samples from rhizosphere of an allelopathic plant species.

ii. Remove course debris and roots from samples with a 2-mm sieve.

iii. Dry the soil in an oven at 90 °C and weigh 1 kg of the sample.

iv. Mix the soil into a substantial volume of water. After allowing the heavy soil particles to settle for a few second, decant the water through a series of sieves. Roots and course debris are collected on a coarse (750 mm) screen, while spores are captured on finer screens. A small amount (±100 mg) of finally powdered kaolin clay can be added to water with the sievings to help to form a stable pellet during centrifugation.

v. Add sievings to water and centrifuge for 5 min at 2000 rpm.

vi. Suspend pellet in 50 % sucrose. Then, centrifuge for 1 min at 2000 rpm.

vii. Immediately after centrifugation, wash the spores from sucrose supernatant on a 50 mm sieve to remove sucrose.

viii. After rinsing the spores, wash them onto a pre-wetted filter paper in a Buchner funnel before vacuum filtration (see *Observation* 1).

ix. Transfer the spores to a microscopic slide with the help of a paintbrush or a wooden dowel with very fine tip. Semi-permanent microscopic slides can be prepared using PVLG with or without Melzer's reagent.

x. Observe the prepared slide under different powers of compound microscope.

xi. Identify the spores of different mycorrhizal species by referring to the Synoptic Keys to the genera and species of Zygomycetous mycorrhizal fungi by Trappe (1982).

Calculations

Spore density data can be obtained by counting spores in a given soil sample.

Observations

1. Glass fiber filters are superior to paper ones. These filters can be stamped with parallel lines 5–7 mm apart to separate microscope fields for spore counting.

Experiment 2: Extraction, staining and measurement of AM fungal hyphae from rhizosphere of allelopathic plants

The AM fungal hyphae also extend from the roots out into the soil where they interface with soil particles. These extraradical hyphae function as absorptive structures for mineral elements and water. Since they can extend out several centimeters away from the roots, they can effectively bridge over the zone of nutrient depletion around roots and absorb immobile elements from the bulk soil (Bethlenfalvay and Linderman, 1992). Several methods have been used to measure hyphae in soil samples. The efficacy of the different soil hyphae extraction procedures that have been used may vary and is likely to be influenced by factors as soil texture and organic matter content. The method given by Miller and Jastrow (1992) is presented here with some modifications (Brundrett *et al.*, 1994).

Principle

Soil is incubated in a dilute solution of sodium hexametaposphate that helps to break up soil aggregates, thereby releasing hyphae into a soil suspension. Aliquots of this suspension are passed over mesh filters that trap hyphae while allowing smaller colloidal particles to pass through. The hyphae trapped on these mesh filters can be easily stained by placing the filter in the desired stained solution.

Materials and equipments required

Sieve with 2-mm pore size, drying oven, weighing boats, glass stirring rods, 600 mL beakers (2 for each soil sample), magnetic stirrer plate, 10 mL pipette, sidearm vacuum filter flask, filters with 20 mm pore size, white cellulose nitrate filters (1.2 mm pore size) with grid markings, wash bottle, fine tipped forceps, glass slides, cover slips, microscope.

Reagents

Sodium hexametaphosphate (35.7g L^{-1}), trypan blue stain in 1:2:2 lactic acid:glycerol:deionized water (0.6g trypan blue per litre), Cargille Type A non-drying immersion oil.

Procedure

i. Collect soil samples from rhizosphere of an allelopathic plant. Pass the soil through 2 mm sieve. Discard root samples greater than 1 mm in diameter. Cut smaller roots (less than 1 mm in diameter) into 2–3 mm segments and mix with the sieved soil.

ii. Weigh 5g soil from each sample (also measure dry mass equivalent of soil for calculations).

iii. Place the sample into 600 mL beaker. Add 250 mL deionized water and 31 mL sodium hexametaphosphate to the beaker. Allow to stand for 30 min and then stir the sample vigorously with glass rod. Rinse the glass rod with deionized water.

iv. Place beaker on magnetic stirrer and stir at high speed to allow sand and other heavy particles to settle to bottom. Remove aliquots for dilution.

v. Remove Two aliquotes of 6 mL from the original soil solution and transfer both into a new 600 mL beaker. Add 125 mL deionized water and 15.5 mL sodium hexametaphosphate solution.

vi. Place the beaker on magnetic stirrer and stir at high speed to resuspend the hyphae. While stirring, take two aliquots of 10 mL from the beaker and pour into a filter funnel. Connect sidearm of filter flask to vacuum tubing and turn vacuum on. After pulling down of all the solution, spray 8 times into the filter funnel.

vii. With the vacuum still on, remove the funnel part of the filter holder using a rolling action to leave the mesh filter on the stopper.

viii. Turn vacuum off, and grasp the filter paper near the edge with the help of a fine tipped forceps. Carefully curl it off the stopper and slide it into a plastic vial. Rinse any hyphae adhering to the bottom of the filter funnel into the vial. Add 5 mL trypan blue stain to the vial, then cap, and vortex for 30s making sure that the filters spin with the stain solution. Allow filters to stain for 1.5 h.

ix. Rinse the funnel part of filter holders with deionized water, and reset with white cellulose nitrate filters to collect stained hyphae. Place the cellulose filters on the stoppers with the grid side up.

x. After passing the staining period, vortex each vial for few seconds and decant into filter holders with cellulose filters. Turn on vacuum and after pulling down all the solution, each 8 times, into each vial with mist spray bottle. Cap vial, vortex, and pour into funnel to rinse. Repeat rinsing till no blue color from the staining solution remains in the vial.

xi. Without turning off the vacuum, remove the funnel part of filter holder. Turn the vacuum off and grasp each along the edge with a fine tipped forceps, carefully remove from stopper, and place on a labeled glass slide. Allow it to dry for 30 min.

xii. Clear the filters and mount them on slides using low-viscosity, non-florescent immersion oil.

xiii. The length of hyphae present on a membrane filter can be measured by the line intersect method (Tennant, 1975):

a. Insert a graticule (with 10 × 10 grid of squares) into the eyepiece of a compound microscope.

b. Examine 50 fields of view on the membrane filter at 200× magnification (*i.e.,* a 20× objective with 10× eyepiece) as recommended by Abbott *et al.* (1984).

c. Move the field of view to reside over one corner of one square in the grid of large squares on the membrane filter.

d. Position the eyepiece graticule to rest in the chosen corner.

e. Count the intersections.

f. Repeat for 2 corners per membrane filter square over the entire membrane filter (which will give about 50 fields of view per microscopic slide).

g. Count every intersection between a gridline and hypha, casting the eye along all horizontal and vertical lines in the 10 × 10 grid, except uppermost horizontal line and the vertical line at the extreme right. Record on paper the number of counts "*c*".

h. At some corners of the membrane filter, large squares will fall near the edge of the circular perimeter of the catchment area from the filtration. Record on paper these small squares as "*s*".

Calculations

Calculations are presented as suggested by Brundrett *et al.* (1994) as sum of the number of counts, Σc, and number of small squares Σs, for all fields.

Suppose that microscope has a magnification of 200×, the size of a square (a) is 0.05 mm and size of one side of 10 × 10 grid (g) is 0.5 mm:

$$L\ (\text{mm slide}^{-1}) = [\Sigma c \times (11/14) \times a] \times [\pi r^2] / [\Sigma s/100) \times g^2];$$

where $\Sigma c \times (11/14) \times a$ is the length of hyphae actually seen per slide; $(Ss/100) \times g^2$ is the total area actually examined per slide; πr^2 is the area of the circular filtration catchment on the filter; *L* is the length of hyphae per slide. Using the dilution factor for extraction of hyphae from soil, hyphal length in mm slide^{-1} can be converted to mm hyphae in the soil sample and then to m hyphae g^{-1} of soil.

Experiment 3: Estimation of root length colonized by AM fungi

Spores of AM fungi germinate and produce thick-walled hyphae that penetrate the host root causing internal infection. After penetrating into the root, the hyphae spread inter and/or intra-cellularly in the root cortex without damaging the integrity of the cells (Strack *et al.*, 2003). On passing to the inner cortical layers, specialized haustoria-like structures called arbuscules are formed within the cells where metabolites exchanges take place between the fungus and host cytoplasm (Parniske, 2000). Characteristic vesicles usually also form later as terminal or intercalary swellings in the cortical cells and function as nutrient storage organs or as propagules in root fragments. They are inter and/or intra cellular, depending on the host species. These structures can be stained using different procedures after clearing the roots. These structures are then seen under microscope and percentage of root length colonized by AM fungi can be measured.

Principle

Root segments or whole roots are heated in 10 % KOH. This removes the cytoplasm and most of the nuclei, and the roots become very clear with the vascular cylinder distinctly visible. The stain penetrates readily and there is no stained host cytoplasm to obscure the deep blue fungal tissues. Trypan blue is a differential stain which reacts with the complex carbohydrate called chitin which is largely present in the cell walls of fungi.

Materials and equipments required

Autoclave, test tubes, fine nylon mesh screen, Petri plate, dissecting microscope.

Reagents

10 % KOH (w/v) dissolved in water, 0.03 % (w/v) chlorazol black E (CBE) in lactoglycerol (1:1:1 lactic acid:glycerol:water), 0.05 % (w/v) trypan blue in lactoglycerol.

Procedure

i. Collect fine roots of an allelopathic plant.

ii. Carefully rinse roots with tap water and transfer them to heat resistant test tubes.

iii. Add sufficient quantity of 10 % KOH solution and autoclave for about 30 min at 121 °C to clear the roots (Phillips and Hayman, 1970).

iv. Wash the cleared roots with water.

v. Place the materials in dilute HCl for few min to neutralize the roots.

vi. Bleach the cleared roots with alkaline H_2O_2 (0.5 % NH_4OH and H_2O_2 in water) in case of pigmented roots.

vii. Add 0.03 % CBE in a lactoglycerol solution (Brundrett *et al.*, 1984) to the cleared roots.

viii. Autoclave the samples at 121 °C for 15 min.

ix. Alternatively cleared roots can be stained with 0.05 % trypan blue in lactoglycerol (Phillips and Hayman, 1970).

Calculations

The most frequently used procedure is a modification of the gridline intersect method described by Giovannetti and Mosse (1980):

i. Disperse the stained roots in a 9 cm diameter Petri plate on which a grid is drawn.

ii. Using a dissecting microscope at ×40 magnification cast your eyes along the lines in both the horizontal and vertical directions.

iii. Note that each intersection between line and root is either infected or not infected. The percentage infection is then calculated as:

$$\text{AM infection (\%)} = \frac{\text{Number of infected intersections}}{\text{Total number of intersections}} \times 100\,;$$

Precautions

i. Avoid contact with KOH solution.

ii. CBE and trypan blue are suspected to be carcinogens.

iii. Phenol, which has traditionally been used in stain preparations should not be used as it is a toxic chemical. Its omission will not influence staining quality (Brundrett *et al.*, 1994).

Experiment 4: Measurement of mycorrhizal inoculum in the rhizosphere of allelopathic plants

The total mycorrhizal inoculum potential of soils can be determined either by "most probable number" assay procedures (An *et al.*, 1990) or by growing baits plants in intact cores of soil (Brundrett, 1991). The later method is superior to the former one as it measures infectivity due to intact hyphal networks, as well as that from disturbance propagules such as spores. Results of bioassay experiments vary with the use of different bait plants and environmental conditions, but can be relied upon to provide a relative measure of mycorrhizal fungus inoculum.

Principle

Since AM fungi cannot be cultured on synthetic media, therefore, its population in the soil can only be determined by growing living plants in the sampled soils.

Materials and equipments required

Steel cylinder of 12 cm diameter × 14 cm deep, autoclave, pots of 1 L, test tubes, fine nylon mesh screen, Petri plates, dissecting microscope.

Reagents

10 % KOH (w/v) dissolved in water, 0.03 % (w/v) CBE in lactoglycerol.

Procedure

i. Sample relatively undisturbed 1 L soil cores from rhizosphere of selected allelopathic plants by driving 12 cm diameter × 14 cm deep steel cylinder into the ground.

ii. Push out the soil cores out of the steel ring into plasic bag-lined 1 L pots.

iii. Sow pregerminated seeds of a plant species with rapidly growing root system in the pots.

iv. Allow the plants to grow in the core for 2–4 weeks with regular watering to field capacity.

v. Carefully harvest the plants and wash their roots.

vi. Take fresh weight of roots and shoots.

vii. Clear and stain the roots with chlorazol black E or trypan blue in lactoglycerol.

viii. Measure the total and AM colonized root length of samples by gridline intersect method.

Precautions

i. Avoid exposing soil cores to extremes in temperature during transport or storage.

ii. Do not add additional nutrients.

iii. Soil cores should be used as soon as possible.

iv. Avoid longer growth times.

Experiment 5: Preparation of monoculture inoculum of AM fungi for bioassays

Principle

Inoculum of AM fungi cannot be prepared on synthetic media, because these fungi are unable to grow saprophytically except that their spores can germinate to some extent on non-living media. These fungi are, therefore, usually propagated by growing them with a living host plant in soil pot cultures. These pot cultures, which consist of soil, root pieces, hyphal fragments and AM spores can be used as inoculum for experiments or to introduce fungi into soils.

Materials and equipments required

Sandy soils, pots, stacking sieves with a range of pore size (40–1000 mm), wash bottle, centrifuge (2000 rpm), vacuum filter apparatus, filter papers, Petri dishes, dissecting microscope, slides.

Reagents

10 % KOH (w/v) dissolved in water, 0.03 % (w/v) CBE in lactoglycerol.

Procedure

i. Collect coarse textured, sandy soil with moderate nutrient level.

ii. Sterilize the sandy soil by autoclaving.

iii. Fill pots with 500–1000g of autoclaved soil.

iv. Provide the sterilized soil with nutrient either by incorporating in the soil before experiment or by periodic applications of a dilute hydroponic solution.

v. Extract AM fungal spores from field soil by wet sieving and centrifugation.

vi. Identify and separate the spores of different AM species under dissecting microscope.

vii. Transfer healthy spores of different mycorrhizal species to small filter paper triangles with a very fine brush or sharpened dowel.

viii. Wash 100 or more AM spores on a filter paper triangle into a wide, 3 cm deep hole in pots of sterilized sandy soil watered to field capacity.

ix. Cover the spores with soil and incubate for a week.

x. After 1 week sow the surface sterilized (with 1 % hypochlorite for 2–3 min) seeds of a host plant (maize, mungbean, etc.) over the spores.

xi. Maintain the pots in a glasshouse.

xii. Remove a small sample from the pot cultures with a small corer every 2 months or so:

   a. Study root pieces for AM colonization after clearing and staining.

   b. If desired, isolate spores by wet sieving and centrifugation procedure.

c. Discard the pot cultures without any AM after 4 months.

d. Successful cultures will require 4–8 months (depending on the host species) to form adequate levels of mature spores and other propagules.

e. Place the pot cultures in a sealed container for storage.

f. Store the sealed cultures by refregrating damp inoculum at 4 °C or by air drying inoculum (Jarstfer and Sylvia, 1993).

g. Inoculum can also be cryopreserved in liquid nitrogen for longterm storage (Morton *et al.*, 1993).

h. This inoculum can be used in further experiments to study the effects of AM inoculation under allelopathic stress.

Precautions

Avoid excessive nutrient supply to the pot cultures as it adversely affects the AM development.

Experiment 6: Determination of hydraulic conductance and water potential gradients in leaves of mycorrhizal plants (Auge *et al.*, 2008)

Colonization of plant roots and soils by AM fungi is often accompanied by changes in g_s and transpiration in the host plant (Cho *et al.*, 2006; Querejeta *et al.*, 2006). There are some reports of AM-induced changes in whole-plant and root hydraulic conductance (Allen, 1982; Augé, 2001), but the possibility of AM-induced changes in leaf hydraulic conductance has not been examined.

Principle

The mycorrhizal plants are better tolerant to a variety of environmental stresses, which include drought, salinity, pollution and other. The mycorrhizae enable host plant to enhance its outreach to microclimates in the soil with an increased water availability in addition also helping them to survive in soils with extremely negative water potential.

Materials and equipments required

2-L plastic pots (30), potting medium, AM pot cultures, pressure chamber, porometer, resistive sensor, Epson expression 1680 flatbed scanner and Winfolia software, inductively coupled plasma mass spectroscopy, seeds of *Cucurbita pepo*, turface. *Glomus intraradices* Schenck & Smith isolate IA509 and a non-AM fungus (can be obtained from INVAM culture collection centre, USA).

Reagents

NPK fertilizers, micronutrients, trypan blue, KOH, HCl.

Procedure

i. Thirty 2-L plastic pots are seeded with *C. pepo* L. and thinned after germination to one plant per pot in the potting medium which was calcined montmorillonite clay (Surface).

ii. Pots are inoculated with 200 mL of fresh pot culture, banded beneath seeds: half of the pots with *G. intraradices* Schenck & Smith isolate IA509 and the other half with non-AM pot culture.

iii. Plants are fertilized weekly with a water-soluble fertilizer at 100 ppm N (Peters Professional Dark Weather Feed, N:P:K=15:0:15, Scotts-Sierra Horticultural Product, Marysville, OH) and twice with a micronutrient solution at 0.02 mM Fe (Microplex, Miller Chemical & Fertilizer, Hanover, PA) during the growth phase preceding hydraulic measurements. Phosphorus is supplied once per week as 0.6 mM KH_2PO_4 to AM plants and as 1.2 mM KH_2PO_4 to non-AM plants. Plants were grown in a glasshouse with temperature maintained at 22–25/18–22 °C (day/night) under natural light.

Leaf hydraulic measurements

i. Leaf hydraulic conductance (k_{leaf}) normalized to leaf area is measured in mol m^{-2} s^{-1} MPa^{-1}, following the electrical analogue approach (Tyree and Cheung, 1977; Schulte, 1993) using the procedures of Franks (2006).

ii. To obtain k_{leaf}, the lamina of a detached, fully hydrated leaf is pressurized so that a small amount of water moves in the opposite direction, from the storage tissue to the xylem:

$$k_{leaf} = J_{w0}/\text{Ä}P \;;$$

where J_{w0} (mol m^{-2} s^{-1}) is the initial, maximal rate of water efflux from the tip of the petiole of the pressurized leaf and ÄP (in MPa) is the difference between the initial and final pressure of the step change.

iii. J_{w0} can be estimated as the volume of xylem sap expressed (Äv; in moles) per unit leaf area (A_{leaf}; in m^2) over a time interval t, provided that t is much less than the time required to discharge the leaf's stored water (Franks, 2006). Franks has computed a time constant for discharge as 63 % of the total sap discharge (60–120s for the species used in his study). The time constant for 63 % discharge for squash in this experiment is 45s. J_{w0} it may be estimated using

$$J_{w0} \approx \text{Ä}v / 10\, A_{leaf} \;;$$

where Ä*v* is the volume of sap expressed in the first 10s after the increase in pressure.

iv. Plants are kept amply watered during the experiment. On the morning that a plant's leaf hydraulics is examined, the plant is watered at 9:00 am to bring it to full hydration.

v. The most recently expanded, unshaded leaf is selected from each plant for characterization of k_{leaf} and Ä$Ø_{leaf}$. First, its g_s is measured using a diffusion porometer (AP4, Delta-T Devices, Cambridge, UK). The porometer cuvette is placed parallel to the midrib in the center of the lamina on the abaxial side of the leaf. Stomatal conductance is measured on each side of the midrib, and those two values were pooled, to obtain a more accurate estimate of leaf g_s (g_s varies across the lamina in many species).

vi. After the g_s measurements, the leaf is excised, placed in a pressure chamber (Soilmoisture Equipment, Santa Barbara, CA), and brought to the balance pressure. After allowing the leaf to equilibrate at a balance pressure for 45 s, the chamber pressure is quickly increased by 0.5 MPa. Sap expressed from the leaf is collected for 10 s and weighed on an analytical balance to obtain Ä*v*. Leaf area is measured with an Epson Expression 1680 flatbed scanner and Winfolia software (version 2006a, Regent Instruments, Quebec City, Canada). J_{w0} and k_{leaf} were computed as indicated above. The Ä$Ø_{leaf}$ is calculated as follows::

$$\text{Ä}Ø_{leaf} = g_s \text{Ä}w / k_{leaf};$$

where Ä*w* is the leaf-to-air water vapor concentration difference (kg kg^{-1}) computed from leaf and air temperatures and relative humidities. Ä$Ø_{leaf}$ has been variously referred to as the water potential gradient across the leaf, hydrodynamic (transpiration-induced) water potential drawdown across the leaf, or hydrodynamic pressure gradient (Franks, 2006).

vii. Relative humidity of air is measured with a resistive sensor (U10–003, Hobo datalogger, Onset Computer, Bourne, MA).

viii. Water relations measurements begin 8 weeks after planting. Stomatal conductance of all AM and non-AM plants is compared eight times over 3 weeks, at 9:00–11:30 EST and 1 h after plants are watered. Stomatal conductance, k_{leaf}, J_{w0}, and Ä$Ø_{leaf}$ of unstressed plants is measured for 1 week. Equal numbers of replicates of each of the two mycorrhizal treatments are measured on each day, and measurements of AM and non-AM plants are alternated throughout the day to avoid diurnal bias. A total of 60 leaves are measured from unstressed plants.

ix. Plants may then be subjected to salinity stress and all water relations measurements are repeated.

Shoot, root, and soil characters

i. Shoots were excised in each pot after completing stomatal and hydraulic conductance measurements, and dry weights are determined. Leaves used for hydraulic measurements are pooled for each plant (four leaves per plant, two before and two after exposure to stress), and their elemental analyses are performed using inductively coupled plasma mass spectroscopy (Agilent 7500ce Series, Agilent Technologies, Santa Clara, CA).

ii. Immediately after harvesting shoots, soil is removed from each pot, sealed in a plastic bag, and frozen, for subsequent root and soil measurements.

AM colonization studies

i. Hyphal, arbuscular, and vesicular colonization of roots is determined for each plant, on one grid intersection on each of 50–0.5 cm root pieces, after clearing with boiling 10 % KOH for 10 min, acidifying with 2 % HCl for 1.5 h, staining with 0.05 % trypan blue for 1 h, and destaining in a lactoglycerol solution.

ii. Soil hyphal density is measured as described previously (Augé *et al.*, 2003), on 10g subsamples obtained from the soil after thorough mixing. Roots are carefully excavated from another 25g of soil of each pot, for the measurement of root length, using scanning equipment and imaging software (WinRhizo, Regent Instruments).

Experimental design and statistical analysis

Pots are arranged in a completely randomized design. Analysis of variance is performed Correlations among water relation parameters and other plant and soil variables were tested by computing Pearson correlation coefficients.

Experiment 7: Evaluation of response of transgenic tobacco fluorescent lines to AM inoculation

A number of studies have related the transgenic plants and their response towards mycotrophy. There are numerous examples of cultivars within plant species that differ in the extent of colonization by mycorrhizal fungi and their responsiveness to them (Mercy *et al.*, 1990; Hetrick *et al.*, 1996). There was also a two-fold difference in root colonization among these lines which was not related to mycorrhizal responsiveness. Likewise, the transgenic mutants with different antibiotic resistant or some other markers may be vital to the study of the behaviour of AM and ECM fungi in relation to allelopathy.

Principle

The responsiveness of host plants towards mycotrophy varies from variety to variety and hybrid to hybrid. There are variations between transgenic lines of defferent crops. The probable reasons accountable for these differences may be the imbalance in the production of hormones, proteins or some other phenolic substances by the host or the fungus or both which may prove stimulatory towards the establishment of the relationship.

Materials and equipments required

AM isolates, transgenic seeds, turface, *Glomus intraradices* isolate #IA509 and *G. diaphanum* isolate #wv5789 (can be obtained from INVAM culture collection centre, USA), Transgenic seeds of 10–15 fluorescent lines of tobacco can be obtained from any lab engaged in the research related to fluorescent lines of tobacco.

Reagents

Trypan blue stain, HCl, KOH.

Procedure

i. The roots of pot culture plants with *Glomus intraradices* and *G. diaphanum* are used for inoculating the pots prepared for tobacco experiment. These roots are chopped up into tiny (0.5–1 cm) pieces and about 10 mL of these root pieces along with the turface are used for inoculating one pot (3 × 3"). The inoculum is placed in the above 2 cm of the pot in the form of a layer. After which a thin layer of turface is spread to cover it.

ii. Assessment of AM colonization is carried out after routine staining procedures (see Experiment 2).

iii. Determination of AM propagule in the potting medium is done following the methods of wet sieving and decanting and density gradient centrifugation (see Experiment 1).

v. Soil hyphal density and root density can be determined using a scanner software.

Suggessted Readings

Abbott, L.K., Robson, A.D. and De Boer, G. (1984). The effect of phosphorus on the formation of in soil by the vesicular arbuscular mycorrhizal fungus *Glomus fasciculatum*. *New Phytologist* **97**: 437–446.

Al-Karaki, G.N. (2002). Benefit, cost and phosphorus use efficiency of mycorrhizal field grown garlic at different soil phosphorus levels. *Journal of Plant Nutrition* **25**: 1175–1184.

Allen, M.F. (1982) Influence of vesicular-arbuscular mycorrhizae on water movement through *Bouteloua gracilis* (H.B.K.) Lag ex Steud. *New Phytologist* **91**: 191–196.

Amoo, S.O., Ojo, A.U. and Staden, J.V. (2008). Allelopathic potential of *Tetrapleura tetraptera* leaf extracts on early seedling growth of five agricultural crops. *South African Journal of Botany* **74**: 149–152.

An, Z.Q., Hendrix, J.W., Hershman, D.E. and Henseon, G.T. (1990). Evaluation of the most probable number and wet sieving methods for determining soil-borne populations of endogonaceous mycorrhizal fungi. *Mycologia* **82**: 576–581.

Augé, R.M. (2001). Water relations, drought and VA mycorrhizal symbiosis. *Mycorrhiza* **11**: 3–42.

Augé, R.M., Moore, J.L., Cho, K., Stutz, J.C., Sylvia, D.M., Al-Agely, A.K. and Saxton, A.M. (2003). Relating foliar dehydration tolerance of mycorrhizal *Phaseolus vulgaris* to soil and root colonization by hyphae. *Journal of Plant Physiology* **160**: 1147–1156.

Auge, R.M., Toler, H.D., Sams, C.E. and Nasim, G. (2008) Hydraulic conductance and water potential gradients in squash leaves showing mycorrhiza induced increases in stomatal conductance. *Mycorrhiza* **18**: 115–121.

Bajwa, R., Javaid, A. and Haneef, B. (1999). EM and VAM technology in Pakistan V: Response of chickpea (*Cicer arietinum* L.) to co-inoculation with effective microorganisms (EM) and VA mycorrhiza under allelopathic stress. *Pakistan Journal of Botany* **31**: 387–396.

Bajwa, R., Akhtar, J. and Javaid, A. (2003). Role of VAM in alleviating allelopathic stress of *Parthenium hysterophorus* on maize. *Mycopathology* **1**: 15–30.

Bethlenfalvay, G.J. and Linderman, R.G. (1992). *Mycorrhiza in sustainable agriculture.* ASA special publication number 54, American Society of Agronomy, Inc. Madison, Wisconsin, USA.

Brundrett, M.C., Piche, Y. and Peterson, R.L. (1984). A new method for observing the morphology of vesicular arbuscular mycorrhizae. *Canadian Journal of Botany* **62**: 2128–2134.

Brundrett, M.C. (1991). Mycorrhizas in natural ecosystems. *In*: Advances in Ecological Research. Vol. 21. (*Eds*., A. Macfayden, M. Begon and A.H. Fitter). pp. 171–313. Academic Press, London.

Brundrett, M., Melville, L., and Peterson, L. (1994). *Practical Methods in Mycorrhiza Research.* Mycologue Publications. pp. 24–27.

Cairney, J.W.G. (1999). Intraspecific physiological variation: Implications for understanding functional diversity in ectomycorrhizal fungi. *Mycorrhiza* **9**: 125–135.

Cavagnaro, T.R., Jackson, L.E., Six, J., Ferris, H., Goyal, S., Asami, D. and Scow, K.M. (2006). Arbuscular mycorrhizas, microbial communities, nutrient availability, and soil aggregates in organic tomato production. *Plant and Soil* **282**: 209–225.

Chou, C.H. (1995). Allelopathy and sustainable agriculture *In*: Allelopathy Organisms, Process and Applications (*Eds*., Inderjit, K.M.M. Dakshini and F.A. Einhellig), pp. 211–233. ACS Symposium Series 582, American Chemical Society, Washington, DC.

Cho, K., Toler, H.D., Lee, J., Ownley, B.H., Jean, C., Stutz, J.C., Moore, J.L. and Augé R.M. (2006). Mycorrhizal symbiosis and response of sorghum plants to combined drought and salinity stresses. *Journal of Plant Physiology* **163**: 517–528.

Combier, J.P., Maleyah, D., Raffier, C., Gay, G. and Marmeisse, R. (2003). Agrobacterium tumefaciens-mediated transformation as a tool for insertional mutagenesis in the symbiotic ectomycorrhizal fungus Hebeloma cylindrosporum. *FEMS Microbiology Letters* **220**: 141–148.

Cote, J.F. and Thibault, J.R. (1988). Allelopathic potential of raspberry foliar leachates on growth of ectomycorrhizal fungi associated with black spruce. *American Journal of Botany* **75**: 966–970.

De Groot, M.J.A., Bundock, P., Hooykaas, P.J.J. and Beijersbergen, A.G.M. (1998). *Agrobacterium tumefaciens*-mediated transformation of filamentous fungi. *Nature Biotechnology* **16**: 839–842.

Einhellig, F.A. (1996). Interactions involving allelopathy in cropping systems. *Agronomy Journal* **88**: 886–893.

Franks, P. (2006). Higher rates of leaf gas exchange are associated with higher leaf hydrodynamic pressure gradients. *Plant Cell and Environment* **29**:584–592.

Gazzey, C., Abbot L.K. and Robson, A.D. (1992). The rate of development of mycorrhizas effects the onset of sporulation and production of external hyphae by two species of *Acailospora*. *Mycological Research* **96**: 643–650.

Hetrick, B.A.D., Wilson, G.W.T. and Todd, T.C. (1996). Mycorrhizal response in wheat cultivars: Relationship to phosphorus *Canadian Journal Botany* **74**: 19–25.

Jackson, R.M. and Simson, P.A. (1984). *Mycorrhiza*. Edward Arnold Publishers Ltd London, UK.

Jarstfer, A.G. and Sylvia, D.M. (1993). Inoculum production and inoculation strategies for vesicular arbuscular mycorrhizal fungi. *In*: *Soil Microbial Ecology, Applications in Agriculture and Environmental Management* (*Ed*., F.B. Metting Jr). pp 349–377. Marcel Dekker Inc., New York.

Javaid, A. (2007). Allelopathic interactions in mycorrhizal associations. *Allelopathy Journal* **20**: 29–42.

Javaid, A. (2008). Allelopathy in mycorrhizal symbiosis in the Poaceae family. *Allelopathy Journal* **21**: (in press).

Javaid, A., Hafeez, F.Y. and Iqbal, S.H. (1993). Interaction between vesicular arbuscular (VA) mycorrhiza and *Rhizobium* and their effect of biomass, nodulation and nitrogen fixation in *Vigna radiata* (L.) Wilczek. *Science International (Lahore)* **5**: 395–396.

Javaid, A., Iqbal, S.H. and Hafeez, F.Y. (1994). Effect of different strains of *Bradyrhizobium* and two types of vesicular arbuscular mycorrhizae (VAM) on biomass and nitrogen fixation in *Vigna radiata* (L.) Wilczek var. NM 20–21.. *Science International (Lahore)* **6**: 265–267.

Karandashov, V. and Bucher, M. (2005). Symbiotic phosphate transport in arbuscular mycorrhizas. *Trends in Plant Sciences* **10**: 22–29.

Khaosaad, T., Garcia-Garrido, J.M., Steinkellner, S. and Vierheilig, H. (2007). Take-all disease is systemically reduced in roots of mycorrhizal barley plants. *Soil Biology and Biochemistry* **39**: 727–734.

Koske, R.E. and Tessier, B. (1983). A convenient permanent slide mounting medium. *Mycological Society of America Newsletter* **34**: 59.

Landeweert, R., Hoffland, E., Finlay, R. and van Breemen, N. (2001). Linking plants to rocks: ectomycorrhizal fungi mobilize nutrients from minerals. *Trends in Ecology and Evolution* **16**: 248–254.

Mallik, A.U. and Zhu, H. (1995). Overcoming allelopathic growth inhibition by mycorrhizal inoculation. *In*: *Allelopathy*: *Organisms, Processes and Applications*. (*Eds*., Inderjit, K.M.M. Dakshini and F.A. Einhellig). pp. 39–57. ACS Symposium Series No. 582, American Chemical Society, Washington, D.C.

Mercy, M.A., Shivashankar, G. and Bagyaraj, D.J. (1990). Mycorrhizal colonization in cowpea is host dependent and heritable. *Plant and Soil* **121**: 292–294.

Miller, R.M. and Jastrow, J.D. (1992). Extraradical hyphal development of vesicular arbuscular mycorrhizal fungi in a chronosequence of prairie restoration. *In*: Mycorrhizas in Ecosystems (*Eds*., D.J. Read). pp. 171–176. C.A.B. International, Wallingford.

Morton, J.B., Bentivenga, S.P. and Wheeler, W.W. (1993). Germplasm in the international collection of arbuscular and vesicular arbuscular mycorrhizal fungi (INVAM) and procedures for culture development, documentation and storage. *Mycotaxon* **48**: 491–528.

Narwal, S.S. (1994). *Allelopathy in Crop Production*. Scientific Publishers, Jodhpur, India. p. 288.

Olsson, P.A., Thingstrup, I., Jakobsen, I. and Baath, E. (1999). Estimation of the biomass of arbuscular mycorrhizal fungi in a linseed field. *Soil Biology and Biochemistry* **31**: 1879–1887.

Parniske, M. (2000). Intracellular accommodation of microbes by plants: A common developmental program for symbiosis and disease? *Current Opinion in Plant Biology* **3**: 320–328.

Phillips, J.M. and Hayman, D.S. (1970). Improved procedure for clearing roots and staining parasitic and VA mycorrhizal fungi for rapid assessment of infection. *Transactions of Brtish Mycological Society* **5**: 158–161.

Popa, V.I., Dumitru, M., Volf, I. and Anghel, N. (2008). Lignin and polyphenols as allelochemicals. *Industrial Crops and Products* **27**: 144–149.

Querejeta, J.I., Allen, M.F., Caravaca, F. and Roldan, A. (2006). Differential modulation of host plant delta C-13 and delta O-18 by native and nonnative arbuscular mycorrhizal fungi in a semiarid environment. *New Phytologist* **169**: 379–387.

Renne, I.J., Rios, B.G. Fehmi, J.S. and Tracy, B.F. (2004). Low allelopathic potential of an invasive forage grass on native grassland plants: A cause for encouragement? *Basic and Applied Ecology* **5**: 261–269.

Ruiz-Lozano, J.M. (2003). Arbuscular mycorrhizal symbiosis and alleviation of osmotic stress. New perspectives for molecular studies. *Mycorrhiza* **13**: 309–317.

Sampietro, D.A., Sgariglia, M.A., Soberón, J.R., Quiroga, E.N. and Vattuone, M.A. (2007). Role of sugarcane straw allelochemicals in the growth suppression of arrow leaf. *Environmental and Experimental Botany* **60**: 495–503.

Schliemann, W., Ammer, C. and Strack, D. (2008). Metabolite profiling of mycorrhizal roots of *Medicago truncatula*. *Phytochemistry* **69**: 112–146.

Schulte, P.J. (1993). Tissue properties and water relations of desert shrubs. *In*: Smith, J.A.C., Griffiths, H. (*Eds*). *Water deficits*: *plant responses from cell to community* (*Eds*., J.A.C. Smith and H. Griffiths). pp. 177–192. Bios Scientific, Oxford, UK.

Smith, S.E. and Read, D.J. (1997). *Mycorrhizal Symbiosis*. Academic Press, London.

Strack, D., Fester, T., Hause, B., Schliemann, W. and Walter, M.H. (2003). Arbuscular mycorrhiza: Biological, chemical, and molecular aspects. *Journal of Chemical Ecology* **29**: 1955–1979.

Takeda, N., Kistner, C., Kosuta, S., Winzer, T., Pitzschke, A., Groth, M., Sato, M., Kaneko, T., Tabata, S. and Parniske, M. (2007). Proteases in plant root symbiosis. *Phytochemistry* **68**: 111–121.

Tennant, D. (1975). A test of a modified line intersection method of measuring root length. *Journal of Ecology* **63**: 995–1001.

Tommerup, I.C. (1992). Methods for the sudy of the population biology of vesicular arbuscular mycorrhizal fungi. *In*: Methods in Microbiology. Vol. 24. *Techniques for the study of mycorrhiza*. (*Eds*., J.R. Norris, D.J. Read and A.K. Verma). pp. 23–51. Academic Press, London.

Trappe, J.M. (1982). Synaptic keys to the genera and species of Zygomycetous mycorrhizal fungi. *Phytopathology* **72**: 1102–1108.

Tyree, M.T. and Cheung, Y.N.S. (1977). Resistance to water flow in *Fagus grandifolia* leaves. *Canadian Journal of Botany* **55**: 2591–2599.

Wallander, H. (2000). Uptake of P from apatite by *Pinus sylvestris* seedlings colonized by different ectomycorrhizal fungi. *Plant and Soil* **218**: 249–256.

Wu, O.S. and Xia, R.X. (2006). Arbuscular mycorrhizal fungi influence growth, osmotic adjustment and photosynthesis of citrus under well-watered and water stress conditions. *Journal of Plant Physiology* **163**: 417–425.

Wu, O.S., Xia, R.X. and Zou, Y.N. (2008). Improved soil structure and citrus growth after inoculation with three arbuscular mycorrhizal fungi under drought stress. *European Journal of Soil Biology* **44**: 122–128.

Yamane, A., Nishimura, H. and Mizutani, J. (1992). Allelopathy of yellow field cress. *Journal of Chemical Ecology* **18**: 683–691.

Zeng, R.S. and Mallik, A.U. (2006). Selected ectomycorrhizal fungi of black spruce (*Picea mariana*) can detoxify phenolic compounds of *Kalmia angustifolia*. *Journal of Chemical Ecology* **32**: 1473–1489.

CHAPTER 8

Biochemical Aspects of Plant – Pathogen Interactions

DIEGO A. SAMPIETRO[1], MELINA A. SGARIGLIA[1], JOSÉ R. SOBERÓN[1], EMMA N. QUIROGA[1] AND MARTA A. VATTUONE[1]

1. INTRODUCTION

Plants are non-motile organisms subjected to constant attack of microorganisms, vertebrate and invertebrate animals and even other plants. This feature forced plant evolution to the development of an original defence system, much different from that found in animals. Immunological system of vertebrate animals, including human beings, involves the mobilization of specialized cells through a circulatory system to the body's site under attack to kill or restrict the invading organism (Buchanan *et al.*, 2000). In plant defence, each plant cell has to perform immune functions, which can be preformed and/or induced after pathogen attack. The passive defence mechanisms involve structural barriers provided by waxy cuticles or cell wall bonded polyphenols and accumulation of secondary metabolites in specific locations of plant tissues. Active or induced defence occurs in plant cells after pathogen attack and requires host metabolism to function (Hutcheson, 1998). If pathogen is recognized after infection, plant readily initiates a local response into cells in contact or very close to the attacking organism. The first response frequently detected is an increase in active oxygen species (AOS), which can occur within less than 5 min in the infected site. This accumulation of AOS is also known as oxidative burst and triggers several biochemical changes in plant cells such as accumulation of phytoalexins and increase in activity of enzymes (*i.e.,* peroxidases and phenylalanine ammonia lyase). The outcome of this primary response is often the hypersensitive response, which consists in the death of plant cells in contact and in the surroundings of the infected site. Recognition of pathogen attack is then transmitted through the whole plant body through a systemically acquired resistance (SAR). The SAR is hormonally induced and often related with an increase in salicylic and jasmonic acid levels (Dmitriev, 2003). The following paragraphs provide general concepts and

1. Instituto de Estudios Vegetales "Dr. A. R. Sampietro", Facultad de Bioquímica, Química y Farmacia, Universidad Nacional de Tucumán, España 2903, 4000, San Miguel de Tucumán, Tucumán, Argentina, E-mail: dasampietro2006@yahoo.com.ar

criteria for the detection of organic molecules involved in both passive and active plant defence responses. It also comprises the measurement of some enzyme activities related to active defence response.

2. PHYTOANTICIPINS

Phytoanticipins are antimicrobial compounds found in plant tissues prior to infection of pathogenic organisms (Buchanan *et al.*, 2000). In fact, several phytoanticipins are stored in specific plant cell compartments (*i.e.,* cell vacuoles) as inactive precursors. When plant tissues suffer damages or insect attack, cell compartmentalization is broken and phytoanticipin's precursors make contact with plant enzymes that convert them to biological active forms (Dmitriev, 2003). These enzymes are already present in healthy plant tissues but are separated from their substrates by cell compartmentalization. Some secondary compounds such as saponins, glycosides of hydroxamic acids, cyanogenic glycosides and glucosinolates are known precursors of phytoanticipins. Many phytoanticipins may be constitutively expressed in one part of a plant but induced as phytoalexin in other organs. In other situations, the highest levels of phytoanticipins are found in young plant stages (*i.e.,* hydroxamic acids and cyanogenic glycosides) and at low concentrations in mature plants (Morrissey *et al.*, 1999). The experiment below allows estimation of sorghum's phytoanticipins using a colorimetric reaction.

Experiment 1: Detection of hydrogen cyanide (Konig's reaction)

Principle

Cyanogenic glycosides are commonly found in *Sorghum* genus species. They are hydrolyzed by β-glucosidases and lipases after disruption of plant cells, releasing hydrogen cyanide and an aldehyde or ketone. Hydrogen cyanide can be detected through the Konig's reaction (Goldstein and Spencer, 1985). Hydrogen cyanide is reacted with pyridine to produce an intermediate which hydrolyzes to a conjugated dialdehyde, the glutaconic aldehyde (Fig 1). This compound is then coupled with a primary amine or a compound containing reactive methylene hydrogens to form colored compounds that can be read at 580 nm.

Materials and equipments

Assay tubes, Erlenmeyer's flasks, pipettes, analytical balance with accuracy of 0.01 mg, spectrophotometer or filter photometer capable of measurement at 580 nm, equipped with a 1-cm path length glass microcuvette, germination tray, sorghum seeds.

Reagents

1 N NaOH: Dissolve 40g of NaOH in 400 mL of distilled water. Make up to a final volume of 1 L.

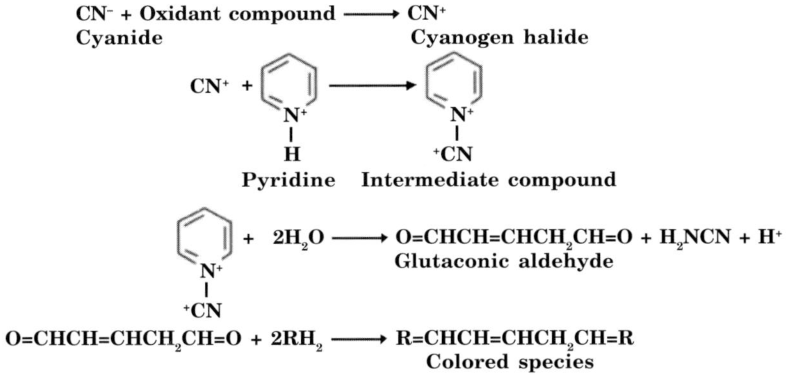

CN^- + Oxidant compound → CN^+
Cyanide — Cyanogen halide

CN^+ + Pyridine (N^+–H) → Intermediate compound (N^+–^+CN)

Intermediate compound (N^+–^+CN) + $2H_2O$ → $O{=}CHCH{=}CHCH_2CH{=}O$ + H_2NCN + H^+
Glutaconic aldehyde

$O{=}CHCH{=}CHCH_2CH{=}O$ + $2RH_2$ → $R{=}CHCH{=}CHCH_2CH{=}R$
Colored species

Fig 1. Reactions involved in detection of cyanide production

2 N acetic acid: Dilute 57.2 mL of acetic acid to a final volume of 1 L.

Oxidant reagent: Dissolve 2.5g of succinimide in 300 mL of distilled water; add 0.25g *N*-chlorosuccinimide and stir to dissolve; dilute the solution to 1 L).

Coupling compound reagent: Dissolve 12g of barbituric acid in 120 mL of pyridine; make up to a final volume of 400 mL with distilled water).

0.5 mg mL^{-1} stock NaCN solution : Dissolve 50 mg of NaCN in 100 mL of NaOH 1 N).

Standard NaCN solution : Dilute 1 mL of stock NaCN solution in 49 mL of 1N NaOH).

Plant extract: Soak seeds of *Sorghum bicolor* in a beaker filled with distilled water; after 24 h, place seeds between two layers of cheese cloth stretched over a metal screen and place in a germination tray at 28 °C in the dark. Leave to grow in the darkness for 3 days and then harvest seedling shoots. Grind shoots with the help of mortar and pestle adding a little known volume of water (see *Observation* 1).

Procedure

i. **Preparation of calibration curve:** Dilute standard NaCN solution with 1 N NaOH to prepare solutions of 20, 40, 60 and 80 µM in test tubes. Label tubes from 1–5. Then, add 1 mL of 0, 20, 40, 60 and 80 µM NaCN to tube 1, 2, 3, 4 and 5, respectively (see *Observation* 2).

ii. Set up your experimental assay tubes. Prepare tubes 6 and 7 adding 0.50 and 1 mL of plant extract, respectively. Complete volume of tube 6 to 1 mL with 1 N NaOH.

iii. Add 0.5 mL of acetic acid 2 N and 5 mL of the oxidant reagent to each tube.

iv. Add 1 mL of the coupling compound reagent and vortex. Stand tubes for 10 min.

v. Read absorbance at 580 nm.

Precautions

NaCN is extremely poisonous. Eating or inhaling it is harmful or fatal.

Calculations

Step 1: Graph the calibration curve relating absorbance at 580 nm to NaCN concentration. Fit a line function to the measured data applying regression analysis with $r^2 > 0.80$. You should obtain a calibration curve as shown in Fig 2.

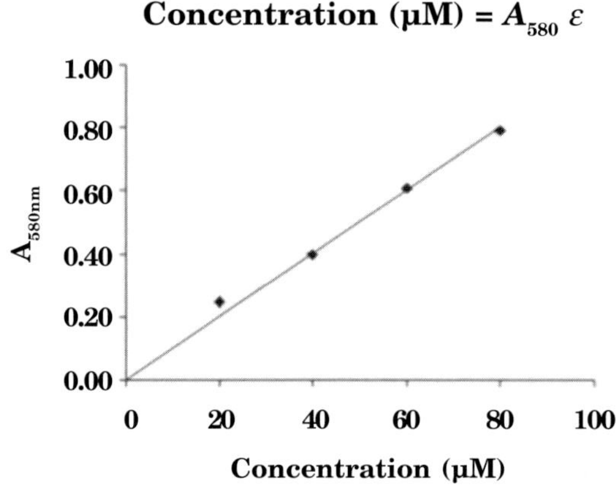

Fig 2. Absorbance at 580 nm vs NaCN concentration. ε is extinction coefficient and $A_{580\ nm}$ is absorbance at 580 nm.

Step 2: Calculate cyanide concentration (C) produced from plant extract expressed in equivalent concentrations of NaCN. For example, if absorbance at 580 nm (A_{580} = 0.5) and extinction coefficient (*å*) obtained from linear regression = 0.89, concentration (C) should be for tube 6:

$$C\ (\text{mM}) = (A_{750} \times å) \times 2 \qquad \text{Hence,}\ \ C = (0.5 \times 0.89) \times 2$$

$$C = 0.72\ \text{mM}$$

In the example above, if calculations were on tube 7, concentration (C) should be:

$$C\ (\text{mM}) = A_{750} \times å \qquad \text{Hence,}\ \ C = 0.5 \times 0.89$$

$$C = 0.36\ \text{mM}$$

Observations

1. If sorghum seeds are not available, an alternative is to use cassava extracts: collect root peels of cassava; grind 1g of root peel in a mortar; stand for 2 h. Add 7 mL of NaOH 1 N; grind again in the mortar; filter and use for the assay.
2. Make at least three independent experiments ($N = 3$).

3. PHYTOALEXINS

Phytoalexins are low molecular weight metabolites with antimicrobial activity which are not found in healthy plant tissue and are synthesized in response to pathogen attack or environmental stresses, probably as a result of *de novo* synthesis of enzymes. Although both disease-resistant and susceptible plants may respond to pathogen attack by producing phytoalexins, these compounds generally accumulate more rapidly and to higher levels in the infected tissues of resistant plants. Chemical nature of phytoalexins is very diverse. They may be derived from one or several primary biosynthetic pathways. In general, they are neither specific to pathogen species nor with a particular potent activity. Simple phenylpropanoid derivatives, flavonoid- and isoflavonoid-derived

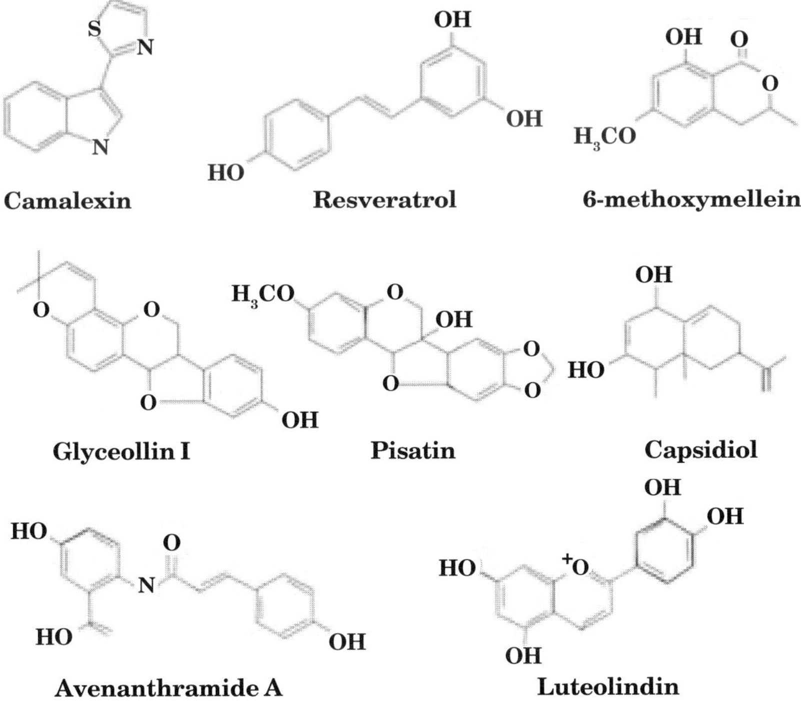

Fig 3. Representative phytoalexin structures

phytoalexins, sesquiterpenes, and polyketides are good known examples (Fig 3). Despite diversity in chemical structure, many plant families produce phytoalexins that fall into the same chemical class, which sometimes allow the use of these compounds for chemotaxonomic studies.

Experiment 2: Induction of pisatin biosynthesis in pea leaves

Principle

Pisatin is an isoflavonoid phytoalexin produced by species from *Pisum* and *Lathyrus* genus (*Leguminosae*). High amounts of pisatin accumulate after fungal infection of pea (*Pisum sativum* L.) leaves, which can be detected through the drop diffusate technique described in this experiment (Harborne, 1998).

Materials and equipments

Pea (*Pisum sativum* L.) seeds, greenhouse, conidial suspensions of *Helminthosporium* sp., clear plastic boxes (*ca* 15 × 25 cm) with tapes, rotary evaporator, plates of silica gel 60 F-254, short wave UV lamp, scanning spectrophotometer capable to register the range from 290–325 nm, equipped with a 1-cm path length glass microcuvette, pencil, tubes.

Reagents

0.05 % Tween 20, ethyl acetate, chloroform, methanol, concentrated HCl, ethanol.

Procedure

i. Excise leaves from pea plants grown in a greenhouse. Float leaves on water with adaxial side uppermost, in the plastic boxes. Inoculate leaves with the conidial suspension (*i.e.,* a 2.5 × 10^5 spores mL^{-1}) in 0.05 % Tween 20. Leaves without fungal inoculation will serve as controls (see *Observation* 1).

ii. Cover the boxes with plastic tapes and incubate at room temperature under continuous diffuse a lighting for up to 72 h.

iii. Separately collect the droplets and extract into ethyl acetate in a ratio 1:3, evaporate the organic extracts to dryness *in vacuo* at 40 °C.

iv. Dissolve the two extracts (control and infected) separately in 0.2–0.4 mL ethanol. Develop the plate by ascent in chloroform containing 2 % methanol for 60–90 min (see *Observation* 2).

v. On examining in UV light, the phytoalexins should be clearly visible as a dark absorbing band which can be marked with a pencil. For the control, mark a corresponding blank area of silica gel.

vi. Remove phytoalexin band and the control. Elute with ethanol (2 mL aliquots), concentrating the eluates *in vacuo*. Dissolve the two residues in 2 mL ethanol, register UV spectrum using the control eluate as blank and annotate absorbances at 310 and 350 nm. Then, add 5 drops of concentrated HCl and record the spectral shift (see *Observation* 3).

Precautions

Concentrated HCl is highly corrosive. Ethyl acetate, chloroform and methanol are toxic solvents. Avoid exposure. Obtain special instruction before use.

Calculations

The amount of phytoalexin formed is calculated according to spectral absorbances at 310 (A_{310}) and 350 nm (A_{350}):

$$\mu\text{g pisatin mL}^{-1}\text{ diffusate} = [D \times (A_{310} - A_{350}) \times B] / C \,;$$

where D is the dilution of sample; B is the pisatin constant which is equal to 47; C is the volume of diffusate (mL).

Observations

1. Make at least three independent experiments ($N = 3$).
2. TLC development is achieved by adding a small amount of mobile phase into 1 L beaker. Cover the beaker with a glass and leave for 30 min. Then, open the beaker and place the TLC plate so that the solvent is touching the plate but below the spotted samples. Close the beaker again and allow the solvent to rise. After the solvent front has stopped rising or approaches the top of the TLC plate, remove the plate, mark the solvent front with a pencil and allow the plate to dry in the hood. After drying, visualize under UV light.
3. Pisatin has maxima at 213, 280, 286 and 309 nm and in the presence of HCl characteristically undergoes dehydration to dehydropisatin, which has maxima at 248, 342, and 358 nm.

4. ELICITORS

Elicitors are chemicals (*e.g.,* organic compounds) or physical factors *e.g.,* wounding, UV-radiation that act as initiation signals for plant defence responses (Stout *et al.*, 1998). Organic elicitors may be from pathogen or plant origin and are very diverse in their chemical structures. Some elicitors participate only in specific plant-pathogen interactions, while, others lack specificity. They can be unspecific such as proteins (including glycoproteins), polyenoic fatty acids, or fragments of chitin and β-1,3-glucans (Dmitriev,

2003). Resistance mechanisms can also be triggered by specific elicitors such as heptaglucan and oligopeptide produced by *Phytophthora sojae* which initiate defence responses in soybean (*Glycine max* L.) and parsley (*Petroselinum crispum* L.), respectively. The recognition of microbial elicitors is accomplished by putative plant-cell receptors in plasmalemma. Although none of these receptors have been isolated, it is believed that elicitor interaction with a receptor activates the enzymatic mediated processes triggering a complex network of signal transduction, which allows defence responses.

Experiment 3: Increased phenolics metabolism in mung bean seeds induced by microbial polysaccharides

Principle

Polysaccharides are water soluble compounds with complex structures made up of multiple sugar molecules. The ability to produce polysaccharides is wide spread in live beings. Bacteria produce extracellular polysaccharides (exopolysaccharides; EPS) with molecular masses varying between 10 and 104 kDa (approximately 50–50000 glycosil units). These molecules protects the microbial cell against stress conditions (*i.e.,* desiccation, exposure to antibiotic or toxic compounds, phagocytosis and phage attack). Some EPS were identified with elicitor activity such as Phytagel™ (P), a gellan gum isolated from *Pseudomonas elodea* and Cp Kelco xanthan, a gum isolated from *Xanthomonas campestris* (McCue and Shetty, 2002). These gums are used as viscosityling, stabilizing, emulsifying, gelling or water binding agents in food and non-food industry (Bueno and Garcia-Cruz, 2006). Stimulation of phenolic metabolism occurs in mung bean seeds due to EPS elicitor activity, which may be used to produce food with enhanced phenolic antioxidant contents (McCue and Shetty, 2002). In the following experiment, content of phenolic compounds is measured in seedlings of mung bean. Seedlings are produced from seeds pretreated with commercially available bacterial gums.

Materials and equipments

Erlenmeyer's flasks, assay tubes, beakers, pipettes, analytical balance with accuracy of 0.01 mg, mung bean seeds, filter paper, aluminum foil, plastic flats, germination chamber, spectrophotometer or filter photometer capable of measurement at 725 nm, equipped with a 1-cm path length glass microcuvette, microwave oven, centrifuge capable of 15000g, vortex mixer.

Reagents

Gellan gum solution: Add 1g of gellan gum per 100 mL of distilled water; heat in the microwave oven for 30–60s (see *Observation* 1).

Xanthan gum: Add 1g of xanthan gum per 100 mL of distilled water; heat in the microwave oven for 30–60s, (see *Observation* 1); 2 N stock solution of Folin & Ciocalteu's reagent (a volume of this solution is half diluted before use).

Gallic acid solution: Dissolve 2g of gallic acid in 10 mL of distilled water).

5 % Na_2CO_3: Dissolve 50g of Na_2CO_3 in 500 mL of distilled water); 95 % ethanol.

Procedure

i. **Priming of mung bean seeds:** Place mung bean seeds in 200 mL Erlenmeyer's flasks. Each treatment consists of 25g seeds which are embedded overnight in 50 mL of gellan or xanthan gum solution. A water treatment is used as control.

ii. **Seedling growth:** Place seeds of mung bean between two wet filter paper towels, inside empty plastic trays. Cover with aluminum foil and place in a dark germination chamber for 5 days. Then, collect seedlings from the trays and dissect plant material in cotyledons, shoots and roots immediately before use.

iii. **Extraction of phenolic compounds:** Add 2.5 mL of 95 % ethanol to 50 mg of plant material (cotyledon, shoot or root). Leave for 48–72 h. Then, homogenize in mortar with a pestle. Centrifuge at 16000g for 8–10 min. Assay total phenolic content in 1 mL of supernatant from each sample (see *Observations* 2 and 3). Prepare each tube by adding:

   a. 1 mL of supernatant or 1 mL of ethanol (blank).

   b. 1 mL of ethanol.

   c. 5 mL of distilled water.

   d. 0.5 mL of 1 N Folin & Ciocalteu's reagent.

   Vortex tubes and then incubate at room temperature for 5 min. Then, add 1 mL of 5 % Na_2CO_3 to each sample, vortex and place in dark for 1 h at room temperature. Vortex again and readings at 725 nm against a blank (see *Observation* 4).

iv. **Standard curve:** Assay different concentrations of gallic acid (25, 50, 75, 100, 150 and 200 mg mL^{-1}) following the procedure described in step (iii).

Calculations

Graph the calibration curve relating absorbance at 725 nm to gallic acid concentration. Then, proceed as indicated for step 1, 2 and 3 to determine the hydrogen cyanide content.

Observations

1. Heat in a microwave oven to increase polysaccharide solubility. This is observed as a slight yet noticeable thickening of the solution.

2. Make at least three independent experiments ($N = 3$).

3. Prepare a duplicate of each tube.

5. GUAIACOL PEROXIDASE (GPX)

Guaiacol peroxidases (EC 1.11.1.7) are glycosylated hemoproteins of about 50 kDa found as multiple isozymes in plant tissues. Activity of guaiacol peroxidases often increase in response to pathogen attack. Enhanced guaiacol peroxidase activity leads to biosynthesis of lignin, a complex polymer consisting in phenyl propanoid units interconnected by a variety of carbon-carbon bonds and ether linkages. Lignification of cell wall acts as a barrier to fungal infections. These peroxidases can use guaiacol as substrate oxidizing it (Fig 4), in presence of H_2O_2, to dimmers, which can be detected at 470 nm (McCue *et al.*, 2000).

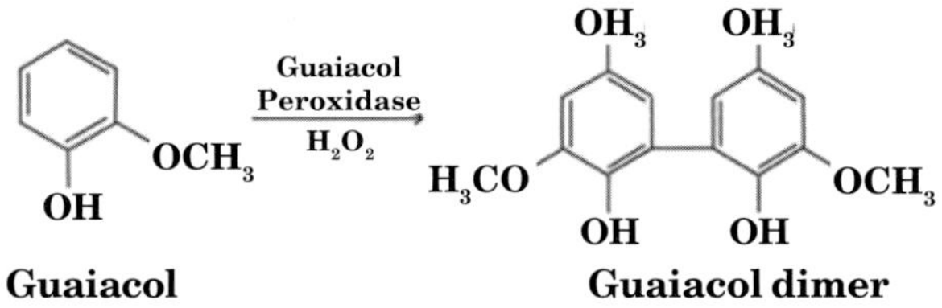

Fig 4. Guaiacol peroxidase reaction

Experiment 4: GPX activity

Materials and equipments

Assay tubes, Erlenmeyer's flasks, beakers, pipettes, analytical balance with accuracy of 0.01 mg, spectrophotometer or filter photometer capable of measurement at 470 nm, equipped with a 1-cm path length microcuvette, centrifuge capable of 3000g, ice bath, vortex mixer.

Reagents

1 mM H_2O_2 (add 60 µL of 20 v/v H_2O_2 to 100 mL of distilled water); 25 mM guaiacol (add 310 mg of guaiacol to 100 mL of 20 % aqueous glycerol); 0.5 % polyvinyl polypirrolidone, PVPP, (dissolve 0.5g of PVPP in 100 mL of water); sodium phosphate buffer (pH 6.8).

Procedure

i. Take tissue samples of 0.5g (fresh weight). Tissues should be obtained from plants infected and non-infected with pathogen (see *Observation* 1).

ii. Grind each sample in 2 mL of extraction buffer (tissue extract), using a mortar and a pestle kept on ice.

iii. Centrifuge at 3000g for 10 min at 2–5 °C and then keep in ice.

iv. Prepare reaction mixture containing:

   a. 0.2 mL of 1 mM H_2O_2

   b. 0.2 mL of 0.5 M sodium phosphate buffer

   c. 0.2 mL of guaiacol 25 mM

   d. 0.4 mL of tissue extract

v. Read increase in absorbance (guaiacol oxidation) at 470 nm over a period of 5 min at room temperature (*i.e.,* 25 °C).

vi. Obtain at least four tissue samples per treatment and repeat the experiment twice.

Calculations

Step 1: Calculate concentration of oxidized guaiacol (C) in each sample.

$$C = A_{470} / \mathring{a}_{470}\ (2), \qquad \text{where } \mathring{a}_{470} = 26.6\ \text{mM}^{-1}\ \text{cm}^{-1}$$

For example, if A_{470} is 0.66, concentration should be:

$$C = 0.66 / 26.6\ \text{mM}^{-1}\ \text{cm}^{-1}$$

$$C = 0.025\ \text{mM (or } 0.025\ \mu\text{mol mL}^{-1})$$

Step 2: Considering that amount Total Px (tPx) content in the assayed volume of tissue extract, considering that total volume (Vte) of tissue extract is known and the assayed volume is 0.4 mL.

$$\text{tPx } (\mu\text{mol}) = \text{C } (\mu\text{mol / mL}) \times \text{Vte} / 0.4\ \text{mL}$$

Following example in step 1 and considering Vte is 2.5 mL, tPx amount should be:

$$\text{tPx} = 0.025\ \mu\text{mol mL}^{-1} \times 2.5\ \text{mL} / 0.4\ \text{mL}$$

$$\text{tPx} = 1.25\ \mu\text{mol}$$

Step 3: GPX activity is expressed as amount of oxidized guaiacol (μmol min^{-1} g^{-1} fw^{-1}):

GPX activity (μmol min^{-1} g^{-1} fw^{-1}) = tPx (μmol) / Δt (min)] / tissue weight (g fw)

Following the example in step 2, if Δt is 5 min and tissue weight is 0.5g (fw):

GPX activity = (1.25 µmol/5 min) / 0.5g

GPX activity = 0.5 µmol min^{-1} g^{-1} fw

Observations

Make at least three independent experiments ($N = 3$).

6. LIPID PEROXIDATION

The activation of oxygen (O_2) is an important aspect of metabolism in plant cell (Elstner, 1982). Oxygen activation involves formation of superoxide anion free radical (O_2^-), the hydroxyl free radical (OH^-), or singlet oxygen which are all intermediates of hydrogen peroxide production. These oxygen species are highly reactive and their levels are maintained under non-toxic levels by plant cells. The action of AOS in cells is called oxidative stress. Oxidative stress is an inescapable feature of life in an oxygen atmosphere. In plant defence, resistance against pathogens is associated with a fast increase of AOS which lead to lipid peroxidation. Lipid peroxidation is a chemical reaction between polyunsaturated fatty acids and AOS. Oxygen is incorporated into the lipid molecule to form reactive hydroperoxides. These molecules easily react with cell components, such as lipids and other chemicals, generating an oxidative chain reaction. Oxidative process only terminates, if antioxidant system of plant cells are active enough to scavenge the formed free radicals. Malondialdehyde (MDA) is a secondary end product of lipid autooxidation or enzymic degradation of cell lipids, which can be measured as an indicator of lipid peroxidation. This is possible through reaction with TBA (Fig 5). Each MDA molecule reacts with two TBA molecules *via* an acid-catalyzed nucleophilic-addition reaction yielding a pinkish-red chromogen with an absorbance maxima at 532 nm (Hodges *et al.*, 1999). Classic protocols of MDA available in the literature cannot be applied on colored samples or plant tissues. The following protocol overcomes these problems, allowing reliable measurement of MDA (Sampietro *et al.*, 2007).

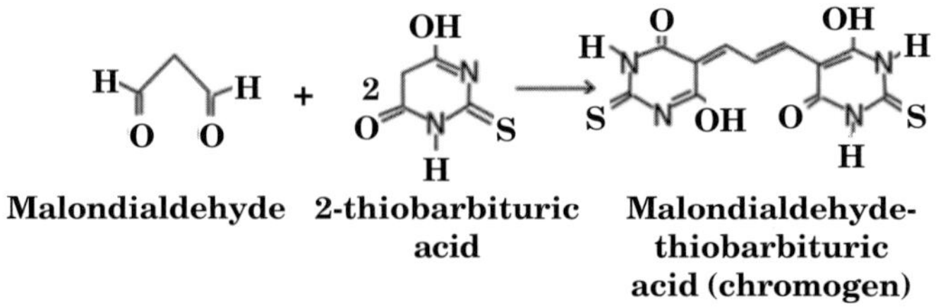

H O O H + 2 OH N O N S H → H N O S N OH OH N H O N S H

Malondialdehyde **2-thiobarbituric acid** **Malondialdehyde-thiobarbituric acid (chromogen)**

Fig 5. Malondialdehyde – thiobarbituric acid reaction

Experiment 5: MDA production

Materials and equipments

Assay tubes, Erlenmeyer's flasks, pipettes, analytical balance with accuracy of 0.01 mg, spectrophotometer or filter photometer capable of measurement at 440, 532 and 600 nm; equipped with a 1-cm path length microcuvette; centrifuge capable of 3000g; vortex mixer; boiling water bath.

Reagents

–TBA and +TBA solutions [add 40g of trichloroacetic acid (TCA) to 200 mL of distilled water; store 100 mL of this solution which is the –TBA solution; add 0.65g of TBA to the remaining 100 mL of TCA solution; this is the +TBA solution], 96° ethanol, distilled water.

Procedure

i. Take fresh tissue samples of 0.5g (fresh weight). Tissues should be obtained from plants subjected and not subjected to biotic and/or abiotic stress (see *Observation* 1).

ii. Add 3 mL of 80:20 (v/v) ethanol:water to each sample and centrifuge at 3000g for 15 min at 4 °C.

iii. Take 1 mL of each supernatant. Prepare two tubes per sample as under:

   a. **Tube 1:** Add 1 mL of +TBA solution.

   b. **Tube 2:** Add 1 mL of –TBA solution.

iv. Vortex each tube and heat at 95 °C for 25 min.

v. Cool and centrifuge at 3000g for 10 min.

vi. Read absorbance at 440, 532 and 600 nm using a 1-cm-path-length (1 mL) microcuvette.

vii. Obtain at least four tissue samples per treatment and repeat the experiment twice.

Precautions

TBA has a strong mercaptan odor. Do not breathe dust. Avoid contact with skin and eyes. TCA is caustic acid, so handle with care.

Calculations

Step 1. Calculate concentration of MDA (C) in each sample as under:

$$\text{coef. } A = [(A_{532\text{+TBA}} - A_{600\text{+TBA}}) - (A_{532\text{-TBA}} - A_{600\text{-TBA}})]$$

$$\text{coef. B} = [(A_{440+TBA} - A_{600+TBA})\ 0.0571]$$

$$\text{MDA (nmol·mL}^{-1}) = (\text{coef. A} - \text{coef. B} / 157000) \times 10^{6}$$

Step 2. Calculate total MDA (tMDA) considering total volume of supernatant (tVol):

$$\text{tMDA (µmol)} = \text{C (µmol mL}^{-1}) \times \text{tVol (mL)}$$

Step 3. Express MDA content per gram of fresh weight:

$$\text{MDA content (µmol g}^{-1}\text{ fw)} = \text{tMDA (µmol) / weight of tissue sample (g fw)}$$

Observations

Make at least three independent experiments ($N = 3$).

7. CATALASE

Catalase (EC 1.11.1.6) is one of the main H_2O_2–scavenging enzymes in plants. It decomposes H_2O_2 to H_2O and O_2 in the glyoxysomes of fatty seeds and in leaf peroxisomes. Subcellular compartmentalization of catalase suggests that it scavenges the photorespiratory H_2O_2. However, H_2O_2 is able to diffuse freely through the membranes, being possible that catalase also reduces other H_2O_2 sources (Willekens *et al.*, 1997). Several studies indicate that catalase is critical for maintaining the redox balance during oxidative stresses. Current evidences suggest that catalase activity is suppressed during the responses of plants against pathogen attack avoiding H_2O_2 sequestration, which is needed for local responses (*i.e.,* crosslinking of hydroxyproline proteins and proline proteins of cell walls or increasing peroxidase activity) and as signal in SAR response (Chen *et al.*, 1993). The following experiment may be used to measure the catalase activity in plant tissues through disappearance of H_2O_2 at 240 nm (Fig 6).

$$2H_2O_2 \xrightarrow{\textbf{Catalase}} 2H_2O + O_2$$

Fig 6. Catalase reaction

Experiment 6: Catalase activity

Materials and equipments

See *Materials and equipments* of Experiment 4, spectrophotometer or filter photometer must be capable of measurement at 240 nm, equipped with a 1-cm-path-length microcuvette.

Reagents

See *reagents* for Experiment 4.

Procedure

i. Obtain plant tissue and prepare enzyme extracts as indicated to measure peroxidase activity.

ii. Prepare reaction mixture containing:

  a. 0.2 mL of 50 mM H_2O_2

  b. 0.2 mL of 0.5 M sodium phosphate buffer

  c. 0.2 mL of guaiacol 25 mM

  d. 0.2 mL of tissue extract

iii. Record the decrease in absorbance (H_2O_2 decomposition) at 240 nm after 5 min at room temperature (see *Observations* 1 and 2). Control tube should contain 0.2 mL water, instead of H_2O_2 in the reaction mixture.

iv. Record the absorbance at 240 nm using a 1-cm-path-length (1 mL) microcuvette.

Calculations

Catalase activity is expressed as the rate constant (k) of a first-order enzymatic reaction (Aebi, 1983):

$$k = (2.3/\text{Dt})(\log A1/A2)$$

where $A1$ and $A2$ are the absorbance at 240 nm of H_2O_2 at the start and the end of the time interval (Δt), respectively.

For example, if $A1$ is 0.45, $A2$ is 0.40 and Δt is 15 s, then:

$$k = (2.3 / 15 \text{ s}) (\log 0.45 / 0.40)$$

$$k = 0.15 \times 0.05 \quad \text{Hence,} \quad k = 7.68 \times 10^{-3} \text{ s}^{-1}$$

Catalase activity should be measured at 15 s intervals within the first minute of reaction. Then plot graph k vs time for each treatment (see *Observation* 3).

Observations

1. Readings at different times may be appropriate to follow catalase activity.

2. Make at least three independent experiments ($N = 3$).

3. Catalase also can be related to the disappearance of H_2O_2, considering that $\varepsilon = 40 \text{ mM}^{-1} \text{ cm}^{-1}$.

8. PHENYLALANINE AMMONIA-LYASE (PAL)

Phenyl alanine ammonia-lyase (PAL, EC 4.3.1.5) is a key enzyme of phenylpropanoid pathway for synthesis of many derivatives of cinnamic acid. PAL catalyzes the non-oxidative deamination of l-phenylalanine to *trans*-cinnamic acid (Fig 7) which is the precursor of many phenylpropenoids, viz., lignins, flavonoids and coumarins (Dixon *et al.*, 2002). An increase in PAL activity is often observed in plant defence, which may imply the synthesis of antimicrobial compounds (*i.e.,* phytoalexins of flavonoid nature) and/or structural components needed for lignin synthesis and crosslinking with cell wall polysaccharides (*i.e.,* ferulic and *p*-coumaric acids). In this experiment, PAL activity is measured through production of *trans*-cinnamate, the content of *trans*-cinammate is measured spectrophotometrically at 290 nm.

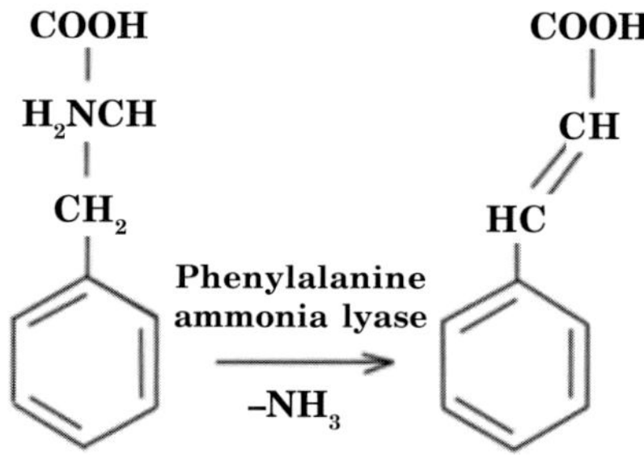

L-Phenylalanine ***trans*-cinnamate**

Fig 7. Phenylalanine ammonia-lyase reaction

Experiment 8: PAL activity

Materials and equipments

Bean seeds, analytical balance, beakers, pipettes, assay tubes, centrifuge tubes, dark bottle, germination paper, mortar and pestle, pH meter (or pH indicator sticks), polypropylene flasks, refrigerated preparative centrifuge capable of speed of 2000g, water bath capable to incubate at 40 °C, ice bath.

Reagents

Nutrient solution: [Prepare a nutrient solution containing 6 mM KNO_3, 3 mM $Ca(NO_3)_2$, 3 mM $MgSO_4$, 1.5 mM KH_2PO_4, 125 µM Fe–EDTA, 10 µM $MnSO_4$, 1 µM $CuSO_4$, 1 µM $ZnSO_4$, 30 nM $(NH_4)_6(Mo7O_2)_4$ and 0.1 µM $CoCl_2$; adjust pH to 5.5]; methanol; 1 N NaOH (dissolve 40 g of NaOH in 500 mL of distilled water and make up to 1 L); 0.1 M sodium borate buffer (dissolve 3.8 g of $Na_2B_4O_7$ in 80 mL of distilled water; adjust to pH 8.8 with 1 N NaOH and make up to 100 mL); 50 mM l-phenylalanine (dissolve 0.8g of l-phenylalanine in 100 mL of distilled water).

Procedure

i. **Bean seedlings:** Place the bean seeds in plastic trays (*i.e.,* 55 × 40 × 15 cm) filled with sterile sand. Irrigate with nutrient solution. Place trays in a greenhouse at 25 °C, relative humidity of 80 % and 14 h photoperiod. Germinated seeds are left to grow until plantlet stage at which the primary leaves become expanded (about 9 days). Irrigate again during the growth period if needed.

ii. **Induction of PAL activity:** Spray leaves with 10 mM salicylic acid with a hand atomizer. Plants sprayed with water will serve as control. After 3 days, collect leaves from each treatment at random. Weigh the tissue, freeze on dry ice, and store at –80 °C for later analysis.

iii. **Measure of PAL activity:** Procedure for each sample is as under:

   i. Grind 2g of leaves in a mortar with a pestle, placed on an ice bath, with 4 mL of 0.1 M sodium borate buffer (see *Observation* 1). Centrifuge homogenates at 2200g for 15 min, containing 0.05g of polyvinylpyrrolidone, in a blender. Filtrate through a double cheesecloth layer and centrifuge (25000g) for 15 min.

   ii. Prepare the reaction mixture adding to a tube:

      a. 1 mL of sodium borate buffer

      b. 0.25 mL of enzyme extract

      Preincubate for 5 min at 37 °C (see *Observation* 2).

   iii. Add 0.3 mL of l-phenylalanine to the reaction mixture and incubate for 2 h at 37 °C.

   iv. Read absorbance of reaction mixtures at 290 nm, against a control consisting in the reaction mixture plus 0.3 mL of distilled water.

Statistical analysis

Experiment is designed in a split plot arrangement in a randomized complete design with factors A (cultivars) × B (treatments) and 5 replicates. The sample analyses are done in triplicate and the enzyme specific activities are subjected to analysis of variance. Correlation coefficients may be calculated and means are compared using the Duncan test.

Calculations

Step 1. Calculate concentration of *trans*-cinnamic acid (C) in each sample.

$$C = A_{290} / \varepsilon_{290} \,(2), \quad \text{where } \varepsilon_{290} = 9630 \text{ M}^{-1} \text{ cm}^{-1}$$

For example if A_{290} is 0.35, concentration should be:

$$C = 0.35 / 9630\ M^{-1}\ cm^{-1}$$

$$C = 0.036\ mM\ (\text{or } 36.34\ \mu mol\ mL^{-1})$$

Step 2. Considering that amount total *trans*-cinnamic acid (tta) content in the assayed volume of tissue extract, considering that total volume (Vte) of tissue extract is known and the assayed volume is 0.25 mL.

$$tta\ (\mu mol) = C\ (\mu mol / mL) \times Vte / 0.25\ mL$$

Following example in step 1 and considering Vte is 1.55 mL, tta amount should be:

$$tta = 36.34\ nmol\ mL^{-1} \times 1.55\ mL / 0.25$$

$$tta = 225.3\ nmol$$

Step 3. PAL activity is expressed as amount of formed *trans*-cinnamic acid (nmol h^{-1} g^{-1} fresh weight):

$$\text{PAL activity (nmol } h^{-1}\ g^{-1}\ fw^{-1}) = \frac{[\text{tta (nmol } mL^{-1}) / \Delta t]}{\text{weight of tissue (g fw)}}$$

Following the example in step 2, if Δt is 2 h and tissue weight is 2g:

PAL activity = (225.3 nmol /2 h)/2 g; Hence, PAL activity = 56.33 nmol h^{-1} g^{-1} fw^{-1}

Observations

1. All material (pestle and mortar, centrifuge tube) must be refrigerated before enzyme extraction.
2. Make at least three independent experiments ($N = 3$).

9. LECTINS

Lectins are proteins or glycoproteins which recognize and bind the specific carbohydrates forming bonds very similar to those of antigen-antibody or substrate-enzyme. They can be defined as “plant proteins possessing at least one non-catalytic domain, which binds reversibly to a specific mono- or oligo-saccharide”. This definition states both main properties of lectins: bond reversibility and specificity. Lectins are profusely found in seeds of *Leguminosae* family, but they are not exclusive of them. Viruses, bacteria, fungi, mammals and animals such as nematodes and eels also contain lectins. Several functional role were proposed for lectins. Perhaps the most consistent function proposed for lectins is plant defence, because they allow specific recognitions and sometimes plant lectins bind to carbohydrates other than those synthesized by plants (*i.e.,* chitine often found in insects' exoskeleton or fungi cell walls and syalic acid which is the main carbohydrate of animals' glycoproteins).

Lectins can be classified according to their structures as under:

a. **Merolectins:** These are proteins with a single carbohydrate-binding domain. They are small and due to their monovalent nature, are not able to precipitate glycoconjugates or agglutinate cells.

b. **Hololectins:** These also consist in only carbohydrate-binding domains but contain two or more identical or homologous domains that bind either the same or very structurally similar sugars. As they are multivalent, they can agglutinate cells and/or precipitate glycoconjugates. Superlectins are a special type of hololectins, which have two or more carbohydrate-binding domains able to recognize structurally unrelated sugars.

c. **Chimerolectins:** They are proteins consisting of one or more carbohydrate-binding domains tandemly arrayed to an unrelated domain, which can possess a well-defined enzymatic or another biological activity, acting independently from carbohydrate-binding domain.

8.2 Agglutination and agglutination inhibitors

When a hololectin is in contact with many cells (*i.e.,* red cells and lymphocytes), it agglutinates them (Fig 8). This is because each hololectin molecule can bind to carbohydrates on the surface on different cells producing their grouping. In red cells, this is macroscopically observed as red clumps. Adding enough amount of a carbohydrate together with the hololectin, red clumps will not be observed. This is known as the hemoagglutination inhibition.

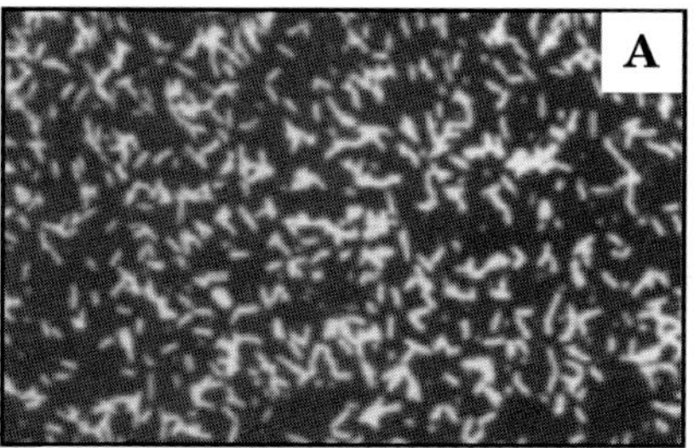

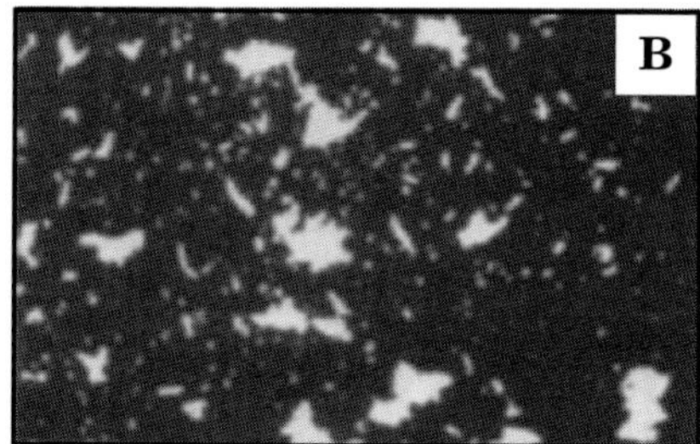

Fig 8. Agglutination reaction: red cell suspension (A) before and (B) after addition of a plant lectin (agglutination reaction)

Experiment 8: Partial purification of a lectin from lentil seeds

Materials and equipments

Assay tubes, Erlenmeyer's flasks, pipettes, blender, lentil seeds, centrifuge capable of 17500g, Pasteur pipette, cheese cloth.

Reagents

10 mM phosphate buffered saline, PBS (add 0.23g of NaH_2PO_4, 1.15g of NaH_2PO_4 and 5.8g of NaCl to 900 mL of distilled water. Adjust pH to 7.2 with 1 M NaOH or 1 M HCl and take up to final volume of 1000 mL); red cell suspension (Add 0.5 mL of blood collected with anticoagulant to 4.5 mL of 0.9 % NaCl (see *Observation* 1); homogenise carefully inverting the tube and centrifuge at 400g, eliminate the supernatant with a Pasteur pipette, avoiding re-suspension of red cells; this washing procedure is twice repeated; re-suspend red cells in 4.5 mL of 0.9 % NaCl to obtain a suspension with 4 % of red cells); Bradford reagent (dissolve 100 mg Coomassie Brilliant Blue G-250 in 50 mL 95 % ethanol; add 100 mL 85 %, w/v phosphoric acid; dilute to 1 L with distilled water when the dye has completely dissolved and filter through Whatman #1 paper just before use); stock solution of commercial bovine serum albumin, BSA (dissolve 100 mg of BSA in 100 mL of distilled water); 0.1 mg/mL standard BSA solution (dilute 1 mL of BSA stock solution in 9 mL distilled water; this solution can be stored at –20 °C and used whenever is needed).

Procedure

i. Add 50g of lentil to 1000 mL of distilled water. Leave in the refrigerator overnight.

ii. Grind seeds in a blender with 50 mL of 10 mM PBS. Filter through double cheese cloth.

iii. Centrifuge at 17500g for 10 min at 4 °C. Add solid $(NH_4)_2SO_4$ to the supernatant until 50 % saturation (see *Observation* 2).

iv. Centrifuge at 17500g for 10 min at 4 °C.

v. Dissolve the precipitate in a small volume of 0.9 % NaCl.

vi. Add NaCl solution in a dialysis tube previously hydrated and clipped at one of its ends. Then, clip the other end and place in a beaker filled with 0.9 % NaCl. Leave dialyzing overnight.

vii. Centrifuge the dialyzed extract. Content of the dialysis tube is then added to a 2 × 17 cm column filled with Sephadex G-150 pre-equilibrated with 0.9 % NaCl. The lectin is adsorbed to the dextran matrix of the Sephadex. Then, wash the column with 0.9 % NaCl to eliminate no-adsorbed proteins and then elute lectin from the column with 0.1 M glucose. Receive fractions in assay tubes.

viii. Measure total protein content in column fractions by Bradford method (see *Observation* 3).

ix. Evaluate agglutination of red cells in the collected fractions. To do it, mix in a tube 0.1 mL of each fraction with 0.1 mL of the red cell suspension. Agglutination reaction is visualized as red clumps.

x. Titer each fraction as follows:

   a. Label 8 tubes and add 0.1 mL of 0.9 % NaCl in each tube.

   b. Add 0.1 mL of the column fraction to tube 1 and vortex (1/2 dilution). Take 0.1 mL from tube 2 and vortex (1/4 dilution). Proceed as indicated until tube 8 (1/256 dilution). Discard 0.1 mL from tube 8. Add 0.05 mL of red cell suspension and 0.1 mL of PBS to each tube.

   c. Prepare control adding 0.2 mL of NaCl 0.9 % to 0.05 mL of red cell suspension (red cell suspension control). Clumps should be absent in this tube.

   d. Add 0.1 mL of red cell suspension from tube 1–8. Leave 30 min at room temperature and observe if red clumps are formed.

Calculations

Step 1. Calculate titration value (TV): The hemagglutination titer is the inverse of highest dilution of a fraction causing agglutination of red cells. For example, considering a fraction if hemagglutination is observed at dilution 1/4 but not at 1/8, then hemoagglutination titer is 4.

Step 2. Calculate the hemoagglutination unit (HAU): The HAU is the minimum amount of lectin needed to agglutinate red cells and is calculated as follow:

$$\text{HAU } (\mu g\ mL^{-1}) = \text{TPC } (\mu g\ mL^{-1}) / \text{TV}$$

where TPC is total protein concentration and TV is titration value. Following the example in step 1, if TPC is 200 $\mu g\ mL^{-1}$:

$$\text{HAU} = 200 / 4 \qquad \text{then, HAU} = 50\ \mu g\ mL^{-1}$$

Lectin specificity to carbohydrate binding can be evaluated adding glucose, mannose or galactose on dilutions containing 2 HAU. To do it, add 0.1 mL of a fraction containing two HAU, 0.1 mL of 0.2 M glucose, galactose or mannose and 0.05 mL of the red cell suspension at 4 %. Leave 30 min and watch for red clump formation. If inhibition of agglutination occurs, no red clumps will be observed.

Observations

a. Blood of any group can be used because lectin from lentil is not specific to a group. For this experiment, it was assumed that 100 mL blood has 40 mL of red cells. Hence, 0.5 mL should have 0.2 mL of red cells.

b. Label tubes from 1 to 5. Dilute 0.1 mg mL^{-1} BSA solution with distilled water to prepare solutions 25, 50, 75 µg mL^{-1}. Then, add 2 mL of 0, 25, 50, 75 and 100 µg mL^{-1} BSA to tube 1, 2, 3, 4 and 5, respectively. Prepare tubes 6 and 7 adding 0.10 and 0.20 mL of the protein homogenate preparation. Complete volume of tube 6–0.20 mL with distilled water. Add 1 mL Coomassie Brilliant Blue solution and vortex. Allow to stand 5 min at room temperature and read absorbance at 595 nm. Graph the calibration curve relating absorbance at 595 nm to BSA concentration. Calculate total protein concentration in the protein homogenate (µg mL^{-1}) expressed as BSA equivalent concentration, considering absorbance at 595 nm (A_{595}) from experimental tubes and extinction coefficient (ε) obtained from linear regression.

c. Saturation until 50 % is achieved adding 35.3g of $(NH_4)_2SO_4$ per 100 mL of supernatant.

10. SALICYLIC ACID (SA)

Salicylic acid (SA) is benzoic acid derivative, with signalling functions in local and SAR resistance responses (Fig 9). An increase of SA concentration in plant tissues has been related to expression of pathogenesis-related (PR) proteins and other protective factors induced following pathogen challenge. When SAR is activated, SA concentration increases in plant organs both inoculated and non-inoculated with the pathogen (Dmitriev, 2003). This increase of SA can be followed through the plant, collecting phloem sap from different plant organs. Local increases of SA in plant tissues has been related to transient higher levels in PAL activity suggesting that SA is synthesized *de novo* in such tissues but not translocated through the plant (Rasmussen *et al.*, 1991). This also means that an unknown mobile molecule

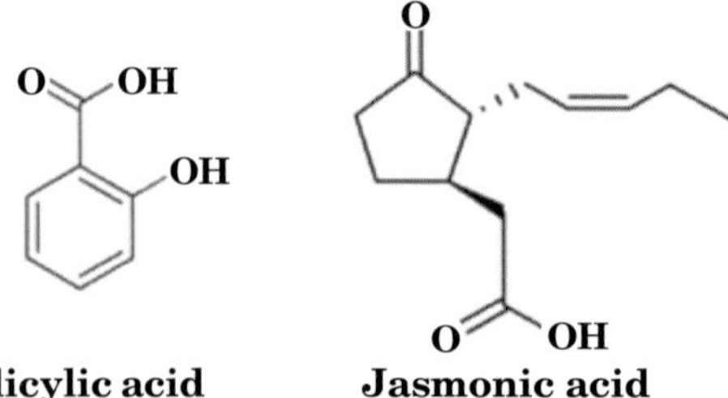

Fig 9. Chemical structure of salicylic and jasmonic acids

other than SA is responsible for SAR signalling through the non-inoculated plant parts (Becker *et al.*, 1998). A protocol is proposed in the experiment below to follow induction of SA synthesis in cucumber seedlings using Thin Layer Chromatography (TLC).

Experiment 9: Induction of SA biosynthesis

Materials and equipments

Sterile syringe and needle, suspension of *Pseudomonas syringae*, cucumber seedlings, 50 µL capillary pipette, centrifuge capable of 10000g, plates of Silica gel 60 F-254, 1 L glass beaker and a glass piece to cover it (or a TLC tank), long wave UV lamp, analytical balance with accuracy of 0.01 mg.

Reagents

Ethanol, dioxane, ethyl acetate, acetic acid, benzene, salicylic acid solution (add 1 mg of salicylic acid to 10 mL of ethanol).

Procedure

i. Inoculate cucumber plants with two leaves, by injection into one leaf with cells of a wild type of *Pseudomonas syringae* (recommended level of inoculum: 1×10^8 CFU mL^{-1}).

ii. Obtain samples of phloem from cucumber seedlings at different times (6, 12, 16, and 24 h). To do it, cut the ends of petioles of cucumber seedlings. Collect phloem exudates from these cuts with a 50 µL capillary pipette.

iii. Place known volumes of phloem exudates into three volumes of ethanol to precipitate proteins and other high molecular weight materials.

iv. Centrifuge at 10000g for 5 min. Extract pellets with ethanol and combine ethanol extracts.

v. Evaporate ethanol extract to dryness in a rotary evaporator at 40 °C and dissolve in 50 % ethanol, in a volume equal to that original collected as phloem exudates.

vi. Spot ethanol extract onto a silica gel plate and develop it (see *Observation* 1). Solvents suggested as mobile phase are toluene: dioxane: acetic acid (90:25:4; v/v), acetic acid: chloroform (1:9; v/v) and ethyl acetate: benzene (9:11; v/v). A control consisting in a salicylic acid standard solution with known concentration is also developed.

vii. Visualize salicylic under long wave UV light (365 nm, see *Observations* 2 and 3).

Precautions

Dioxane, ethyl acetate, acetic acid and benzene are toxic solvents. Avoid exposure. Obtain special instructions before use.

Observations

1. See Experiment 2, *Observations* 1.

2. Content of salicylic acid can be measured using a fluorometer. When fluorometer is not available, an alternative is to take a picture of the developed plate and then scan it in a computer for image analysis. NIH image (http://rsb.info.nih.gov/nih-image/) and Scion image (http://www.scioncorp.com/) are public domain image processing programs often used for TLC analysis.

3. Salicylic acid can also be visualized spraying with Folin & Ciocalteu's reagent. Dilute the commercial source of this reagent with distilled water in a ratio 1:3 and spray onto the plate. After drying, the plate is also exposed to ammonia fuming to increase intensity. Salicylic acid is visualized as a blue spot.

11. JASMONIC ACID (JA)

Jasmonic acid (JA) is an oxylipin-like hormone synthesized from α-linolenic acid (Fig 9) through octadecanoid pathway (Vick and Zimmerman, 1984). Increases in JA have been observed in response to pathogen attack, wounding and osmotic stress both at local and systemic levels. For this reason, JA is considered a member of the signal transduction pathway in plant defence. JA has been involved in the induction of several genes encoding pathogenesis related proteins (*i.e.,* chitinases and glucanases – enzymes that degrade polysaccharides from fungal cell walls) and also of low-molecular weight proteins with antimicrobial activity known as defensins (Dmitriev, 2003). The following protocol allows the evaluation of JA content in plant tissues.

Experiment 10: Induction of JA in plant tissues.

Materials and equipments

Sterile syringe and needle, suspension of *Pseudomonas syringae*, cucumber seedlings, plates of Silica gel 60 F-254, 1 L glass beaker and a glass piece to cover it (or a TLC tank), analytical balance with accuracy of 0.01 mg, rotary evaporator.

Reagents

p-anisaldehyde reagent (dissolve 1 mL of *p*-anisaldehyde and 1 mL of H_2SO_4 in 18 mL of ethanol); methanol; ethyl acetate; saturated $NaHCO_3$ solution (add solid $NaHCO_3$ to 100 mL of distilled water until incipient sediment is observed); chloroform; anhydrous Na_2SO_4; 2 N HCL (dilute 16.6 mL of concentrated HCl, 12.06 M, in 83.4 mL of distilled water and store in a dark bottle); a JA standard solution.

Procedure

i. Proceed as indicated in step 1 of *Procedure* for SA biosynthesis.

ii. Add 10 g (fresh weight) of plant tissue to 50 mL of methanol cooled at 4 °C. Leave for 10 min, recuperate the solvent and add again 50 mL of cool methanol to the plant tissue.

iii. Filtrate the methanolic extractions with filter paper. Evaporate the methanolic extract to dryness in a rotary evaporator.

iv. Dissolve the residue in 50 mL of ethyl acetate. Then, extract components of the organic phase adding 30 mL of the saturated $NaHCO_3$ solution to the separation funnel and the ethyl acetate. Insert the stopper, invert the separation funnel and shake vigorously for 3 min. Place the separating funnel with the stopper in the upside and leave 10 min to separate in two phases. Carefully recover the lower phase ($NaHCO_3$ solution) transferring it to an Erlenmeyer's flask. Discard the upper phase.

v. Acidify the $NaHCO_3$ phase with 2 N HCl. Add the acidified phase to the separation funnel. Then, add 10 mL of chloroform. Proceed as indicated in step (iii), but transferring the upper phase (chloroform) to a 20 mL assay tube. Add anhydrous Na_2SO_4 to eliminate water from the organic phase.

vi. Spot the chloroform phase and JA standard solution on a TLC plate (see *Observation* 1). Develop the plate in chloroform-methanol-water (140: 20:1; v/v/v).

vii. After development, leave mobile phase to evaporate. Then, spray the *p*-anisaldehyde solution and heat for 5–10 min at 120 °C in an oven. In this solvent, JA has Rf < 0.5 (see *Observation* 2).

Observations

1. See Experiment 2, *Observation* 1.

2. See Experiment 8, *Observation* 2.

Suggested Readings

Aebi., H.E. (1983). Catalase. *In*: *Methods of enzymatic analysis* (*Ed.*, H.U. Burgmeyer). pp. 273–286. Academic Press, New York.

Becker, J.S., Marois, E., Huguet, E.J., Midland, S.L., Sims, J.J. and Keen, N.T. (1998). Accumulation of salicylic acid and 4-hydroxybenzoic acid in phloem fluids of cucumber during systemic acquired resistance is preceded by a transient increase in phenylalanine ammonia-lyase activity in petioles and stems. *Plant Physiology* **116**: 231–238.

Buchanan, B.B., Gruisem, W. and Jones, R.L. (2000). *Biochemistry and Molecular Biology of Plants*. American Society of Plant Physiologists. Rockville, USA.

Bueno, S.M. and Garcia-Cruz, C.H. (2006). Optimization of polysaccharides by bacteria isolated from soil. *Brazilian Journal of Microbiology* **37**: 296–301.

Chen, Z., Silva, H. and Klessig, D.F. (1993). Oxygen species in the induction of plant systemic acquired resistance by salicylic acid. *Science* **262**: 1883–1886.

Dmitriev, A.P. (2003). Signal molecules for plant defence responses to biotic stress. *Russian Journal of Plant Physiology* **50**: 417–425.

Elstner, E.F. (1982). Oxygen activation and oxygen toxicity. *Annual Review of Plant Physiology* **33**: 73–96.

Goldstein, W.S. and Spencer, K.C. (1985). Inhibition of cyanogenesis by tannins. *Journal of Chemical Ecology* **11**: 47–58.

Harborne, J.B. (1998). *Phytochemical Methods: A Guide to Modern Techniques of Plant Analysis*. Chapman and Hall, London.

Hodges, D.M., DeLong, J.M., Forneym C.F. and Prange, R.K. (1999). Improving the thiobarbituric acid-reactive-substances assay for estimating lipid peroxidation in plant tissues containing anthocyanin and other interfering compounds. *Planta* **207**: 604–611.

Hutcheson, S.W. (1998). Current concepts of active defence in plants. *Annual Review of Phytopathology* **36**: 59–90.

McCue, P. and Shetty, K. (2002). A biochemical analysis of mung bean (*Vigna radiata* L.) response to microbial polysaccharides and potential phenolic-enhancing effects for nutraceutical applications. *Food Biotechnology* **16**: 57–79.

Morrissey, J.P. and Osbourn, A.E. (1999). Fungal resistance to plant antibiotics as a mechanism of pathogenesis. *Microbiology Molecular Biology Review* **63**: 708–724.

Rasmussen, J.B., Hammerschmidt, R. and Zook, M.N. (1991). Systemic induction of salicylic acid accumulation in cucumber after inoculation with *Pseudomonas syringae* pv *syringae*. *Plant Physiology* **97**: 1342–1347.

Sampietro, D.A., Sgariglia, M.A., Soberón, J.R., Quiroga, E.N. and Vattuone, M.A. (2007). Role of sugarcane straw allelochemicals in the growth suppression of arrowleaf sida. *Environmental and Experimental Botany* **60**: 495–503.

Stout, M.J., Workman, K.V., Bostock, R.M. and Duffey, S.S. (1998). Stimulation and attenuation of induced resistance by elicitors and inhibitors of chemical induction in tomato (*Lycopersicon esculentum*) foliage. *Entomologia Experimentalis et Applicata* **86**: 267–279.

Vick, B.A. and Zimmerman, D.C. (1984). The biosynthesis of jasmonic acid by several plant species. *Plant Physiology* **75**: 458–461.

Willekens, H., Chamnongpol, S., Davey, M., Schraudner, M., Langebartels, C., Van Montagu, M., Inzé, D. and Van Camp, W. (1997). Catalase is a sink for H_2O_2 and is indispensable for stress defence in C_3 plants. *EMBO Journal* **16**: 4806–4816.

CHAPTER 9

Plant Growth Promoting Rhizobacteria Assays

FABRICIO CASSÁN[1], OSCAR MASCIARELLI[1] AND VIRGINIA LUNA[1]

1. INTRODUCTION

Soil is a natural system colonized by various microorganisms including bacteria, fungi, and actinomycetes (Foster, 1988). The portion of soil in which microbial proliferation is induced by plant roots is called "rhizosphere" (Garate and Bonilla, 2000). Bacteria are most common soil microorganisms and those growing in the rhizosphere are called "rhizobacteria". Rhizobacteria possessing direct capacity to promote plant growth are referred as Plant Growth Promoting Rhizobacteria, (PGPR; Kloepper *et al.*, 1989) and those that promote plant growth through indirect mechanism are called biocontrol Plant Growth Promoting Bacteria, (biocontrol-PGPB; Bashan and Holguin, 1998). Recently a new group of free-living microorganisms involved in direct regulation of the plant's abiotic stress response has been reported and called Plant Stress Homeo-regulating Rhizobacteria, (PSHR; Cassán *et al.*, 2005). Direct promotion of growth occurs when rhizobacteria provide compounds that affect plant metabolism, or when they facilitate non-available nutrient uptake from soil. In PGPR, the most important direct plant growth promoting mechanism besides biological nitrogen fixation is synthesis of phytohormones or plant growth regulating compounds. Examples are the production of indole-3-acetic acid (IAA) by *Acetobacter diazotrophicus* and *Herbaspirillum seropedicae* (Bastian *et al.*, 1998); zeatin and ethylene by *Azospirillum* sp. (Strzelczyk *et al.*, 1994); gibberellic acid (GA_3) by *Azospirillum lipoferum* strain op33 (Bottini *et al.*, 1989); and abscisic acid (ABA) by *Azospirillum brasilense* strains Cd and Az39 (Perrig *et al.*, 2007; Boiero *et al.*, 2006). Another important PGPR mechanism is biosynthesis and release of siderophores, acidic compounds that supply iron or phosphates to the plant when not available in the soil (Seshadri *et al.*, 2000). PGPR's comprises free-living bacteria that colonize the rhizosphere or certain tissues within the plant (Kloepper *et al.*, 1989). Besides PGPR, there are many symbiotic rhizobacteria that establish a complex, intimate, mutual beneficial relationship with the plant (Bohlool, 1990). Presently, the most common explanation for the effect of rhizobacteria on plants is

1. Plant Physiology Lab, University of Río Cuarto, Ruta 8 Km 601, 5800, Río Cuarto, Córdoba, Argentina. E-mail: fcassan@exa.unrc.edu.ar

based on production of phytohormones that modify the plant metabolism and morphology, leading to improved mineral and water absorption. In contrast, the additive hypothesis of Bashan and Levanony (1990) proposes that rhizobacteria operate through multiple mechanisms which act simultaneously or in succession to promote plant growth. Inoculant produced with symbiotic and free-living PGPR, are being used in modern agriculture in several countries (Bashan and Holguin, 1998). An inoculant consists of not only microorganisms but also their metabolites released into the growing medium and they influences early germination and seedling growth. Thus, it is important to evaluate all the inoculant components and bacterial mechanisms that promote plant growth to make accurate quality control or strain selection for inoculants´ formulation. This chapter describes the experiments to evaluate bacterial mechanisms involved in plant growth promotion like production of phytohormones, nitrogen fixation, production of siderophores, phosphate solubilization and polyamine or amino acid biosynthesis and describes two experimental approaches: (a) the bacterial culture in chemically defined medium (see Experiments 1, 3, 5, 7, 9 and 10); and (b) the bacterial plant growth promotion under hydroponics culture conditions (see Experiments 2, 4, 6 and 8).

Experiment 1: Identification and quantification of indol-3-acetic acid (IAA), zeatin (Z), abscisic Acid (ABA) and gibberellic acid (GA_3) in a chemically defined medium

Principle

The effects of PGPR on plant growth are commonly explained through their production of phytohormones such as auxins (IAA), gibberellins (Bashan *et al.*, 2004), cytokinins (Tien *et al.*, 1979) and abscisic acid (ABA). Production of phytohormones by microbial isolates in chemically defined media varies among different species and strains of the same species and is also influenced by culture conditions, growth stage and availability of substrates and precursors (Boiero *et al.,* 2007; Perrig *et al.,* 2007). *In vitro* studies have demonstrated that some microbial strains can produce small amounts of phytohormones in absence of physiological precursors. Nevertheless, the presence of precursors (*e.g.,* tryptophan for auxins and mevalonic acid for gibberellins) often increase the microbial production of phytohormones. The procedure proposed below allows identification and quantification of phytohormones in a chemically defined medium. Experiments are conducted to investigate the production and release of phytohormones in the medium that is modified by inoculation with different strains.

Materials required

Bacterial cultures in late exponential growth phase, pH meter (or pH indicator sticks), refrigerated preparative centrifuge capable of 8000g, deuterated internal standards ($^2H_6$–ABA, $^2H_5$–IAA, or $^2H_2$–GA_3), orbital evaporator, high performance liquid chromatography system coupled to a

UV diode-array spectrometer (HPLC-UV) provided with a reverse phase C_{18} HPLC column (300 × 3.9 mm), gas chromatography-mass spectrometer system (GC-MS) provided with a DB1-15N (15 m × 0.25 mm, 0.25 µM methyl silicone) capillary column.

Reagents

Solutions: 1 % acetic acid in water, 1 % acetic acid in ethyl acetate, 1 % acetic acid and 30 % methanol in water, 1 % acetic acid and 15 % acetonitrile in water, 30 % methanol in water, 50 % pyridine in BSTFA [bis (trimethylsilyl) trifluoroacetamide plus 1 % trimethyl-chlorosilane] solution: mix an equal volume of pyridine and BSTFA solutions in a dark glass flask before the use.

Procedure

i. The bacterial cultures are separated into several fractions of 20 mL to determine IAA, ABA, GA_3, and zeatin.

ii. Fractions are centrifuged at 8000g for 20 min at 4 °C and supernatants are acidified at pH 2.5 with 1 % acetic acid in water.

iii. 100 ng of corresponding $^2H_6$–ABA, $^2H_5$–IAA, or $^2H_2$–GA_3 deuterated internal standard are separately added to individual acidified fractions and kept at 4 °C for 2 h. No deuterated internal standard is used for zeatin.

iv. Each sample is partitioned 4 times with the same volume of 1 % acetic acid in ethyl acetate solution. After the last partition, acidic ethyl acetate is evaporated to dryness at 36 °C. Dried samples are diluted in 100 µL of 1 % acetic acid and 30 % methanol solution for ABA determination, 1 % acetic acid and 15 % acetonitrile solution for IAA determination and 30 % methanol solution for GA_3 and zeatin determination.

v. Samples are injected into the HPLC–UV for identification and quantification of zeatin or for pre-purification of IAA, ABA and GA_3. Elution is done at 1 mL min^{-1} flow rate and fractions eluting at the corresponding retention time from each pure standard are collected. Zeatin is identified by HPLC–UV at 254 nm (Tien *et al.*, 1979) and quantified by comparison with a pure external standard.

vi. IAA, ABA and GA_3 are identified and quantified by GC–MS with a selective ion monitoring. To do it, HPLC–UV–absorbing fractions obtained at 254, 262 and 220 nm grouped for IAA, ABA, and GA_3, respectively, are evaporated to dryness at 36 °C and then methylated for 30 min at 90 °C by addition of 100 µL of 50 % pyridine in BSTFA solution, to obtain methyl-trimethylsilyl derivatives (MeTMSi) of IAA, ABA, and GA_3. The GC temperature program is 60–195 °C at 20 °C min^{-1}, then 4 °C min^{-1} to 260 °C. Helium is used as carrier gas at a flow rate of 1 mL min^{-1} with an interface temperature of 280 °C. Data acquisition is controlled by a computer.

Observations

1. Make at least three independent experiments ($N = 3$).

2. Bacterial supernatant should not be exposed for direct light. The use of 0.5 mg of 2,6-di-tert-butyl-*p*-cresol in 20 mL of bacterial supernatant is strongly recommended for ABA or IAA determination.

3. Pyridine: BSTFA solution should be prepared immediately before use.

4. Using the isocratic HPLC method, identification of zeatin is possible by comparison of its peak retention time (Rt) in the sample with pure external standard. The compound is quantified comparing the peak area obtained with the pure external standard at the same retention time. Based on peak areas, zeatin quantification is determined as described in calculations.

5. Culture medium used for the bacterial fermentation could be supplemented with 1 µg mL^{-1} of l-tryptophan before inoculation. This fact, make it possible to evaluate if the microorganisms ability to produce IAA depend of the presence of its natural precursor.

Calculations

The amount of free ABA is calculated by comparing the peak areas of ion at a mass/charge (*m/z*) 196 (molecular ion for [$^2H_6$]ABAMeTMSi) and the ion at *m/z* 190 (molecular ion for [^{1}H]ABAMeTMSi) at the corresponding time (Kovats, 1958). Similarly, we calculate the amount of free IAA by comparing the peak areas for the parent ion (*m/z*) 194 and (*m/z*) 189 and amount of free GA_3 by comparison of peak areas for parent ion (*m/z*) 508 and (*m/z*) 506 (Fig 1). Zeatin is quantified by comparing the peak area obtained at the same retention time (Rt) for internal standard and the sample. The general equation for quantification is described below and the concentration of phytohormone is expressed in ng of ABA/AIA/Z/GA_3. mL^{-1} of culture medium.

$$[C_P\,(\mu L\ L^{-1})]\cdot[A_S]^{-1} = [K_P\,(\mu L\ L^{-1})]\cdot[A_I]^{-1}$$

where C_P is phytohormone concentration in a sample; K_P is phytohormone concentration in an internal standard; A_S is peak area in sample and A_I is peak area in internal standard.

Statistical analysis

Data from three independent experiments are analyzed by ANOVA followed by *post hoc* Tuckey's test.

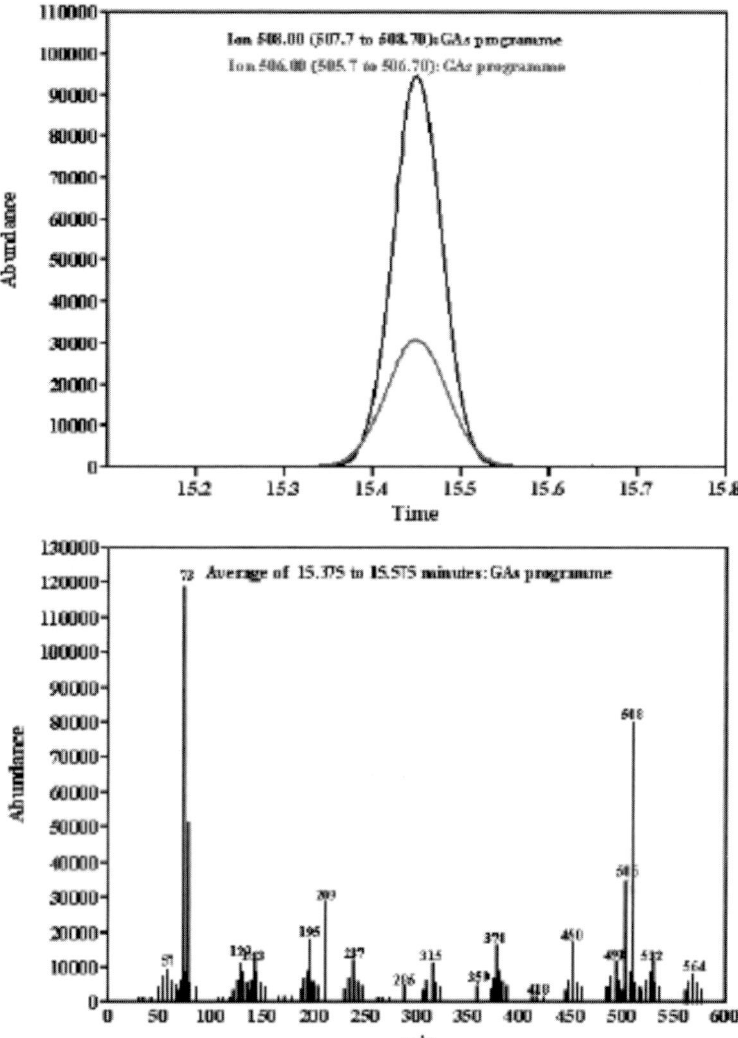

Fig 1. GC–MS–SIM chromatogram for a hypothetic PGPR sample obtained after growth for 48 h in chemically defined medium. Superior part show the retention time range (15.375 to 15.575 min) for a peak of the deuterated external standard [$^2H_2$]GA_3MeTMSi. The ion population as well as the abundance of the peak is confirmed in the inferior part according to the retention time selected

Precautions

Methanol is flammable, keep away from ignition sources. It is toxic by inhalation, direct contact or swallowing.

Ethyl acetate is toxic by inhalation and if swallowed. Irritating to the eyes and skin.

Experiment 2: Physiological effects of bacterial gibberellic and indole-3-acetic acid on plant growth

Principle

PGPR-based inoculants are not only a carrier of microorganisms, but also a source of microbial metabolites synthesized from precursor compounds added to the culture medium. Biological activity of microbial phytohormones depends on their production by microbes and their concentrations in culture supernatants. These secondary metabolites influences the early seed germination, early seedling growth, plant colonization, and bacterial establishment. Another group of plant growth regulators, called retardants, are a synthetic group of compounds, which inhibit the biosynthesis of certain bioactive phytohormones, like gibberellins and auxins. Uniconazole and other triazole-type plant growth regulators inhibit the internodes growth in many plant species (Rademacher, 1991) by blocking the oxidation steps of GAs biosynthesis and impeding the formation of bioactive molecules. Lucangeli and Bottini (1996) reported that GAs produced by PGPRs play an important role in plant growth and development (Fig 2). IAA inhibitors, like chloroindole, are structural analogues of the indolic ring system that can not only be useful to understand the IAA physiological role, but also have potential for studies on the importance of a particular pathway during development (Ilic *et al.*, 1999). The following methodology determines the production and release of bacterial phytohormones (IAA and GA_3) in inoculated plants treated with the inhibitors of phytohormone biosynthesis uniconazole and chloroindole.

Materials required

Maize (*Zea mays* L.) and rice (*Oryza sativa* L.) seeds, bacterial cultures in late exponential growth phase, pH meter (or pH indicator sticks), 200 mL individual pots (200 mL volume), sterile perlite–sand (1:1) mixture, plant growth chamber supporting 16 h at 25–30 °C at a light intensity of 450 µmol $m^{-2} s^{-1}$, and 8 h at 20 °C in darkness, 100 µL microsyringe and 1000 µL micropipette.

Reagents

80 µM of uniconazole solution [S-3307, (±) (*E*)-1-(4-clorophenyl)-4,4-dimethyl-2-(1,2,4-triazol-1-yl)-1-penten-3-ol], 60 µM of 5-chloroindole solution, gibberellic acid (GA_3) solutions at concentrations 0.1, 0.5, 1.0, 5.0 and 10.0 µg mL^{-1}, indole-3-acetic acid (IAA) solutions at concentrations 0.1, 5.0, 1.0 and 5.0 µg·mL^{-1}, 50 % of complete Hoagland's solution (Table 1), 0.06 M potassium phosphate buffer, pH 7.0.

Table 1. Modified nutrient solution of Hoagland and Arnon (1950)

	$g \cdot L^{-1}$	$mL \cdot L^{-1}$ of standard solution	$mL \cdot L^{-1}$ of P deficient solution	$mL \cdot L^{-1}$ of N deficient solution
Macronutrient				
KH_2PO_4	136.0	2.0	–[c]	2.0
KNO_3	101.0	10.0	10.0	–[d]
$Ca(NO_3)_2 \cdot H_2O$	236.0	10.0	10.0	–[d]
$MgSO_4 \cdot 7H_2O$	246.5	4.0	4.0	4.0
Micronutrient	–[a]	2.0	2.0	2.0
Fe-EDTA	–[b]	5.0	5.0	5.0

[a] Micronutrient solution: 2.86g H_2BO_3, 1.81g $MnCl_2 \cdot 4H_2O$, 0.22g $ZnSO_4 \cdot 7H_2O$, 0.08g $CuSO_4 \cdot 5H_2O$ and 0.02g $H_2SMoO_4 \cdot H_2O$ are dissolved in 1 L of deionized water.

[b] Fe-EDTA solution: 60.25g $FeCl_3 \cdot 6H_2O$ and 65.10g EDTA are separately dissolved in deionized water at 50 °C. After, the solutions are mixed and the volume completed to 1 L of deionized water. To store in a dark glass flask. The solution should be maintained in refrigerator.

[c] Add 2 mL of 1 M KCl solution.

[d] Add 10 mL of 1 M of KCl and 10 mL of 1 M of $CaCl_2$.

Procedure

i. Maize (*Zea mays* L.) or rice (*Oryza sativa* L.) seeds are surface disinfected by soaking for 3 min in 1 % NaClO, and then washed with sterile distilled water to eliminate traces of this compound.

ii. Seeds are pre-germinated for 48 h at 30 °C and 80 rpm shaking in an Erlenmeyer flask containing sterile 80 µM uniconazole [S-3307, (±) (*E*)-1-(4-clorophenyl)-4,4-dimethyl-2-(1,2,4-triazol-1-yl)-1-penten-3-ol].

iii. Almost five seeds are planted at 1-cm depth in individual pots, each containing 200 mL of sterile perlite–sand (1:1) mixture. After full emergence, seedlings are thinned to 3 plants per pot. Plant pots are kept at field capacity by daily irrigation with sterile distilled water and once a week fertilized with 50 % of Hoagland's solution (Hoagland and Boyer, 1936).

iv. Plants (3 plants per pot, 5 replicates per treatment) are cultured in a growth chamber with daily cycles of 16 h at 25–30 °C with a light intensity of 450 $\mu mol\ m^{-2}\ s^{-1}$ and 8 h at 20 °C in darkness for 7 days. At first leaf stage, seedlings are irrigated with nutrition solution modified by adding 60 µM of 5-chloroindole (Ilic *et al.*, 1999).

v. Seedlings are grouped in the following treatments:

   a. **Inoculated seedlings:** Each seedling is inoculated with 1 mL of bacterial culture medium with up to 10^9 $cells \cdot mL^{-1}$.

b. **GA_3 treated seedlings:** GA_3 is dissolved in pure ethanol to obtain 0.1, 0.5, 1.0, 5.0 and 10.0 µg mL^{-1} solutions. 10 µL of these solutions are applied individually at the first leaf of each seedling with the microsyringe.

c. **IAA treated seedlings:** IAA is dissolved in fresh culture medium to obtain 0.1, 0.5, 1.0, 5.0 and 10.0 µg mL^{-1} solutions and then 1 mL are applied to each seedling individually at the level where the root and the shoot join, with a micropipette.

d. **Control seedlings:** Seedlings individually treated with pure ethanol.

vi. At the end of the experiment the plants are carefully pulled out from the pots and rinsed for 5s in the potassium phosphate buffer.

vii. The following parameters are measured from each plant: number of roots, first internodes length, total root and shoot length, second leaf sheath, fresh and dry weight of roots, and aerial shoots. Also, it is strongly recommended quantification of GA_3 and IAA in roots and shoots (see Experiment 1).

Observations

1. Make at least three independent experiments ($N = 3$).

2. There are some dwarf mutants deficient in the production of physiologically active gibberellins, like the dwarf-1 (*d1*) described in maize (*Zea mays* L.) by Fujioka *et al.* (1988) and the dwarf-x (*dx* or cv. Tanginbozu) described in rice (*Oryza sativa* L.) by Kobayashi *et al.* (1989). These mutants express the dwarf phenotype since young seedling stage, but the exogenous application of gibberellic acid (GA_3) or the inoculation with gibberellins producer PGPRs, allows expression of the normal phenotype. For this reason, the use of the described dwarf mutants is strongly recommended.

3. Do not overexpose the maize or rice seeds to NaClO. Wash them with sterile distilled water to eliminate traces of this compound before use in bioassays.

Statistical analysis

Data from three independent experiments are analyzed by ANOVA followed by *post hoc* Tuckey's test.

Precautions

Uniconazole is toxic by ingestion.

Fig 2. Six week–old plants of *d1 Zea mays* L. from left to right: treated with GA_3 10, 1 and 0.1 µg plant^{-1}; inoculated with *A. lipoferum* op 33 and control (*Source:* Lucangelli and Bottini, 1996)

Experiment 3: Biological nitrogen fixation in chemically defined medium

Principle

Many PGPRs are capable of fixing nitrogen from the atmosphere (bacterial diazotrophs) in a chemically defined medium or interacting with plant roots through an associative, non-obligated interaction (James, 2000). The biological nitrogen fixation in a defined medium is estimated through apparent nitrogenase activity by measuring by the acetylene-reducing activity (ARA). Measurement of ARA requires the use of a gas chromatograph to detect microbial reduction of acetylene to ethylene under controlled conditions. Acetylene reduction of a pure bacterial culture is done as per Christiansen-Weniger and van Veen (1991). Bacterial ethylene production is measured according to Strzelczyk *et al.* (1994) by gas chromatography-flame ionization detection (GC-FID). The major advantage of this method is the high sensitivity obtained by gas chromatography (GC), which detects nitrogenase activity in as few as 2 or 3 cells of diazotrophic microorganisms (Hardy *et al.*, 1968). The following experiment estimates bacterial nitrogen fixation in a chemically defined medium with or without supply of mineral nitrogen by the acetylene reduction method.

Materials required

Bacterial culture in late exponential growth phase, obtained from N-free chemically defined medium, gas chromatograph with flame ionization detector (GC-FID) equipped with a Kromxpek column (2 m length, 3 cm outer diameter and 2 mm inner diameter) provided with a chromosorb (W-AW-DCMS) column, microsyringe.

Reagents

Pure ethylene N50 and pure acetylene gas for chromatographic use.

Procedure

i. 1 mL of the bacterial culture in late exponential growth phase (OD_{600} ~1) is transferred to 250 mL. Erlenmeyer flasks containing 100 mL of chemically defined (N-free) liquid medium. Flasks are fitted with rubber plugs tightened with metal cowls.

ii. Inject acetylene gas with microsyringe to obtain a final concentration of 10 % (v/v).

iii. Erlenmeyer flasks are incubated at 30 °C and 80 rpm until late exponential growth phase.

iv. Air samples of 500 µL are taken with microsyringe from the flasks after incubation for 24, 48 and 72 h and ethylene production is analyzed with a GC system operated isothermally at 110 °C, using nitrogen as gas carrier and a FID (170 °C).

Observations

1. Make at least three independent experiments ($N = 3$).
2. To determine the influence of mineral N on biological nitrogen fixation by PGPR, make a control treatments cultured in chemically defined medium formulated with NO_3^- and NH_4^+ mineral salts.

Calculations

The nitrogenase reaction is described below:

$$N_2 + 6H^+ + 6e^- \longrightarrow 2NH_3$$

If enzyme is exposed to acetylene in absence of N_2 the reaction is substituted by:

$$C_2H_2 + 2H^+ + 2e^- \longrightarrow 2C_2H_4$$

Using a GC, the samples that contain ethylene and acetylene are detected and registered by the FID. Each component of the gas mixture is shown as a single peak. The area of this peak is proportional to component quantity and each compound has a proportionality factor determined by the using the external standard. The acetylene reduction is measured as under:

$$[\mathrm{CE}\ (\mu\mathrm{L\ L^{-1}})]\ [E_\mathrm{S}\ (\mathrm{mm})]^{-1} = [\mathrm{KE}\ (\mu\mathrm{L\ L^{-1}})]\ [E_\mathrm{E}\ (\mathrm{mm})]^{-1};$$

where CE is ethylene concentration in a sample, KE is ethylene concentration in an external standard, E_S is the peak length in sample and E_E is the peak length in external standard.

Ethylene concentration is changed to molar units as under:

$$\mathrm{CE} = [E_\mathrm{S} \cdot \mathrm{KE}]\ [E_\mathrm{E} \cdot 24450]^{-1}$$

The first approach to the method supposes that all samples were injected in a GC system at 25 °C (298 °K), where the volume of any gas is equal to 24450 mL at 25 °C. The samples are obtained from a recipient volume F (mL) with a time T (h) of incubation between samples, the rate of ethylene production is X (µmol h^{-1}):

$$X = [CE_2 - CE_1 F] T^{-1};$$

where CE_1 is the initial ethylene concentration and CE_2 is the final ethylene concentration.

Integrating the previous equations, we get

$$X = [E_2 - E_1 F \, KE] [24450 \times E_E T]^{-1};$$

where E_1 is initial ethylene peak length and E_2 is final ethylene peak length.

The value X is considered acetylene reduction rate (µmol C_2H_4 h^{-1}) by a number of microorganisms cultured and measured in a specific growth phase. In this sense, we strongly recommend to determine the number of viable microorganisms in the recipient before the incubation to correlate the specific acetylene reduction to a number of microorganisms (*e.g.,* 1×10^9 CFU mL^{-1}).

Statistical analysis

Data from three independent experiments are analyzed by ANOVA followed by *post hoc* Tuckey's test.

Precautions

Ethylene and acetylene gas are flammables, keep away from ignition sources.

Experiment 4: Bacterial nitrogen fixation and plant growth promotion

Principle

Field studies provide the most important evidence for N_2 fixation, although greenhouse and laboratory studies can provide useful information to explain growth increases and nitrogen accumulation associated with nitrogenase activity on inoculated plants at field conditions (Baldani *et al.*, 1983). The method to evaluate *in vivo* nitrogen fixation is the measurement of acetylene-reducing activity (ARA). Two techniques can be applied:

i. Soil rhizospheric evaluation, which consists in collection of soil samples and their incubation and evaluation under laboratory conditions (Stewart *et al.*, 1967) and

ii. Evaluation of plant systems by inoculation with a PGPR (Christiansen-Weniger and van Veen, 1991). *In vivo* systems have the advantage that they consider the fixation activity of rhizobacteria interacting with other organisms in soil or interacting with the host plant.

Materials required

Seeds of maize (*Zea mays* L.) or wheat (*Triticum aestivum* L.), bacterial cultures in late exponential growth phase, pH meter (or pH indicator sticks), individual growth container (2000 mL volume), 2 L containers filled with sterile perlite–sand (1:1) mixture, paraffin-lanolin seal, plant growth chamber that supports 16 h at 20–30 °C at a light (450 µmol m^{-2} s^{-1}) and 8 h at 15–20 °C in darkness, aeration pump and membrane filter (0.2 µm), water bath at 25 °C, gas chromatography system with flame ionization detector (GC-FID).

Reagents

Pure ethylene N50, pure acetylene, and N and air mixture gas for chromatographic use, N-deficient Hoagland's solution (see Table 1), draining solution: 2 mM $MgSO_4$, 0.5 mM KH_2PO_4, 0.4 mM $Ca(HPO_4)_2$ and 0.2 % Hoagland's micronutrients solution (see Table 1).

Procedure

i. Seeds of wheat (*Triticum aestivum* L.) or maize (*Zea mays* L.) are surface disinfected by soaking for 3 min in 1 % NaClO and then washed with sterile distilled water to eliminate traces of this compound. Five seeds are planted at 1-cm depth in individual pots containing a 200-mL volume of sterile perlite–sand (1:1) mixture.

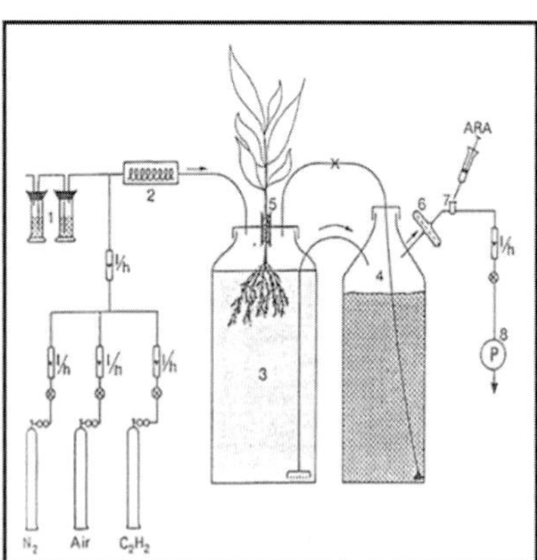

Fig 3. Functional design of a plant growth system for *in situ* acetylene reduction activity (ARA) under controlled environmental conditions, according to Christiansen-Weniger and van Veen (1991) with some modifications. 1, Communicating tubes; 2, cotton filter; 3, growth container; 4, nutrient solution; 5, lanolin seal; 6, membrane filter; 7, sampling port; 8, vacuum pump, (1/h) $L{\cdot}h^{-1}$

ii. Plant pots are kept at field capacity by daily irrigation with 50 % of *N*-free Hoagland's solution (Hoagland and Boyer, 1936) and cultured in a plant growth chamber with photoperiod of 16 h at 20–30 °C at a light, intensity of 450 µmol m^{-2} s^{-1} and 8 h at 15–20 °C in darkness. At the third day, each seedling is inoculated with 1 mL of bacterial culture medium containing up to 10^9 cells mL^{-1} and on 5th day individual seedlings are transferred to plant–soil systems.

iii. The plant–soil system (Fig 3) consists of a 2 L container filled with carefully washed and sterile perlite–sand mixture (1:1). The mixture is sealed airtight around the stem with a 5 % paraffin and lanolin seal to keep the root compartment sterile. Sterilized mineral solution is supplied to the plants by flooding and draining of the growth container. Aeration is done with an aspiration pump *via* a membrane filter (0.2 µm).

iv. Inoculated seedlings are cultured in a plant growth chamber for 7 days with photoperiod of 16 h at 20–30 °C at a light with an intensity of 450 µmol m^{-2} s^{-1} and 8 h at 15–20 °C in darkness.

v. After this, the plant–soil systems are transferred to a water bath at 25 °C to maintain physiological conditions and the nutritive solution is removed by flooding and draining the pot with 2 mM $MgSO_4$, 0.5 mM KH_2PO_4, 0.4 mM $Ca(HPO_4)_2$ and 0.2 % Hoagland's micronutrients solution.

vi. Before the assay, the entire system is flushed for 30 min with a gas mixture of air and N_2 at flow rate of 2.5 L h^{-1} pot^{-1} and the O_2 tension is chosen from a range of 0–20 kPa.

vii. Then, acetylene is added by positive pressure to a final concentration of 10 % (v/v). The ethylene formed as a product of the nitrogenase activity is measured separately for each pot after 1 h of incubation by GC-FID, as per the method described in Experiment 3.

Observations

1. Make at least three independent experiments ($N = 3$).

2. To determine the influence of mineral N on biological nitrogen fixation mediated by PGPRs, make a control treatment irrigated with 50 % of complete Hoagland's solution and control with N-free Hoagland's solution without inoculating. In both cases, a physiological evaluation of nitrogen deficiency symptoms on plants is strongly recommended.

3. Do not overexpose maize or rice seeds to NaClO and wash them with sterile distilled water to eliminate traces of this compound before the use in bioassays.

Calculations

Ethylene and acetylene-reducing activity (ARA) like (nmol C_2H_4 h^{-1} $plant^{-1}$) calculations are as per Experiment 3. We strongly recommended to isolate and determinate the number of viable microorganisms inoculated that colonize the plant tissues (root and shoot), to estimate the specific acetylene reduction for a number of microorganisms (*e.g.,* 1×10^5 CFU·$plant^{-1}$ evaluated in root system).

Statistical analysis

Data from three independent experiments are analyzed by ANOVA followed by *post hoc* Tuckey's test.

Precautions

Ethylene and acetylene gas are flammables, keep away from ignition sources.

Experiment 5: Phosphate solubilization in a chemically defined medium

Principle

Phosphate (P)-solubilizing microorganisms are a group of prokaryotes with relevant beneficial effects on plant growth and development (Vassilev *et al.*, 2006). Visual detection and even semi-quantitative estimation of the phosphate solubilization by microorganisms is done using plate screening methods, which show clearing zones around the microbial colonies in media containing insoluble mineral phosphates, as the single phosphorus source (Katznelson and Bose, 1959). In some cases, there have been contradictory results between the plate halo detection and phosphorus solubilization on liquid cultures. However, it is reliable method for isolation and preliminary characterization of phosphate-solubilizing microorganisms. *In vitro* studies on phosphate solubilisation dynamics by bacterial strains have been done based on the measurement of soluble phosphate (P) released into culture broth, from cultures developed using an insoluble compound as the only phosphorus source (Rodríguez and Fraga, 1999).

Materials required

Bacterial culture in late exponential growth phase; bacteriological culture chamber at 30 °C; orbital shaker (capable of 40 rpm); refrigerated preparative (capable of 8000g) centrifuge and visible light (400–700 nm) spectrophotometer.

Reagents

Trypticase phosphate medium (g L^{-1}): soybean trypticase broth (10.0), $Ca_3(PO_4)_2$ or hydroxyapatite (4.0) at pH 6.8. Trypticase phosphate agar (TPA): plus agar (15.0); 0.025 N HCl solution, 0.03 N NH_4F solution, 5 mM $SnCl_2$ solution and 12 mM ammonium paramolybdate solution.

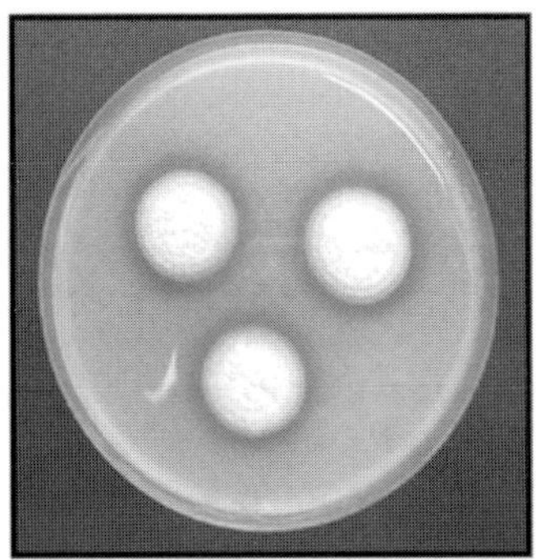

Fig 4. Phosphate solubilization development on trypticase phosphate agar (TPA) plates containing microbial colonies with transparent halos around each one, incubated at 30 °C for up to 4 days on trypticase soya agar (TSA) medium with Ca_3PO_4 (*Picture credit*: Bioq. Florencia Olivieri. Nitragin Argentina)

Procedure

i. **Semi-quantification in agar plate:** Petri dish plate containing Trypticase soya agar (TSA) medium with $Ca_3(PO_4)_2$ or hydroxyapatite, are inoculated with 10 µL of bacterial culture in late exponential growth phase, in halfway points of a plate. Petri dishes are incubated at 30 °C up to 4 days and colonies and diameter of transparent halo formed around each colony are daily measured (Fig 4) as per Katznelson and Bose (1959). The results are expressed as solubilization efficiency (SE; see *Calculations*).

ii. **Phosphorus quantification in liquid culture medium:** Three Erlenmeyer flasks (100 mL) containing 20 mL of TSA liquid medium with $Ca_3(PO_4)_2$ or hydroxyapatite is inoculated with 1 mL of bacterial culture in late exponential growth phase and incubated at 40 rpm and 30 °C for 2, 3, and 4 days, as per Nguyen *et al.* (1992). After incubation, a single culture is processed daily. To do it, the cells are harvested by centrifugation at 8000g and supernatant is used to estimate the phosphorus concentration by the paramolybdate blue method (Olsen and Sommers, 1982). Supernatants are filtered through Whatman #42 paper and 15 µL of filtrate is added to a 50 mL Erlenmeyer flask containing 1.95 mL extracting solution (0.025 N HCl and 0.03 N NH_4F). A 5-mL distilled water and 2 mL ammonium paramolybdate solution (12 mM) are added to the solution, followed by 1 mL $SnCl_2$ dilute solution (5 mM) and well mixed. After 5 min, but before 30 min, phosphorus concentration is measured reading absorbance at 660 nm. Phosphorus quantification is described in *Calculations*.

Observations

1. Positive controls for agar plate method are bacterial strains obtained from international cell collections that inoculated on TPA at the same conditions described above, produces transparent halos around the colonies.

2. Negative control for paramolybdate blue method is an autoclaved noninoculated culture medium.
3. Additional control for paramolybdate blue method is a culture medium after 30 min of bacterial addition.

Calculations

1. **Semi-quantification in agar plate:** Expressed as solubilisation efficiency (SE) is calculated as under:

   $$SE = [\text{solubilization diameter (mm)} \times 100][\text{growth diameter (mm)}]^{-1}$$

2. **Phosphorus quantification in liquid culture:** Expressed as solubilization rate (SR) is calculated as under:

   $$SR = [(Pf - Pc) - Pi]\, T^{-1};$$

where Pf is phosphorus concentration after incubation (2, 3 and 4 days), Pc is phosphorus concentration after inoculation, Pi is theoretical concentration of phosphorus supplied in the medium and T is time (h).

Statistical analysis

Data from three independent experiments are analyzed by ANOVA followed by *post hoc* Tuckey's test.

Experiment 6: Phosphate solubilization and plant growth promotion

Principle

Several reports indicate the ability of different bacterial species to solubilize the insoluble inorganic phosphate compounds such as [tricalcium phosphate, dicalcium phosphate, hydroxyapatite, and rock phosphate (Goldstein, 1986)]. These bacteria often are species from *Pseudomonas, Bacillus, Rhizobium, Burkholderia, Achromobacter, Agrobacterium, Micrococcus, Aereobacter, Flavobacterium* and *Erwinia*. There are considerable populations of phosphate-solubilizing bacteria in soil and in plant rhizosphere (Sperberg, 1958). A considerably higher concentration of phosphate-solubilizing bacteria is commonly found in the rhizosphere in comparison to non-rhizosphere soil. Although several phosphate solubilizing bacteria occur in soil, usually their numbers are not high enough to compete with other bacteria commonly established in the rhizosphere. Inoculation of plants by target microorganisms at a much higher concentration than those normally found in soil is necessary to take advantage of bacterial phosphate solubilisation to increase plant yield. The experiment proposed below allows the estimation of phosphate solubilization bacteria in inoculated plants.

Materials required

Maize (*Zea mays* L.) or bean (*Phaseolus vulgaris* L.) seeds, bacterial cultures in late exponential growth phase, pH meter (or pH indicator sticks),

Fig 5. Phosphorus deficiency in mono- and dicotyledonous plants. Inoculation with phosphate solubilizer's bacterial strain could revert the syntomatology in hydroponical plant culture conditions. The major symptom is the anthocyanins accumulation in superior leaves

individual growth container (200 mL volume), sterile perlite–sand (1:1) mixture, plant growth chamber that support 16 h at 25–30 °C at a light period of (450 µmol $m^{-2} s^{-1}$) and 8 h at 20 °C in darkness.

Reagents

0.06 M potassium phosphate solution, Hoagland's solution: (i) standard formulation (S) or (ii) phosphate deficient formulation (PD) modified by addition of insoluble phosphate source [$Ca_3(PO_4)_2$]. (Table 1; see phosphorus-deficient solution).

Procedure

i. Maize (*Zea mays* L.) or bean (*Phaseolus vulgaris* L.) seeds are surface disinfected by soaking for 3 min in 1 % NaClO and then washed with sterile distilled water. Almost five seeds are planted 1-cm depth in individual pots containing 200 mL of sterile perlite–sand (1:1) mixture, which is kept at field capacity by daily irrigation with sterile distilled water for 3 days.

ii. Then, pots are fertilized every 3 days with 50 % of standard (S) or phosphate deficient (PD) Hoagland's solution (Hoagland and Boyer, 1936) and inoculated with the PGPR, according to the following treatments:

   a. *Control (S)*: noninoculated pot with sterile perlite–sand (1:1) mixture and irrigated with (S) Hoagland's solution.

   b. *Control (PD)*: noninoculated pot with sterile perlite–sand (1:1) mixture and irrigated with (PD) Hoagland's solution.

c. *Inoculated (S)*: inoculated pot with bacterial culture medium containing up to 10^9 cells mL^{-1} to obtain a population of 10^6 cells g^{-1} of sterile perlite–sand (1:1) mixture irrigated with (S) Hoagland's solution.

d. *Inoculated (PD)*: inoculated pot with bacterial culture medium containing up to 10^9 cells mL^{-1} to obtain a population of 10^6 cells g^{-1} of sterile perlite–sand (1:1) mixture, irrigated with (PD) Hoagland's solution.

iii. The experimental design allows to grow 3 plants per pot, after full emergence. The pots (3 replicates per treatment) are cultured in a plant growth chamber for 21 days with photoperiod of 16 h at 20–30 °C at a light with an intensity of 450 µmol m^{-2} s^{-1} and 8 h at 15–20 °C in darkness.

iv. At the end of the experiment, plants are carefully pulled out from the pots and rinsed for 5 s in a 0.06 M potassium phosphate buffer pH 7.0. Then, root and shoot fresh and dry weight (g) is measured for each plant. A physiological evaluation of P-deficiency symptoms is strongly recommended (Fig 5).

Observations

1. Make at least three independent experiments ($N = 3$).
2. The pH is monitored daily, corrected when necessary to give 7.3 ± 0.1 pH units.
3. It is strongly recommended to isolate and determine the number of viable microorganisms inoculated that colonize the soil system to determine the specific phosphorus solubilization for standard number of microorganisms (*e.g.,* 1 × 10^5 CFU g of $soil^{-1}$).
4. Do not expose the seeds of maize or pea to NaClO for more than 2 min and wash with sterile distilled water to eliminate traces of this compound before use in bioassays.

Statistical analysis

Data from three independent experiments are analyzed by ANOVA followed by *post hoc* Tuckey's test.

Experiment 7: Siderophores production in a chemically defined medium

Principle

Many microorganisms, including PGPR, sequester and solubilize the ferric iron incorporating it in low molecular mass compounds (< 1000 Da), which are called siderophores (Neilands, 1984). The most common detection method to determine siderophore production is the universal assay of Schwyn and

Neilands (1997), which is based on the competition for iron between the ferric complex of an indicator dye, chrome azurol S (CAS), and a chelator or siderophore produced by microorganisms. The iron is removed from CAS by a siderophore, which apparently has a higher affinity for iron (III). The positive reaction results in a color change of CAS reagent, usually from blue to orange (Fig 6). The blue dye can be applied as a solution test, or alternatively, it can be incorporated in the solid growth medium and used for screening of several microbial siderophore producers by direct plating procedures (Schwyn and Neilands, 1987).

Materials required

Bacterial cultures in late exponential growth phase, bacteriological cultures chamber at 30 °C, orbital shaker (capable of 40 rpm), a spectrophotometer capable of measurements between 400 and 700 nm, a ruler.

Reagents

CAS liquid medium: 30g of PIPES buffer, 3 mL of (0.05M) tryptophan, 0.16g of triptone and H_2O for 800 mL. Autoclave at 121 °C for 15 min and add the following sterile solutions: 100 mL of (10×) MM9, 100 mL of (10×) CAS, 10 mL of (1 M) mannitol, 1 mL of (1 M) $MgSO_4 \cdot 7H_2O$, 10 mL of (0.01 M) $CaCl_2$ and 0.5 mL of vitamins. Sterilize solutions with (0.2 µm) filters in aseptic conditions. To prepare MM9 (10×) solution: 60g of Na_2HPO_4, 30g of KH_2PO_4, 5g of NaCl, 10g of NH_4Cl and H_2O for 800 mL.

CAS (10×) solution: 60.5g of chrome azurol S, 10 mL of (1mM) $FeCl_3 \cdot 6H_2O$, 0.0729g of HTMD and deionizated H_2O for 50 mL.

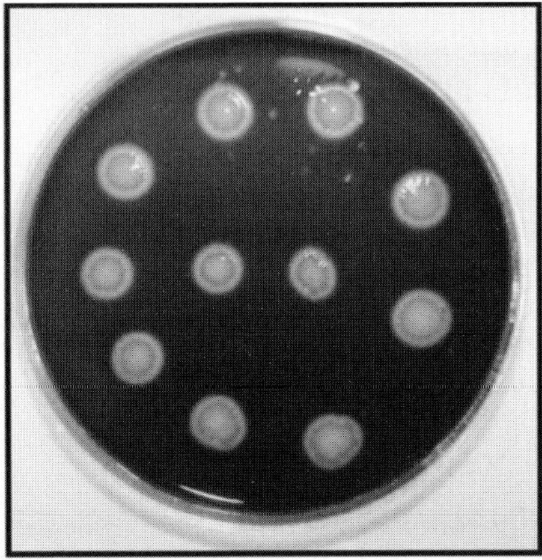

Fig 6. Siderophores production screened on chrome azurol S (CAS) agar plate containing positive colonies incubated at 30 °C for up to 4 days that changed the medium from blue to orange around each colony

Vitamins solution: 20 mg of riboflavine, 20 mg of *p*-aminobenzoic acid, 20 mg of nicotinic acid, 20 mg of biotin, 20 mg of calcic pantothenate, 20 mg of hydrochloride piridoxal, 200 mg of hydrochloride thiamine and H_2O for 100 mL. CAS agar medium: plus 15g of agar each 1000 mL of liquid CAS.

Procedure

i. **Semi-quantitative method:** A Petri dish plates containing CAS medium is inoculated with 1 µL pure bacterial culture in halfway points. These are incubated at 30 °C and observed daily for changes in color from blue to orange, purple, or dark purplish-red (magenta) (Fig 6). After the second day, the diameter of the halos formed around each colony are measured with the rules.

ii. **Quantitative method:** Erlenmeyer flask (100 mL) containing 20 mL of chemically defined medium is inoculated with 1 mL of bacterial culture in late exponential growth phase and incubated at 80 rpm and 30 °C for up to 4 days. After incubation, the cells are harvested by centrifugation at 8000g and 10 mL of supernatant is incubated at 80 rpm and 30 °C with 10 mL of CAS solution for 1 h. After incubation, the absorbance at 630 nm is read in the bacterial supernatant against the blank of culture medium without inoculation.

Observations

Positive controls for qualitative method are bacterial strains obtained from international cell collections (*e.g.,* ATCC) that inoculated on CAS at the same conditions described above, produces color change in CAS reagent (usually from blue to orange).

Calculations

1. **Semi-quantitative method:** Expressed as CAS reaction rate (CR) is described below:

$$CR = [\text{halo diameter (mm)} \times T]^{-1}$$

where T is Time (h).

2. **Quantitative method:** The percentage of iron-binding like-siderophores compounds is calculated bellow:

$$[RA_{630}\ SA_{630}\ (RA_{630})^{-1}] = 100\ \%\ SI$$

where RA_{630} is reference absorbance, SA_{630} is sample absorbance and SI is siderophore units.

Percentages of siderophores units less than 10 are considered as negative and in this case no change in the blue color of CAS solution is observed.

Statistical analysis

Data from three independent experiments are analyzed by ANOVA followed by *post hoc* Tuckey's test.

Experiment 8: Bacterial siderophores production and plant growth promotion

Principle

Plants use strategies to obtain iron from the environment. They may solubilize extracellular iron secreting reductor agents (protons, organic acids, and phenolic compounds). The grasses produce and secrete phytosiderophores to chelate soil iron in a manner similar to microbial siderophores (Sugiura *et al.*, 1981). Plants also use siderophores produced by rhizosphere microorganisms. The ability to take up and use iron-bound siderophores secreted by PGPRs enhances the plant survival by increasing the iron availability, but also enables the opportunistic organisms to effectively compete with other soil organisms for iron (Glick *et al.*, 1999). Promotion of plant growth by siderophore-producing rhizobacteria occurs through direct supply of iron for plant use and/or by removing the iron from the environment of phytopathogens, thereby reducing their competitive ability. This experiment demonstrates that PGPRs use siderophores to obtain iron and how these bacterial siderophores benefit plant growth in hydroponics culture systems.

Materials required

Maize (*Zea mays* L.) and bean (*Phaseolus vulgaris* L.) seeds, bacterial cultures in late exponential growth phase, pH meter (or pH indicator sticks), 200 mL growth containers, sterile perlite–sand (1:1) mixture, plant growth chamber that supports 16 h at 25–30 °C at light (450 $\mu mol\ m^{-2} s^{-1}$) and 8 h at 20 °C in darkness.

Fig 7. Iron deficiency in mono- and dicotyledonous plants. Inoculation with siderophores producer bacteria strain could revert these symptoms in hydroponical plant culture conditions. The major symptom is the high chlorosis located in the interveinal spaces, mainly in upper leaves

Reagents

Half-strength Hoagland's solution modified by addition of 200, 20 or 2 µM of FeEDTA, 0.06 M potassium phosphate buffer at pH 7.0.

Procedure

i. Maize (*Zea mays* L.) and bean (*Phaseolus vulgaris* L.) seeds are selected as experimental tools, because they use different strategies to obtain iron from the environment (Sugiura *et al.*, 1981). Seeds are surface disinfected by soaking for 3 min in 1 % NaClO and then washed with sterile distilled water to eliminate traces of this compound.

ii. Five seeds are planted (1-cm depth) in individual pots containing a 200 mL of sterile perlite–sand (1:1) mixture. Pots are kept at field capacity by irrigation with 50 % of Hoagland's solution (Hoagland and Boyer, 1936) containing 200 µM (Fe-normal); 20 µM (Fe-deficient) FeEDTA or 2 µM (Fe-critical) FeEDTA. Pre-inoculation with bacterial culture medium (to obtain a population of 10^6 cells g^{-1} of hydroponics mixture) is done in each pot, under 4 day before the seed sowing.

iii. The experimental design is as under:

   1. **Control (Fe-normal):** Noninoculated pots irrigated with Hoagland's (Fe-normal) solution.

   2. **Control (Fe-deficient)**: Noninoculated pots irrigated with Hoagland's (Fe-deficient) solution.

   3. **Control (Fe-critical):** Noninoculated pots irrigated with Hoagland's (Fe-critical) solution.

   4. **Inoculated (Fe-normal):** Inoculated pots irrigated with Hoagland's (Fe-normal) solution.

   5. **Inoculated (Fe-deficient):** Inoculated pots irrigated with Hoagland's (Fe-deficient) solution.

   6. **Inoculated (Fe-critical):** Inoculated pots irrigated with Hoagland's (Fe-critical) solution. Pots with plants (3 replicates per treatment) are cultured in a plant growth chamber for 21 days with a photoperiod of 16 h at 25–30 °C at light intensity of 450 µmol m^{-2} s^{-1} and 8 h at 20 °C in darkness.

iv. At the end of experiment, the plants are carefully pulled out from the pots and rinsed for 5 s in a 0.06 M potassium phosphate buffer pH 7.0. Then, root and shoot fresh and dry weight (g) are measured as per plant basis. A physiological evaluation of Fe-deficiency symptoms on plants is strongly recommended (Fig 7).

Observations

1. Make at least three independent experiments ($N = 3$).
2. It is strongly recommended to isolate and determinate the number of viable microorganisms inoculated that colonize the soil system and the plant tissue (root or shoot) to estimate the specific siderophores production for standard number of microorganisms (*e.g.*, 1×10^5 CFU·g^{-1} of hydroponic mixture and 1×10^5 CFU·g^{-1} of root).
3. Do not overexpose the seeds of maize or pea to NaClO and wash with sterile distilled water to eliminate traces of this compound before their use in bioassays.

Statistical analysis

Data from three independent experiments are analyzed by ANOVA followed by *post hoc* Tuckey's test.

Experiment 9: Amino acids production in chemically defined medium

Principle

Application of diazotrophic bacteria inoculants improve the crops development and yield. It was first assumed that inoculation with nitrogen-fixing bacteria enhanced plant growth mainly by their ability to fix N_2, but, the amount of fixed N in several cases meets only a fraction of N_2 demand. Therefore, growth promotion may be attributed to other mechanisms. Many genera of soil and rhizosphere bacteria were investigated for the production of biologically active substances (Thuller *et al.,* 2003), mainly amino acids by these microorganisms in many processes that stimulated the plant growth. Many soil and rhizosphere microorganisms (*Azotobacter, Azospirillum, Rhizobium, Mesorhizobium,* and *Sinorhizobium*) produce amino acids (Salmeron-López *et al.,* 2004). The aim of this experiment is to determine the quantity and quality of amino acids produced by PGPRs in chemically defined media (Marur *et al.*, 1994). It is possible to design a bioassay to probe the beneficial effect of microbial amino acid production, adding an exogenous pattern of amino acids equal to those provided by bacteria, to maize or pea seedlings growing in hydroponics culture system.

Materials required

Bacterial cultures in late exponential growth, pH meter, refrigerated centrifuge capable of 14,000 and 20,000g, internal standards: alanine (Ala), aspartic acid (Asp), glutamic acid, asparagines, serine (Ser), arginine, glutamine (Glu), histidine, glycine, threonine, tyrosine, methionine, tryptophan, valine, phenylalanine, isoleucine, leucine, lysine and ornithine, orbital evaporator, high performance liquid-chromatography system coupled to fluorescence detector (HPLC-F), mini-column Sep-Pak C_{18}, membranes (0.2 µm).

Reagents

Solvent solution: sodium acetate 50 mM and sodium phosphate 50 mM, 0.3 mL of *O*-phthaldialdehyde (OPA), 35 and 65 % of aqueous methanol solution, acetic acid, and tetrahydrofuran.

Procedure

i. Bacterial cultures in exponential growth phase are separated into several 20 mL fractions, for amino acid determination. Samples are centrifuged at 14,000g for 30 min, at 4 °C and supernatants filtered through 0.2 µm membranes and concentrated at room temperature under nitrogen flow.

ii. Samples are resuspended in 3 mL water and centrifuged at 20,000g for 2 min. Supernatant is passed through mini-column Sep-Pak C_{18}, previously conditioned with 35 % of aqueous methanol and sequentially eluted with 1 mL of 35 and 65 % of methanol solution.

iii. Amino acids are derivatizated with *O*-phthaldialdehyde (OPA). Aliquots of 0.1 mL are mixed with 0.3 mL of OPA solution for 10 s before injection. The amino acids in each sample are analyzed by high-performance liquid chromatography (HPLC) on a C_{18} reverse-phase column. The solvent 1 is 65 % methanol and the solvent 2 is a solution of 50 mM sodium acetate, 50 mM sodium phosphate, 1.5 $mL{\cdot}L^{-1}$ of acetic acid, 20 $mL{\cdot}L^{-1}$ tetrahydrofuran, pH 7.25 are combined in the following gradient based on the percent of solvent 2: 35–40 % (0.01–18 min), 40–65 % (18–24 min), 65–68 % (24–35 min), 68–100 % (35–45 min), at flow rate of 1 $mL{\cdot}min^{-1}$. Fluorescence excitation and emission wavelength of 250 and 480 nm, respectively, are used for amino acid detection. Peak areas and retention times are measured by comparison with known quantities of standard amino acids.

Observations

1. Make at least three independent experiments ($N = 3$).
2. Using the HPLC method, identification of amino acids is possible by comparing their peak retention time (Rt) to the pure external standard. Each compound is quantified comparing the peak area obtained with the pure external standard at the same retention time. Amino acids quantification is described in calculations.
3. To determinate the influence of different carbon sources in the culture medium over the bacterial amino acid production, make treatments cultured in different organic compounds as per Gonzalez-Lopez *et al.* (2005). Depending on the strain, add malate, gluconate, fructose or sucrose to the culture medium as unique carbon source.

Calculations

Amino acid quantification is made comparing the peak area obtained at the same retention time (Rt) according to external standard method for each amino acid in the sample. The general equation is described below and the concentration of amino acid is expressed in ìg of amino acid mL^{-1} of culture medium or µM.

$$[C_A\ (\mu L\ L^{-1})] \cdot [A_S]^{-1} = [K_A\ (\mu L\ L^{-1})]\ [A_I]^{-1}$$

where C_A is amino acid concentration in a sample; K_A is amino acid concentration in internal standard, A_S is peak area in sample, and A_I is peak area in internal standard.

Statistical analysis

Data from three independent experiments are analyzed by ANOVA followed by *post hoc* Tuckey's test.

Precautions

Methanol is flammable, keep away from ignition sources. It is toxic by inhalation, direct contact or swallowing.

Experiment 10: Identification and quantification of cadaverine (CAD) and other polyamines in a chemically defined medium.

Principle

Polyamines, particularly cadaverine (CAD), may be correlated with root growth promotion (Niemi *et al.,* 2002) and osmotic stress responses in plants (Aziz *et al.,* 1997). The physiological role of bacterial cadaverine in plant–microbe interaction is unclear, but indirect evidence suggests that this compound promotes or regulates the plant growth. However, in soils with adverse effects on plant growth, microbial cadaverine may contribute to other bacterial molecules such as ABA, to maintain the internal homeostasis. This is new research area, particularly in free-living microorganisms, now classified as third group of beneficial bacteria in plants, and called "Plant Stress Homeo-regulating Rhizobacteria (PSHR)" (Cassán *et al.,* 2005). The objective of this methodology is to demonstrate, if different PGPR strains produce cadaverine in a chemically defined medium and if the production is dependent of the presence on the precursor l-lysine. It is possible that the design of bioassay to probe the beneficial effects of the microbial cadaverine production is similar to bacteria, *i.e.* adding doses of exogenous cadaverine to seedlings growing in a hydroponics culture system.

Materials required

Bacterial cultures in late exponential growth phase, pH meter (or pH indicator sticks), refrigerated preparative centrifuge capable of 8000g, external standards, solvent evaporator, high-performance liquid chromatography system coupled to fluorescence detector (HPLC-F).

Reagents

0.1 mM heptadiamine solution, 2.0 % (w/v) dansyl chloride acetone solution, 10.0 % (w/v) aqueous proline solution, 0.1 mM (w/v) of putrescine, spermine, spermidine, and cadaverine solutions, pure perchloric acid, sodium bicarbonate, toluene, acetonitrile.

Procedure

i. Bacterial cultures are collected from late exponential growth phase by centrifugation at 15,000g, 4 °C for 10 min. An aliquot of 0.1 mM heptadiamine (HTD) is added and supernatants are treated with pure perchloric acid to obtain a 5 % (v/v) solution and centrifuged at 8000g.

ii. Supernatants are neutralized with sodium bicarbonate, added with 400 mL of 2 % (w/v) dansyl chloride acetone solution and kept in darkness for 16 h. Each sample is then added with 100 mL of 10 % (w/v) aqueous proline solution and kept in darkness for 30 min.

iii. Dansyl-amides are extracted in 500 mL toluene and the organic phase is evaporated and resuspended in 100 mL acenonitrile.

iv. Samples are injected in a HPLC-Fluorescence and polyamines identification is made at 510 nm. Elution is done with a gradient of 70 % acetonitrile and 30 % water pumping at 1.5 $mL \cdot min^{-1}$ flow rate.

v. Quantification is done by injection of individual external standard of 0.1 mM (w/v) of putrescine, spermine, spermidine, and cadaverine solutions. Relative abundance of each polyamine was compared with HTD at the same retention time.

Observations

1. Make at least three independent experiments ($N = 3$).

2. Compare the cadaverine biosynthesis with or without $0.1 g \cdot L^{-1}$ l-lysine added in chemically defined medium to determine if diamine production depends of the presence of a biological precursor.

3. Using the isocratic HPLC method, identification of polyamine is possible by comparing its peak retention time (Rt) in the sample with pure external standards. Each compound is quantified according to its peak area. Based on peak areas and relative abundance of each polyamine compared with HTD at the same retention time quantification is determined as described in the calculations.

Calculations

Polyamines identification in the sample is made by comparing the retention time (Rt) with each external standard previously detected. Quantification is made comparing the peak area obtained for a specific retention time (Rt) with the peak area to HTD. The general equation is described below and the concentration of phytohormone is expressed in ìg of polyamine.mL^{-1} of culture medium or µM.

$$[C_A\ (\mu L\ l^{-1})]\ [A_S]^{-1} = [K_{PHTD}\ (\mu L\ l^{-1})] \cdot [A_{HTD}]^{-1}$$

where C_P is Polyamine concentration in a sample; K_{PHTD} is HTD concentration in the sample, A_S is peak area in sample, and A_{HTD} is peak area to HTD.

Statistical analysis

Data from three independent experiments are analyzed by ANOVA followed by *post hoc* Tuckey's test.

Precautions

1. Toluene is flammable, keep away from ignition sources. It is toxic by inhalation, direct contact or swallowing.
2. Dansyl chloride acetone solution and acetonitrile are toxic by inhalation and if swallowed.

Suggested Readings

Aziz, A., Martin-Tanguy, J. and Larher, F. (1997). Plasticity of polyamine metabolism associated with high osmotic stress in rape leaf discs and with ethylene treatment. *Plant Growth Regulation* **21**: 153–163.

Baldani, V., Baldani, J. and Döbereiner, J. (1983). Effects of *Azospirillum* inoculation on root infection and nitrogen incorporation in wheat. *Canadian Journal of Microbiology* **29**: 924–929.

Bashan, Y. and Holguín, G. (1998). Proposal for the division of plant growth-promoting rhizobacteria into two classifications: biocontrol–PGPB (plant growth promoting bacteria) and PGPB. *Soil Biology and Biochemistry* **30**: 1225–1228.

Bashan, L. (2004). *Azospirillum*–plant relationships: physiological, molecular, agricultural, and environmental advances (1997–2003). *Canadian Journal of Microbiology* **50**:521–577.

Bashan, Y. and Levanony, H. (1990). Current status of *Azospirillum inoculation* technology: *Azospirillum* as a challenge for agriculture. *Canadian Journal of Microbiology* **36**: 591–608.

Bastian, F., Cohen, A., Piccoli, P., Luna, V., Baraldi, R. and Bottini, R. (1998). Production of indole-3-acetic acid and gibberellins A_1 and A_3 by *Acetobacter diazotrophicus* and *Herbaspirillum seropedicae* in chemically-defined culture media. *Plant Growth Regulation* **24**: 7–11.

Bohlool, B. (1990). Introduction to nitrogen fixation in agriculture and industry: contribution of BNF to sustainability of agriculture. *In*: *Nitrogen Fixation: Achievement and Objectives (Eds*., P. Gresshoff, L. Roth, G. Stacey and W. Newton). pp. 613–616. Chapman and Hall, New York.

Boiero, L., Perrig, D., Masciarelli, O., Penna, C., Cassán, F. and Luna, V. (2007). Phytohormone production by three strains of *Bradyrhizobium japonicum* and possible physiological and technological implications. *Applied Microbiology and Biotechnology* **74**: 874–880.

Bottini, R., Fulchieri, M., Pearce, D. and Pharis, R. (1989). Identification of gibberellins A_1, A_3, and Iso-A3 in cultures of *A. lipoferum*. *Plant Physiology* **90**: 45–47.

Cassán, F., Paz, R., Maiale, S., Masciarelli, O., Vidal, A., Luna, V. and Ruíz, O. (2005). Cadaverine production by *Azospirillum brasilense* Az39. A new Plant Growth Promotion Mechanism. *XV Annual Meeting Cordoba Biology Society*. Córdoba. Argentina. p. 103.

Christiansen-Weniger, C. and van Veen, J. (1991). Nitrogen fixation by *Azospirillum brasilense* in soil and rhizosphere under controlled environmental conditions. *Biology and Fertility of Soil* **12**: 100–106.

Foster, R. (1988). Micro-environments of soil microorganisms. *Biology and Fertility of Soil* **6**: 189–203.

Fujioka, S., Yamame, H., Spray, S., Phinney, B., Gaskin, P., Mac Millan, J. and Takahashi, N. (1988). Qualitative and quantitative analyses of gibberellins in vegetative shoots of normal, dwarf1, dwarf2, dwarf3 and dwarf5 seedlings of *Zea mays* L. *Plant Physiology* **88**: 1367–1372.

Garate, A. and Bonilla, I. (2000). Nutrición mineral y producción vegetal. *In*: *Fundamentos de Fisiología Vegetal* (*Eds*. A. Azcón-Bieto and Talón, E.). pp. 113–130. McGraw-Hill Interam. Madrid.

Glick, B., Patten, C., Holguin, G. and Penrose, D. (1999). *Biochemical and genetic mechanisms used by plant growth promoting bacteria*. pp. 1–267. Imperial College Press. London. England.

Goldstein, A. (1986). Bacterial solubilization of mineral phosphates: historical perspective and future prospects. *American Journal of Alternative Agriculture* **1**: 51–57.

González-López, J. Rodelas, B., Pozo, C., Salmerón-López, V., Martínez-Toledo, M. and Salmerón, V. (2005). Liberation of amino acids by heterotrophic nitrogen fixing bacteria. *Amino Acids* **28**: 363–367.

Hardy, R., Hosten, R., Jackson, E. and Burns, R. (1968). The acetylene-ethylene assay for N_2 fixation: Laboratory and field evaluations. *Plant Physiology* **43**: 1185–1207.

Hoagland, D. and Boyer, T. (1936). General nature of the process of salt accumulation by roots with description of experimental methods. *Plant Physiology* **11**: 477–507.

Ilic, N., Östin, A. and Cohen, J. (1999). Differential inhibition of indole-3-acetic acid and tryptophan biosynthesis by indole analogues. I. Tryptophan dependent IAA biosyntesis. *Plant Growth Regulation* **27**: 57–62.

James, E. (2000). Nitrogen fixation in endophytic and associative symbiosis. *Field Crops Research* **65**: 197–209.

Katznelson, H. and Bose, B. (1959). Metabolic activity and phosphate-dissolving capability of bacterial isolates from wheat roots, rhizosphere, and non-rhizosphere soil. *Canadian Journal of Microbiology* **5**: 79–85.

Kloepper, J., Lifshitz, R. and Zablotowicz, R. (1989). Free-living bacteria inocula for enhancing crop productivity. *Trends in Biotechnology* **7**: 39–44.

Kobayashi, M., Sakurai, A., Saka, A. and Takahashi, N. (1989). Quantitative analysis of endogenous gibberellins in normal and dwarf cultivars of rice. *Plant Cell Physiology* **30**: 963–969.

Kovats, E. (1958). Gas chromatography characterization of organic compounds: Retention indices of aliphatic halogenides, alcohols, aldehydes and ketones. *Helvetica Chimica Acta* **241**: 1915–1932.

Lester, D., Ross, J., Davies, P. and Reid, J. (1997). Mendel's stem length gene (Le) encodes a gibberellin 3-beta-hydroxylase. *Plant Cell* **9**:1435–1443.

Lucangelli, C. and Bottini, R. (1996). Reversion of dwarfism in *dwarf-1* Maize (*Zea mays* L.) and *dwarf-x* Rice (*Oryza sativa* L.) mutants by endophytic *Azospirillum* spp. *Biocell* **20**: 221–226.

Marur, C., Sodek, L. and Magalhães, A. (1994). Free amino acids in leaves of cotton plants under water deficit. *Revista Brasileira de Fisiologia Vegetal* **6**: 103–108.

Neilands, J. (1984). Methodology of siderophores. *Structure and Bonding* **58**: 1–24.

Nguyen, C., Yan, W., Le Tacon, F. and Lapeirye, F. (1992). Genetic variability of phosphate solubilizing activity by monocaryotic and dicaryotic mycelia of the ectomycorrhizal fungus *Lacaria bicolour* (Maire) P.D. Orton. *Plant Soil* **143**: 193–199.

Niemi, K., Häggman, H. and Sarjala, T. (2002). Effects of exogenous diamines on the interaction between ectomycorrhizal fungi and adventitious root formation in Scots pines *in vitro*. *Tree Physiology* **22**: 373–381.

Olsen, S. and Sommers, L. (1982). Phosphorus. *In*: *Methods of Soil Analysis, Part 2* (*Eds*. A. Page, R. Miller and D. Kennedy). pp. 403–430. American Society of Agronomy, Madison, Winsconsin.

Perrig, D., Boiero, L., Masciarelli, O., Penna, C., Cassán, F. and Luna, V. (2007). Plant growth promoting compounds produced by two agronomically important strains of *Azospirillum brasilense,* and their implications for inoculant formulation. *Applied Microbiology and Biotechnology* On-line first: 10.1007/s00253-007-0909-9.

Rademacher, W. (1991). Biochemical effects of plants growth retardants. In: *Plant Biochemical Regulators*. (*Ed*. Gaussman, H.). pp. 245–263. Marcel Dekker, New York, USA.

Rodríguez, H. and Fraga, R. (1999). Phosphate solubilizing bacteria and their role in plant growth promotion. *Biotechnology Advances* **17**: 319–339.

Salmeron-Lopez, V., Martínez-Toledo, M., Salmeron-Miron, V., Pozo, C. and González-Lopez, J. (2004). Production of amino acids by *Rhizobium, Mesorhizobium* and *Sinorhizobium* strains in chemically defined media. *Amino Acids* **27**: 169–174.

Schwyn, B. and Neilands, J. (1987). Universal assay for detection and determination of siderophores. *Analitycal Biochemistry* **160**: 47–56.

Seshadri, S., Muthukumarasamy, R., Lakshminarasimhan, C. and Ignacimuthu, S. (2000). Solubilization of inorganic phosphates by *Azospirillum halopraeferans*. *Current Science* **79**: 565–567.

Sperberg, J. (1958). The incidence of apatite-solubilizing organisms in the rhizosphere and soil. *Australian Journal of Agricultural Research* **9**: 778.

Stewart, W., Fitzgerald, G. and Burris, R. (1967). *In situ* studies on N_2 fixation using the acetylene reduction technique. *Proccedings of National Academy of Science* **58**: 2071–2078.

Strzelczyk, E., Kamper, M. and Li, C. (1994). Cytokinin-like-substances and ethylene production by *Azospirillum* in media with different carbon sources. *Microbiological Research* **149**: 55–60.

Sugiura, Y., Tanaka, H., Mino, Y., Ishida, T., Ota, N., Inoue, M., Nomoto, K., Yoshioka, H. and Takemoto, T. (1981). Structure and properties and transport mechanism of iron (III) complex of mugineic acid, a possible phytosiderophore. *Journal of American Chemistry Society* **103**: 6979–6982.

Thuler, D., Flosh, E., Handro, W. and Barbosa, M. (2003). Plant growth regulators and amino acids released by *Azospirillum* sp. in chemically defined medium. *Letters in Applied Microbiology* **37**: 174–178.

Tiburcio, A., Altabella, T., Borrell, A. and Masgrau, C. (1997). Polyamine metabolism and its regulation. *Physiologia Plantarum* **100**: 664–674.

Tien, T., Gaskins, M. and Hubbell, H. (1979). Plant growth substances produced by *Azospirillum brasilense* and their effect on the growth of pearl millet (*Pennisetum americanum* L.). *Applied Environmental Microbiology* **37**: 1016–1024.

Vassilev, N., Vassileva, M. and Nikolaeva, I. (2006). Simultaneous P-solubilizing and biocontrol activity of microorganisms: potentials and future trends. *Applied Microbiology and Biotechnology* **71**: 137–144.

CHAPTER 10

Hypersensitive Response

Lucas D. Daurelio[1], María Laura Tondo[1], Germán Dunger[1], Natalia Gottig[1], Jorgelina Ottado[1] and Elena G. Orellano[1]

1. INTRODUCTION

Plants have evolved to defense themselves against many pathogens (bacteria, viruses, and fungi). Hypersensitive response (HR) is a rapid and localized cell death associated with disease resistance including race-specific (classical R–Avr interaction), race-non-specific and non-host resistance (Flor, 1971). The HR was first described by Stakman (1915). It is characterized by the rapid death of a limited number of cells in the vicinity of the invading pathogen and its function is thought to restrict or delay further pathogen spread and eventually kill the pathogen (Heath, 2000). Then, subsequent signaling events induce local and systemic activation of a set of defence responses that play a role in resistance. One of the most rapid plant responses is associated with the oxidative burst, which constitutes the production and accumulation of active oxygen species (AOS), primarily superoxide (O_2^-), and hydrogen peroxide (H_2O_2), at the site of invasion (Lamb and Dixon, 1997) which can produce toxic effects via DNA damage, protein degradation and modification, and lipid peroxidation (Imlay and Linn, 1988). Other physiological and molecular modifications correlated with the HR are ion leakage from dead cells into the apoplast (Mysore and Ryu, 2004), deposition of lignin and callose into the plant cell wall (Verma and Hong, 2001), and production of phytoalexins, hydrolytic enzymes and pathogenesis-related proteins (PRs) (van Loon *et al*., 2006). Plant cell morphological modifications observed during HR are schematized in Fig 1.

Xanthomonas axonopodis pv. *citri* (*Xac*), the bacterium responsible for citrus canker, induces a non-host resistance with HR in non-host plants such as pepper, tobacco, tomato, bean, and cotton (Dunger *et al*., 2005). This chapter describes many techniques to determine the HR in the interaction between bacteria and non-host plants using *Xac* and

1. Molecular Biology Division, IBR (Instituto de Biología Molecular y Celular de Rosario), Consejo Nacional de Investigaciones Científicas y Técnicas (CONICET), Facultad de Ciencias Bioquímicas y Farmacéuticas, Universidad Nacional de Rosario, Suipacha 531, (S2002LRK) Rosario, Argentina. E-mail: orellano@ibr.gov.ar

tobacco, pepper or cotton. A detailed description of plant growth and inoculation techniques is shown (Experiment 1; Fig 2). Pathogen growth restriction in inoculated plant leaves is evaluated using growth curves (Experiment 1; Fig 3). Also the AOS produced during the HR are biochemical detected. Hydrogen peroxide is detected using DAB (3,3'-diaminobenzidine) staining that is oxidized by endogenous peroxidases in the presence of H_2O_2 (Experiment 2; Fig 4), while superoxide production is characterized by NBT (nitro-blue tetrazolium) reduction and the formation of insoluble blue staining (Experiment 3; Fig 5). Ion leakage that reveals cellular membrane damage is determined by conductivity measurements (Experiment 4; Fig 6) and cell death can be monitored by Trypan blue staining since the dye only passes through membranes of dead cells (Experiment 5; Fig 7). Finally, cell wall alterations revealed by callose deposition as a form of a non-host resistance is determined by staining with aniline blue that selectively stains β-1-3 glycosidic linkages of the glucose units in callose (Experiment 6; Fig 9).

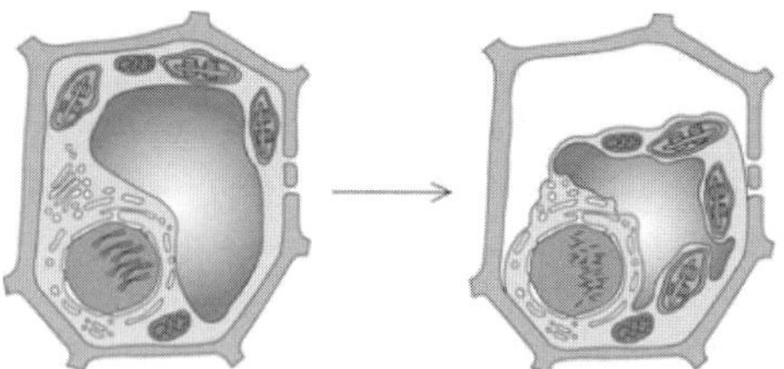

AOS Generation

Vacuolar and plasma membrane blebbing

Nuclear DNA fragmentation

Cell wall modification and callose deposition

Fig 1. Plant cell morphological modifications during hypersensitive response

Experiment 1: Bacterial growth curve *in planta*

Principle

The rapid and localized necrosis of cells at the infection site and the induced defence responses that typically occur within the dying cells and their adjacent live cells during the HR have inhibitory effects on the invading pathogen and restricts the pathogen growth. Thus, pathogen inoculated in resistant or non-host plants will either not multiply significantly or grow slowly in the plant tissue. The bacterial population present in plant tissues can be quantified by a phytopathological technique used to evaluate the growth of bacterial pathogen within specific plant species.

Materials required

Non-host plants: tobacco (*Nicotiana tabacum* cv. Petit Havana), pepper (*Capsicum annuum* cv. Grossum) or cotton (*Gossypium hirsutum* cv. DP 404Bg), *Xac* bacterial suspension, 1.5-mL microfuge tubes, micropipettes, 2.5-mL needleless syringe, cork borer, Petri dishes, Drigalski spatula, autoclave, refrigerated shaker.

Reagents

Silva-Buddenhagen (SB) medium: dissolve 0.5g of sucrose, 0.5g of yeast extract, 0.5g of peptone and 0.1g of glutamic acid in 100 mL of distilled water (if necessary adjust the pH to 7) and sterilize by autoclaving; SB-agar: make up liquid medium according to the formula given above and add 1.5g of bacto-agar just before autoclaving; 10 mM $MgCl_2$: dissolve 0.2g of $MgCl_2 \times 6H_2O$ in 100 mL of distilled water and sterilize by autoclaving; 70 % (v/v) ethanol; 25 mg mL^{-1} ampicillin: dissolve 25 mg of the sodium salt of ampicillin in 1 mL of distilled water, sterilize by filtration and store at –20 °C.

Procedure

Xanthomonas inoculum preparation

i. Bacteria are streaked out from a –80 °C glycerol stock onto a plate of SB medium supplemented with 25 µg mL^{-1} ampicillin and grown for 1 or 2 days at 28 °C. *Xac* is naturally resistant to low concentration of ampicillin.

ii. Bacteria from the fresh streak are transferred to a liquid SB culture with 25 µg mL^{-1} ampicillin and grown with shaking at 28 °C for 12–16 h, when bacterial culture should reach late log phase of growth.

iii. The optical density (OD) of the bacterial cell suspension is quantified using a spectrophotometer set at 600 nm. For *Xac* a late log phase culture usually has an OD_{600} of 3–4 and corresponds to a bacterial concentration of approximately 1×10^9 colony-forming units (CFU) mL^{-1}. The inoculum is made by calculating the proper dilution necessary to obtain a bacterial concentration of 10^7–10^8 CFU mL^{-1} and then diluting that volume of bacteria in sterile 10 mM $MgCl_2$.

iv. A dilution of the inoculum prepared is plated on SB medium supplemented with 25 µg mL^{-1} ampicillin to determine the actual bacterial concentration of the inoculum.

Inoculation of leaves

i. Plants are grown in a greenhouse at 23–25 °C with a photoperiod of 16 h and the inoculum is prepared as described above.

ii. Leaves to be infiltrated are selected and marked so that they can be identified later. The selected leaves are rinsed with distilled water and then with a 70 % ethanol solution.

iii. After washing, the leaves are carefully inverted, exposing the abaxial side, and a 2.5-mL needleless syringe containing the bacterial suspension is used to pressure-infiltrate the leaves intercellular spaces (Fig 2). Only a small amount of inoculum (~100 µL) will infiltrate the leaf. As this occurs an apparent water-soaking of the leaf is observed.

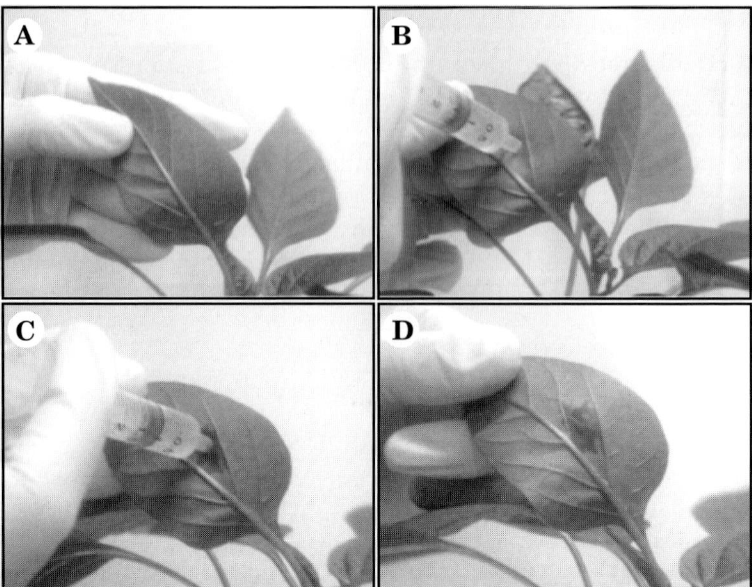

Fig 2. Syringe infiltration of pepper leaves. (A) The abaxial side of a pepper leaf to be syringe-infiltrated. (B) Placement of the syringe on the right side of the leaf, avoiding the midvein. (C) Gentle infiltration of a portion of the leaf's intercellular space. (D) The syringe-infiltrated leaf. Note that the infiltrated area appears water-soaked

Determination of the bacterial population in the inoculated leaf tissue

Infiltrated leaves are harvested at 0, 2, 8, 16, 24 h postinoculation and every 24 h up to 8 days postinoculation. Leaves harvested are surface sterilized as under:

i. Whole leaves are removed from the plant and rinsed with distilled water. After washing, the leaves are placed in a 70 % ethanol solution for 1 min and gently mixed in the solution occasionally. The leaves are then removed, blotted briefly on paper towels and then rinsed in sterile distilled water for 1 min. The leaves are removed and blotted dry on paper towels.

ii. Leaf disks are excised from leaves with a 1-cm-diameter cork borer and placed in a 1.5-mL microfuge tube with 100 µL sterile 10 mM $MgCl_2$.

iii. The tissue samples are ground with a microfuge tube plastic pestle, either by hand or using a small hand-held electric drill. The samples are thoroughly macerated until pieces of intact leaf tissue are not longer visible. The pestle is rinsed with 900 µL of sterile 10 mM $MgCl_2$, with the rinse being collected in the original sample tube so that the sample is now in a volume of approximately 1 mL. Following grinding of the tissue, the samples are thoroughly vortexed to evenly distribute the bacteria within the water/tissue sample.

iv. A 100-µL sample is removed and diluted in 900 µL sterile 10 mM $MgCl_2$. A serial 1:10 dilution series is prepared for each sample by repeating this process. The samples are then plated on SB medium supplemented with 25 µg mL^{-1} ampicillin to select the inoculated bacterial strain. The plates are placed at 28 °C for approximately 2 days and then the colony forming units for each dilution of each sample are counted.

Observations

1. Three or more samples are needed for each time point to generate statistically analyzable data. The standard deviation within the replicate samples for each time point must be calculated and indicated as error bars in the growth curve.

2. Avoid the vascular system of the leaf for injection; mechanical damage of the midrib will have detrimental effects on the viability of the leaf tissue.

3. The number of serial dilutions necessary to get countable colonies must be determined for each sample empirically, but dilutions to 10^{-5} are usually sufficient for any time point.

4. The number of colonies in the plate should be ≥ 30 and ≤ 300 to be reliably countable.

Calculations

Step 1. Calculate the bacterial concentration in the sample plated, based on the number of colonies counted and the volume of the aliquot spread on the plate.

For example, if the number of colonies counted is 150 and the volume used for plating is 100 µL, then the bacterial concentration in the sample should be:

$$\text{Bacterial concentration} = 150 \times 10$$

$$\text{Bacterial concentration} = 1500 \text{ CFU mL}^{-1}$$

Step 2. Calculate the bacterial concentration in the original (not diluted) sample, based on the dilution factor of the corresponding sample.

Following example in step 1, if the dilution factor of the sample is 10^{-4}, then the bacterial concentration in the original sample should be:

$$\text{Bacterial concentration} = 1500 / 10^{-4} \text{ CFU mL}^{-1}$$

$$\text{Bacterial concentration} = 1.5 \times 10^{7} \text{ CFU mL}^{-1}$$

Step 3. Calculate the bacterial population in the inoculated leaf tissue based on the area of the leaf disk processed. Following the above example, if the diameter of the leaf disk is 1 cm, then the area (A) of the disk is:

$$A = \eth r^2$$

$$A = \eth\ (0.5\ \text{cm})^2\ /\ A = 0.785\ \text{cm}^2$$

Then, the bacterial population in the tissue expressed as CFU cm^{-2} should be:

$$\text{Bacterial population} = 1.5 \times 10^7 / 0.785\ \text{CFU cm}^{-2}$$

$$\text{Bacterial population} = 1.9 \times 10^7\ \text{CFU cm}^{-2}$$

Note: Similar procedure should be applied to calculate the bacterial population present within the plant tissue at each time point.

Plotting log (bacterial population) against time (usually in days) after pathogen inoculation produces an unfitted curve usually known as a *growth curve*. A typical growth curve of *Xac* in non-host plants is shown in Fig 3.

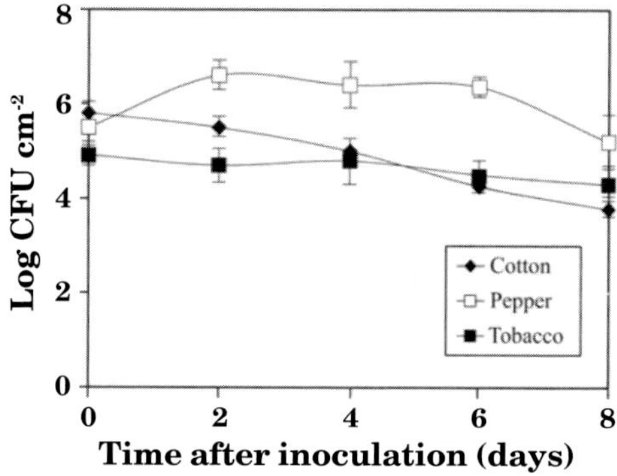

Fig 3. Bacterial growth in non-host plants. *Xanthomonas axonopodis* pv. *citri* is inoculated in cotton, pepper, and tobacco leaves. At different times bacterial growth is measured. Values represent means of three samples; error bars represent standard deviations

Precautions

The inoculation of leaves by the syringe injection method can produce splashing of the bacterial suspension. It is recommended to protect the eyes with goggles.

Experiment 2: Active oxygen species: Hydrogen peroxide detection

Principle

The oxidative burst, characterized by a rapid production of active oxygen species (AOS) released into the apoplast, is described as one of the earliest responses to pathogen infection and is generally associated with HR (Lamb and Dixon, 1997). Several works based on different interactions have described the oxidative burst as a two-phase phenomenon: phase I is an immediate and very transient AOS production, non-specifically stimulated by compatible, incompatible, and even saprophytic bacteria. In contrast, phase II is a delayed (1–3 h) and prolonged AOS production that is specifically stimulated by incompatible HR-causing bacteria and is therefore characteristic of the HR. AOS have direct antimicrobial activities and they have been also implicated in plant programmed cell death as secondary messengers (Venisse *et al.,* 2001). Two principal AOS species are generated in HR response, hydrogen peroxide, and superoxide (see Experiments 2 and 3).

Hydrogen peroxide (H_2O_2) is generated by bivalent reduction of molecular oxygen (O_2). This compound is moderately reactive but it can be converted into more reactive species, especially the hydroxyl radical ($OH^{\bullet}$) (Hammond-Kosack and Jones, 1996). Hydrogen peroxide is generated in plant cells through spontaneous or catalytic dismutation of superoxide (generated in physiological procedures or under stress conditions), or by enzymatic synthesis. In biotic stress, hydrogen peroxide can also be generated by cell-wall-bound peroxidases and amino oxidases in the apoplastic space and then diffuse to the cytoplasm (Mittler, 2002). Hydrogen peroxide can be detected in the sites where it is formed by histochemistry using DAB reagent. DAB is oxidized by endogenous peroxidases in the presence of H_2O_2 rendering an insoluble reddish-brown polymer stable in most solvents. The following experiment is conducted to evaluate the hydrogen peroxide production in the HR produced during the interaction between *Xac* and the non-host plant cotton.

Materials required

Seeds, ground, pots, greenhouse, beakers, plastics tubes, dark bottle, clear bottles, pipettes, measuring cylinder, pH meter (or pH indicator sticks), autoclave, sealing inner-dark box, syringes, Petri dishes, laboratory burner or laminar flow, refrigerated shaker, sterilized toothpicks, little pincers.

Reagents

1 mg mL^{-1} DAB–HCl pH 3.8: dissolve 0.1g of dye in 4 mL of 0.5 N HCl to make a stock solution 25 mg mL^{-1} and store in a dark bottle, dilute in distilled water to make the working solution 1 mg mL^{-1} prior to use and check the pH of the solution; 0.5 N HCl (100 mL): dilute 4.15 mL of concentrated HCl (12.06 N) in 95.85 mL of distilled water, 96% (v/v) ethanol, 10 mM $MgCl_2$, SB medium, SB-agar, 25 mg mL^{-1} ampicillin (see Experiment 1, *Reagents*).

Procedure

i. Cotton seeds are spread uniformly on ground in germination pots. Seeds are germinated in a greenhouse at 25 °C under natural light with a 16 h photoperiod. At two-leaves stage, plants are transferred to individual pots and grown in the same conditions. Two to three months old plants are used for the assays (leaves that are 2-months old are used).

ii. For inoculums preparation, *Xac* is grown as described in Experiment 1 and diluted to a concentration of 10^7 to 10^8 CFU mL^{-1}. Inoculation is done by infiltration of leaves with a needleless syringe (Fig 2). A 10-mM $MgCl_2$ solution is used as negative control.

iii. The detection of hydrogen peroxide production is done using the protocol of Thordal-Christensen *et al.* (1997) with some modifications. Leaves are cut 2 h after infiltration and hanged in a beaker to submerge only the petiole in the DAB solution and incubated for 18 h in darkness. Then leaves are placed in 96 % (v/v) ethanol to stop reaction, remove chlorophyll, and preserve tissue integrity.

iv. Stained tissues are stored in 96 % (v/v) ethanol at room temperature (25 °C) and digitalized using a photographic camera or scanner. Color images may be transformed to black and white 8-bit images (Fig 4) and DAB polymer color intensity can be determined by an image processing software if quantitative results are required (see *Calculations*).

Observations

1. Preferebally, DAB working solution must be prepared before use in the experiments. The solutions are dark-red in color and must not present precipitates or colloids particles (which indicate DAB oxidation).

2. Make at least three independent experiments to confirm results.

3. Alternative DAB staining protocols can be used. In Torres *et al*. (2002), vacuum-infiltrating DAB solution is used. In Thordal-Christensen *et al.* (1997), leaves are placed in DAB solution to stain. Protocol election depends on the plant and tissue used in the experiment.

4. DAB is less soluble in water, hence, prepare the stock solution in 0.5 N HCl and check the pH of working solution.

5. DAB synonyms: 3,3',4,4'-biphenyltetramine; 3,3',4,4'-tetraaminobiphenyl; (1,1'-biphenyl)-3,3',4,4'-tetramine; 3,3',4,4'-diphenyltetramine.

6. DAB is an abbreviation for similar chemical compounds: 3,3'-diaminobenzidine tetrahydrochloride and 4-dimethylaminoazobenzene.

7. Store DAB reagent in a cool dry and well-ventilated area, in a container tightly closed. DAB solutions can be stored at 20 °C, but do not freeze because defrost and refreeze will cause precipitation of the DAB.

Calculations

A quantitative analysis of stain intensity can be done as Torres *et al.* (2002). Briefly, three spots of treated leaves and three spots of control leaves are selected to calculate the index of staining. Then, the index of brown pixels (in color images) or gray pixels (in black and white 8-bit images) must be measured in the selected spots. The index of staining will be calculated as the average of pixels in the spots of treated areas minus the average of pixels in the spots of control areas.

Statistical analysis

Differences between the data of treated and untreated (negative control) leaf regions may be evaluated by Student's *t*-test or non-parametric methods.

Precautions

DAB may be harmful if swallowed, inhaled, or absorbed through the skin. It causes severe skin and eye irritation. Avoid breathing vapors or mist. Wash hand thoroughly after handling. Use with adequate ventilation in a hood. Wear protective goggles, gloves and clothing. In case of contact with eyes, immediately flush with plenty of water and get medical attention. DAB appeared to be inactive in a carcinogenicity-screening test using rats, but some individuals claim that it is a carcinogen. The substance may be toxic to kidneys, lungs, bladder, upper respiratory tract, skin, and central nervous system (CNS). DAB reagent and solutions are sensible to light, hence, place in dark. HCl solution is highly corrosive, handle with care.

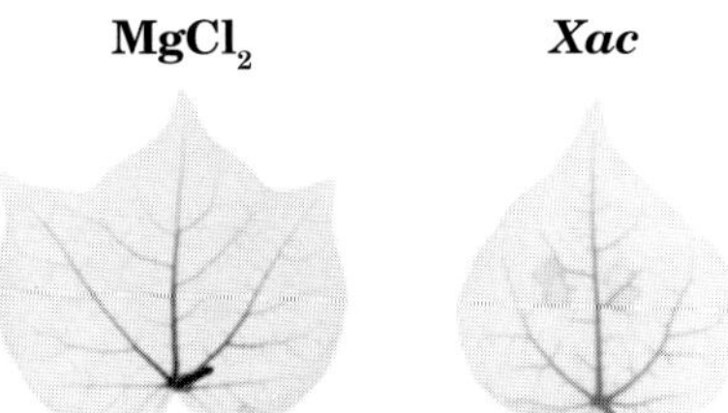

Fig 4. Hydrogen peroxide formation in cotton leaves in the hypersensitive response to *Xanthomonas axonopodis* pv. *citri*. Leaves were inoculated with 10-mM $MgCl_2$ as negative control and with 10^8 CFU mL^{-1} of *Xac* suspension. DAB staining was carried out 2 h postinoculation and gray precipitates in the infiltration region are observed

Experiment 3: Active oxygen species: Superoxide detection

Principle

Superoxide anion ($O_2^{\cdot-}$) is the active oxygen species produced by univalent reduction of molecular oxygen (O_2). Like hydrogen peroxide, superoxide is moderately reactive, but it can be converted into more reactive hydroperoxyl free radical (Hammond-Kosack and Jones, 1996). In plant cells, superoxide is produced in some reactions involved in normal metabolism, such as photosynthesis and respiration, as an undesirable metabolic product of aerobic metabolism. Under stress conditions, it is produced by photorespiration in the photosynthetic apparatus and in the mitochondrial respiration. In addition, biotic stress trigger the active production of cytosolic superoxide by NADPH oxidazes localized in the plasmatic membrane (Mittler, 2002). Superoxide can be detected by cytochemistry using NBT dye (nitro-blue tetrazolium), which is reduced spontaneously in contact with superoxide, producing an insoluble dark blue precipitate called Formazan. The following experiment is conducted to evaluate the superoxide production in the non-host interaction mediated by HR between *Xac* and tobacco or cotton plants, detected by NBT staining.

Materials required

Seeds, soil, pots, greenhouse, beakers, plastics tubes, dark bottle, clear bottles, pipettes, measuring cylinder, pH meter (or pH indicator sticks), autoclave, sealed inner-dark box, syringes, Petri dishes, laboratory burner or laminar flow, refrigerated shaker, sterilized toothpicks, little pincers, vacuum bomb.

Reagents

0.05% NBT: dissolve 0.05g of dye in 100 mL of 0.05 M sodium phosphate buffer (pH 7.5); Phosphate buffer pH 7.5: (100 mL:8 mL of 0.2 M $NaH_2PO_4{\cdot}2H_2O$ (3.12g/100 mL distilled water), 42 mL of 0.2 M $Na_2HPO_4{\cdot}7H_2O$ (5.37g/100 mL distilled water), 50 mL of distilled water, sterilize and check pH; 96 % (v/v) ethanol; 10 mM $MgCl_2$; SB medium; SB-agar; 25 mg mL^{-1} ampicillin (see Experiment 1, *Reagents*).

Procedure

i. Cotton and tobacco seeds are spread uniformly on soil filled pots. Seeds are germinated in a greenhouse at 25 °C under natural light with a 16 h photoperiod. At two-leaves stage, plants are transferred to individual pots and grown in the same conditions. Two to three months old plant are used for assays (preferably leaves that are 2-months old are used).

ii. For inoculum preparation, *Xac* is grown as described in Experiment 1 and diluted to a concentration of 10^7 to 10^8 CFU mL^{-1}. Inoculation is done by infiltration of leaves with a needleless syringe (Fig 2). A 10-mM $MgCl_2$ solution is used as negative control.

iii. Superoxide is detected *in situ* as described by Doke (1983) with some modifications. At the time of detection (4 h in this experiment) leaves are vacuum-infiltrated with 0.05 M sodium phosphate buffer (pH 7.5) containing 0.05 % NBT. After 20 min of staining at room temperature under light, the NBT-treated tissues are placed in 96 % (v/v) ethanol to stop reaction, remove chlorophyll, and preserve tissue integrity.

iv. Stained tissues are stored in 96 % (v/v) ethanol at room temperature (25 °C) and digitalized using a photographic camera or scanner. Color images are transformed to black and white 8-bit images (Fig 5). Formazan color intensity can be determined by an image processing software if quantitative results are required (see *Calculations*).

Observations

1. NBT solution must be prepared in the experiments. The solutions are light yellow color and free from precipitates or colloidal particles.

2. Make at least three independent experiments to confirm results.

3. Alternative NBT staining protocols can be used. For example, the NBT solution can be infiltrated (Rodriguez *et al.*, 2003) or spread on the leaf surface (Cessna *et al.*, 2000). Sometimes leaf disks can be used (Keith *et al.*, 2003). Protocol selection depends of plant and tissue used in the experiment.

4. NBT synonyms: Nitroblue tetrazolium chloride; Nitro BT; 2*H*-(Tetrazolium,3,3'-(3,3'-dimethoxy(1,1'-biphenyl)-4,4'-diyl)bis(4-nitrophenyl)-5-(phenyl-dichloride).

5. Keep NBT reagent in a tightly closed container, stored in a cool, dry, ventilated area. Protect against physical damage. NBT solution can be stored at 20 °C in a dark place.

Calculations

A quantitative analysis of stain intensity can be done as Rodriguez *et al.* (2003). Briefly, three spots of treated leaves and three spots of control leaves are selected to calculate the index of staining. Then, the index of blue pixels (in color images) or gray pixels (in black and white 8-bit images) must be measured in the selected spots. The index of staining will be calculated as the average of pixels in the spots of treated areas minus the average of pixels in the spots of control areas.

Statistical analysis

Differences between data of treated and untreated (negative control) leaf regions may be evaluated by Student's *t*-test or non-parametric methods.

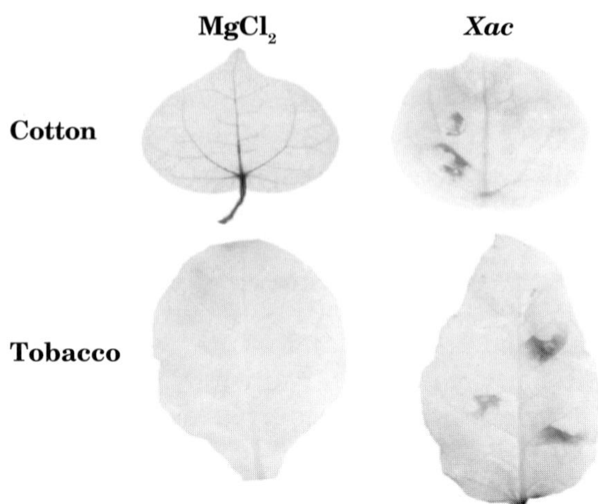

Fig 5. Superoxide formation in tobacco and cotton leaves in the hypersensitive response to *Xanthomonas axonopodis* pv. *citri*. Leaves were inoculated with 10 mM $MgCl_2$ as negative control and with 10^8 CFU mL^{-1} of *Xac* suspension. NBT staining was carried out 4 h postinoculation and dark blue precipitates in the infiltration region are observed

Precautions

NBT, if swallowed, inhaled, or absorbed through the skin, may cause severe irritation to skin, eye and respiratory tract. Avoid breathing vapors or mist. Wash thoroughly after handling. Use with adequate ventilation in a hood. Wear protective goggles, gloves, and clothing. NBT reagent and solutions are sensitive to light so must be stored in the dark.

Experiment 4: Ion leakage

Principle

Damage to cellular membrane is response to a wide variety of stresses. The permeability of the membrane is altered leading to leakage of electrolytes from the cells of stressed tissues, which may result in cell death. Electrolyte leakage from cells may be determined by conductivity. This measurement is a sensitive and reliable indicator of the severity of the stress in plant cells. Measuring solute leakage from plant tissues is a method to estimate the membrane permeability in relation to biotic stresses. Leakage of electrolytes due to stress damage can be determined, after immersing plant tissues, in an aqueous bathing solution where electrical conductivity (EC) is measured. (Dexter *et al.*, 1932). Flint *et al.* (1966) reduced Stuart's calculation to the following equation (Blum and Ebercon, 1981):

$$I_d = 100\,(R_i - R_0) / (1 - R_0)$$

where according to the terminology of Flint *et al.* (1966), I_d is the index of injury, $R_i = EC_{initial}/EC_{total}$ for treated tissues, $R_0 = EC_{initial}/EC_{total}$

for non-treated tissues, and $EC_{initial}$ is the conductivity of the bathing solution following a given period of leakage, EC_{total} is the conductivity of the bathing solution following heat killing to release all ions from the tissue. This calculation of I_d adjusts all values to a scale of 0–100, with higher percentages indicating greater damage. Leaf disks cut with a cork borer are common sample units, although stem or root segments have also been used. Calculation of R_i and R_0 is possible after the measurement of $EC_{initial}$ and EC_{total}. To do it, samples are placed in standardized volumes of distilled water and are allowed to leak solutes into the bathing solution for 2 h. Then, $EC_{initial}$ of each solution is measured. Following measurement, samples are destroyed by autoclaving or rapid freezing in liquid N_2 and allowed to leak for an additional period, typically 2 h. The conductivity is then measured again to obtain EC_{total}. R_0 is calculated from measurements obtained from untreated control samples that have been monitored along with the treated ones.

Materials required

Seeds, soil, pots, plant leaves, syringes, beakers, 12-well polystyrene plate (for leaf disks) or 1.5-mL and 15-mL plastic tubes, conductimeter, autoclave, cork borer, pipettes.

Reagents

10 mM $MgCl_2$, SB medium, SB-agar, 25 mg mL^{-1} ampicillin (see Experiment 1, *Reagents*).

Procedure

i. Cotton seeds are spread uniformly on soil–filled pots. Seeds are germinated in a greenhouse at 25 °C under natural light with a 16 h photoperiod. At two-leaves stage, plants are transferred to individual pots and grown in the same conditions. Two to three weeks old plants are used for assays.

ii. For inoculum preparation, *Xac* is grown as described in Experiment 1 and diluted to a concentration of 10^7 to 10^8 CFU mL^{-1}.

iii. Inoculation is performed by infiltration of leaves with a needleless syringe (Fig 2). A 10-mM $MgCl_2$ solution is used as negative control.

iv. Ion leakage is measured 0, 2, 6, 18 and 24 h after bacterial infiltration. To do it, leaf disks of 1.3-cm-in diameter are removed with a sterile cork borer and floated for 30 min in a 12-well polystyrene plate containing 1 mL of distilled water. EC is measured in an aliquot of this water, at each of the mentioned time intervals. After each measurement, the aliquot is returned to the corresponding plate containing the leaf disks.

v. After ion leakage measurements, the leaf disks in distilled water are transferred to 1.5-mL microfuge tubes, sealed with Parafilm and autoclaved for 10 min. Then, the conductance of boiled leaf disks is measured and this value is taken as the 100 % ion content (Dunger *et al.*, 2005; Baker *et al.*, 1991).

Observations

1. Make at least three independent experiments.

2. Make sure that exactly the same area of leaf disk is used.

Calculations

Electrical conductivity (EC) may be expressed as percentage (%) of control (considered as 100 %) by the equation:

$$EC (\%) = (EC\ Xac\ \text{treatment} / EC\ \text{control}) \times 100$$

or,

$$I_d = 100\,(R_i - R_0) / (1 - R_0)$$

Statistical analysis

Differences between data of treated and untreated (control) leaf disks may be evaluated by ANOVA analysis.

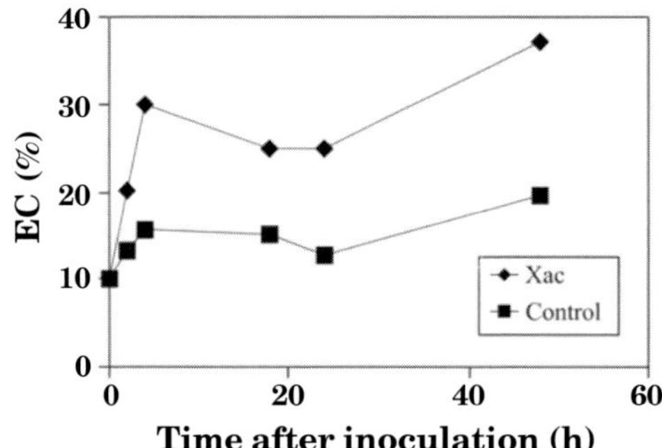

Fig 6. Ion leakage of cotton leaves undergoing HR during the interaction with *Xanthomonas axonopodis* pv. *citri*. Leaves were inoculated with 10-mM $MgCl_2$ as negative control or with 10^8 CFU mL^{-1} of *Xac* suspension. Cotton leaves discs were measured at different times: 2, 4, 18, 24 and 48 h. Values represent means of three samples

Experiment 5: Cell death observation (Trypan blue staining)

Principle

Several plant pathogens are biotrophs and require compounds from living host cells. Recognition of pathogen attack triggers a hypersensitive reaction (HR) in the plant, which includes local cell death (Greenberg, 1997;

Hutcheson, 1998). These dead plant cells can be stained with Trypan blue. Since cells are very selective in the compounds that pass through the membrane, Trypan blue is not absorbed in a viable cell; however, it traverses the membrane of dead cells. Hence, dead cells are shown as a distinctive blue spots in a microscope. Trypan blue is commonly used in microscopy for cell counting and in plants for detection of hypersensitive response. The method cannot distinguish between the necrotic and apoptotic cells. The following experiment is conducted to evaluate the cell death in the non-host interaction mediated by HR between the *Xac* and cotton plants, detected by Trypan blue staining.

Materials required

Seeds, ground, pots, fresh plant leaves, syringes, beakers, pipettes, autoclave, measuring cylinder, glass container, electric heater or laboratory burner.

Reagents

10 mM $MgCl_2$; SB medium; SB-agar; 25 mg mL^{-1} ampicillin (see Experiment 1, *Reagents*); lactic acid; 87 % (w/v) glycerol; 1 mg mL^{-1} Trypan blue: dissolve 1 mg of the dye in 1 mL of distilled water; 2.5 g mL^{-1} chloral hydrate: dissolve 2.5g of chloral hydrate in 1 mL of distilled water.

Procedure

i. Cotton seeds are spread uniformly on soil–filled pots. Seeds are germinated in a greenhouse at 25 °C under natural light with a 16 h photoperiod. At two-leaves stage, plants are transferred to individual pots and grown in the same conditions. Two to three weeks old plant are used for assays.

ii. For inoculum preparation, *Xac* is grown as described in Experiment 1 and diluted to a concentration of 10^7 to 10^8 CFU mL^{-1}. Inoculation is performed by infiltration of leaves with a needleless syringe (Fig 2). A 10-mM $MgCl_2$ solution is used as negative control.

iii. For Trypan blue staining, cotton leaves are boiled for 20 min and shaked very gently in a glass container with a solution containing 10-mL lactic acid, 10 mL glycerol, 10 mL Trypan blue (1 mg mL^{-1}) and 10g phenol. Distaining is performed by four washes with gently shaking in chloral hydrate (2.5g mL^{-1}) or a solution containing 10 mL lactic acid, 10 mL glycerol, and 10g phenol. Leaves are stored in 95 % ethanol at 25 °C (Dunger *et al.*, 2005; Koch and Slusarenko, 1990).

Observations

1. Make at least three independent experiments.

2. Avoid the exposure of cells to Trypan blue for a period longer than 30 min. In this case, it is possible to observe an increase in the population of dead cells (Trypan blue positive) due to the Trypan toxicity.

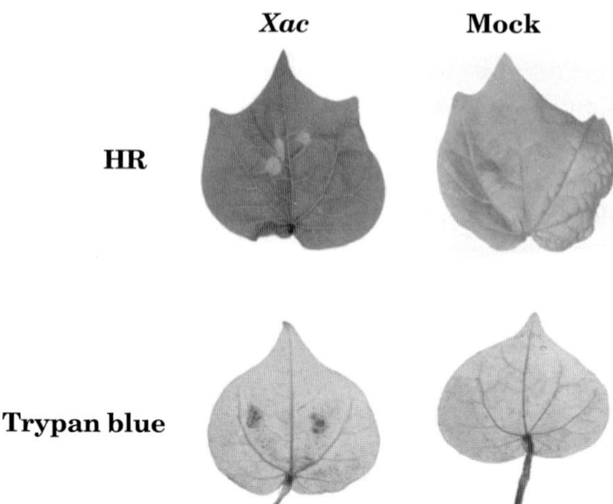

Fig 7. Cell death of cotton leaves in response to *Xanthomonas axonopodis* pv. *citri* infection. Cotton leaves stained with Trypan blue at 18 h postinfiltration are shown. As negative control cotton leaves are infiltrated with 10-mM $MgCl_2$ (mock)

Precautions

Exposure of the skin to concentrated phenol solutions causes chemical burns, which may be severe. It is usually recommended to keep available polyethylene glycol solution for washing off splashes. Chloral hydrate (also known as trichloroacetaldehyde monohydrate or 2,2,2-trichloro-1,1-ethanediol, and under the tradenames Aquachloral, Novo-Chlorhydrate, Somnos) Noctec, and Somnote, is a sedative and hypnotic drug. Higher doses can depress respiration and blood pressure. An overdose is marked by confusion, convulsions, nausea and vomiting, severe drowsiness, slow and irregular breathing, cardiac arrhythmia and weakness. It can intensify various anticoagulants and is weakly mutagenic *in vitro* and *in vivo*. Avoid breathing vapors or mist. Wash thoroughly after handling. Use with adequate ventilation in a hood. Wear protective goggles, gloves, and clothing. In case of contact with eyes, immediately flush with plenty of water and get medical attention.

Experiment 6: Callose deposition

Principle

Cell wall alterations like the formation of minute papillae directly beneath the sites of pathogen penetration, constitute a basal form of resistance to bacterial colonization and are observed during the HR (Keshavarzi *et al.*, 2004). These papillae, which are primarily composed of callose (a β-1,3 glucan; Fig 8) and lignin (a highly complex phenolic polymer), are thought to act as a physical barrier, blocking pathogen penetration into plant cells. Callose deposition may be determined by staining with aniline blue. This selective staining of callose depends upon the structure of this polymer: in callose, the 1,3 glycosidic linkages of the glucose units (Fig 8), causes the polymer to arrange in a helix which could bind the aniline blue and the resulting complex leads to yellow fluorescence. The following assay is based in the staining of callose with aniline blue and cytological observations at the sites of pathogen infection with UV-fluorescence microscopy. The scope of the present experiment is to investigate callose deposition during the non-host interaction of the phytopathogen *Xac* with non-host plants like cotton, pepper and tobacco.

Materials required

Seeds, ground, pots, greenhouse, pipettes, beakers, pH meter (or pH indicator sticks), autoclave, syringes, Petri dishes, laboratory burner or laminar flow, refrigerated shaker, sterilized clear bottles, microscope slides and coverslips, UV-fluorescence microscope.

Reagents

SB medium, SB-agar, 25 mg mL^{-1} ampicillin, 10 mM $MgCl_2$ (see Experiment 1, *Reagents*), sodium phosphate buffer (0.07 M, pH 9): dissolve 12.46g of $Na_2HPO_4{\cdot}2H_2O$ in 1 L of distilled water and adjust the pH to 9 using a solution of 0.07 M $NaH_2PO_4{\cdot}2H_2O$, which is prepared by dissolving 0.966g of $NaH_2PO_4{\cdot}2H_2O$ in 100 mL of distilled water; 0.5 % (w/v) Aniline blue solution: dissolve 0.5g of aniline blue in 100 mL of distilled water. Use this solution to prepare a 0.05 % (w/v) aniline blue solution using sodium phosphate buffer as the diluting agent.

Procedure

i. Plants of cotton, pepper, and tobacco are grown till 2–3 weeks old in a greenhouse at the same conditions indicated in Experiment 1.

ii. For inoculum preparation, *Xac* are grown 48 h on SB-agar supplemented with 25 µg mL^{-1} ampicillin at 28 °C. Then, liquid cultures of SB medium with 25 µg mL^{-1} ampicillin are grown over night with shaking of 220 rpm at 28 °C. Bacterial suspensions for the inoculations are prepared by

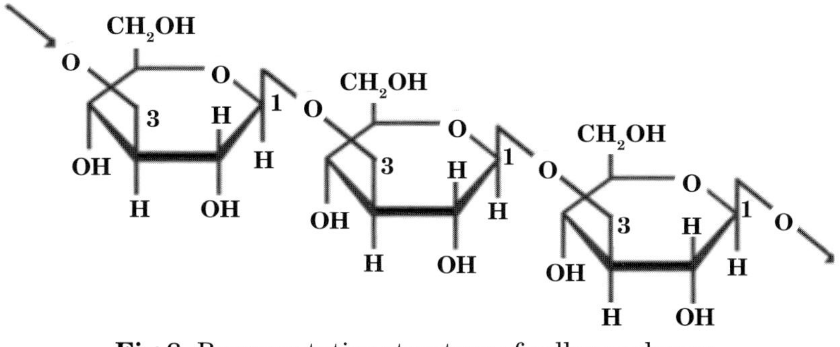

Fig 8. Representative structure of callose polymer

diluting the over night liquid cultures in 10 mM $MgCl_2$ to yield a concentration of 10^8 CFU mL^{-1}.

iii. Inoculations are done by hand-infiltration with a needleless syringe into the intercellular spaces of fully expanded leaves (Fig 2). A 10-mM $MgCl_2$ solution is used as negative control.

iv. The detection of callose deposits is made using the protocol of Adam and Somerville (1996) with some modifications. Leaves are harvested 24 h after bacterial infiltration and cleaned during a whole night in 96 % ethanol in a Petri dishes. Once the leaves are completely distained, they are incubated in the sodium phosphate buffer (0.07 M, pH 9) for 30 min. After this incubation, the buffer is discarded and an appropriate volume of aniline blue solution (0.05 %) to cover the leaves (write down the volume) is added into the dish. The tissue is immersed for 60 min and then the aniline blue solution is discarded.

v. Samples are mounted in 50 % glycerol and examined immediately under a microscope with UV-fluorescence (Nikon Eclipse E800 with a camera Spot RT). In leaves infiltrated with *Xac,* callose deposits are identified as bright-white points, the negative control infiltrated with $MgCl_2$ has not emission of fluorescence indicating the absence of these deposits (Fig 9).

Observations

1. To accelerate the time of leaf distaining, the incubation in 96 % ethanol may be done at 70 °C. Alternatively, the leaf cleaning could be done in methanol (100 %) followed by saturated chloral hydrate (Keshavarzi *et al.*, 2004).

2. Make at least three independent experiments to confirm results.

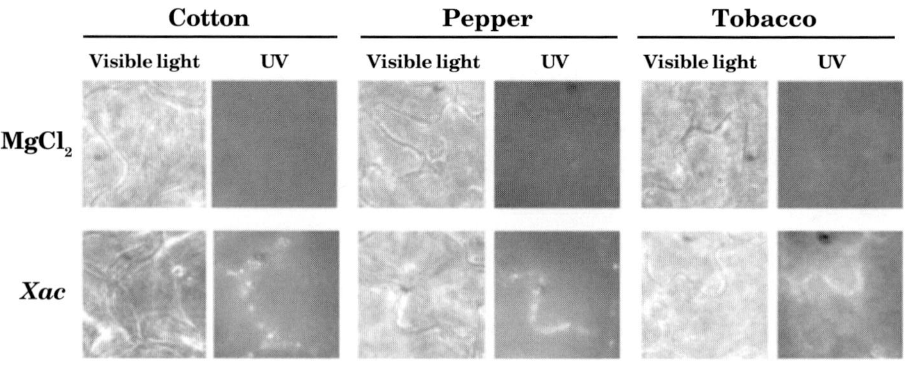

Fig 9. Callose deposition in cotton, pepper, and tobacco is associated with the non-host resistance to *Xanthomonas axonopodis* pv. *citri*. The leaves are inoculated with 10-mM $MgCl_2$ as a negative control and with 10^8 CFU mL^{-1} of *Xac* suspension. The leaves are then stained for callose deposits 24 h postinoculation and observed by light (left sections) and fluorescence microscopy (right sections)

Calculations

A quantitative analysis of stain intensity could be done quantifying the callose depositions with Image Pro Plus software (Media Cybernetics). To do it, more than 10 adjacent fields of view along the length of the leaf may be analyzed and averaged.

Statistical analysis

If a quantitative analysis is done, differences between data of treated and untreated (control) leaves must be evaluated by Student's *t*-test.

Precautions

The inhalation or ingestion of aniline blue may cause irritation to the respiratory or gastrointestinal tract. Skin or eyes contact also may cause irritation with redness and pain. Avoid the inappropriate exposure to the product.

Suggested Readings

Adam, L. and Somerville, S.C. (1996). Genetic characterization of five powdery mildew disease resistance loci in *Arabidopsis thaliana*. *The Plant Journal* **9**: 341–356.

Baker, C.J., O'Neill, N.R., Keppler, L.D. and Orlandi, E.W. (1991). Early responses during plant–bacteria interactions in tobacco cell suspensions. *Phytopathology* **81**: 1504–1507.

Blum, A. and Ebercon, A. (1981). Cell membrane stability as a measure of drought and heat tolerance in wheat. *Crop Science* **21**: 43–47.

Cessna, S.G., Sears, V.E., Dickman, M.B. and Low, P.S. (2000). Oxalic acid, a pathogenicity factor for *Sclerotinia sclerotiorum*, suppresses the oxidative burst of the host plant. *The Plant Cell* **12**: 2191–2199.

Dexter, S.T., Tottingham, W.E. and Graber, L.F. (1932). Investigations of the hardiness of plants by measurement of electrical conductivity. *Plant Physiology* **7**: 63–78.

Doke, N. (1983). Involvement of superoxide anion generation in the hypersensitive response of potato tuber tissues to infection with an incompatible race of *Phytophthora infestans* and to the hyphal wall components. *Physiology Plant Pathology* **23**: 345–357.

Dunger, G., Arabolaza, A.L., Gottig, N., Orellano, E.G. and Ottado, J. (2005). Participation of *Xanthomonas axonopodis* pv. *citri hrp* cluster in citrus canker and nonhost plant responses. *Plant Pathology* **54**: 781–788.

Flint, H., Boyce, B.R. and Beattie, D.J. (1966). Index of injury - A useful expression of freezing injury to plant tissues as determined by the electrolytic method. *Canadian Journal of Plant Science* **47**: 229–230.

Flor, H.H. (1971). Current status of gene-for-gene concept. *Annual Review of Phytopathology* **9**: 275–296.

Greenberg, J.T. (1997). Programmed cell death in plant–pathogen interactions. *Annual Review of Plant Physiology and Plant Molecular Biology* **48**: 525–545.

Hammond-Kosack, K.E. and Jones, J.D. (1996). Resistance gene-dependent plant defense responses. *The Plant Cell* **10**: 1773–1791.

Heath, M. (2000). Hypersensitive response-related death. *Plant Molecular Biology* **44**: 321–334.

Hutcheson, S.W. (1998). Current concepts of active defense in plants. *Annual Review of Phytopathology* **36**: 59–90.

Imlay, J.A. and Linn, S. (1988). DNA damage and oxygen radical toxicity. *Science* **240**: 1302–1309.

Keith, R.C., Keith, L.M.W., Guzmán, G.H., Uppalapati, S.R. and Bender, C.L. (2003). Alginate gene expression by *Pseudomonas syringae* pv. *tomato* DC3000 in host and non-host plants. *Microbiology* **149**: 1127–1138.

Keshavarzi, M., Soylu, S., Brown, I., Bonas, U., Nicole, M., Rossiter, J. and Mansfield, J. (2004). Basal defenses induced in pepper by lipopolysaccharides are suppressed by *Xanthomonas campestris* pv. *vesicatoria*. *Molecular Plant–Microbe Interactions* **17**: 805–815.

Koch, E. and Slusarenko, A. (1990). Arabidopsis is susceptible to infection by a downy mildew fungus. *The Plant Cell* **2**: 437–445.

Lamb, C. and Dixon, R.A. (1997). The oxidative burst in plant disease resistance. *Annual Review of Plant Physiology and Plant Molecular Biology* **48**: 251–275.

Mittler, R. (2002). Oxidative stress, antioxidants and stress tolerance. *Trends in Plant Science* **9**: 405–410.

Mysore, K.S. and Ryu, C.-M. (2004). Non-host resistance: How much do we know? *Trends in Plant Science* **9**: 97–104.

Rodriguez, A.A., Córdoba, A.R., Ortega, L. and Taleisnik, E. (2004). Decreased reactive oxygen species concentration in the elongation zone contributes to the reduction in maize leaf growth under salinity. *Journal of Experimental Botany* **55**: 1383–1390.

Stakman, E.C. (1915). Relation between *Puccinia graminis* and plants highly resistant to its attack. *Journal of Agricultural Research* **4**: 193–199.

Thordal-Christensen, H., Zhang, Z., Wei, Y. and Collinge, D.B. (1997). Subcellular localization of H_2O_2 in plants: H_2O_2 accumulation in papillae and hypersensitive response during the barley-powdery mildew interaction. *The Plant Journal* **11**: 1187–1194.

Torres, M.A., Dangl, J.L. and Jones, J.D. (2002). Arabidopsis gp91phox homologues AtrbohD and AtrbohF are required for accumulation of reactive oxygen intermediates in the plant defense response. *Proceedings of the National Academy of Science USA* **99**: 523–528.

Van Loon, L.C., Rep, M. and Pieterse, C.M.J. (2006). Significance of inducible defense-related proteins in infected plants. *Annual Review of Phytopathology* **44**: 135–162.

Venisse, J.S., Gullner, G. and Brisset, M.N. (2001). Evidence for the involvement of an oxidative stress in the initiation of infection of pear by *Erwinia amylovora*. *Plant Physiology* **125**: 2164–2172.

Verma, D.P. and Hong, Z. (2001). Plant callose synthase complexes. *Plant Molecular Biology* **47**: 693–701.

CHAPTER 11

Plant – Virus Interactions

Gabriella Kazinczi[1], Richard Gáborjányi[1], Erzsébet Nádasy[1], András Takács[1] and József Horváth[1]

1. INTRODUCTION

Plant viruses are obligate parasites containing single or double stranded RNA or DNA surrounded by a coat protein. They produce diseases which cause losses estimated at about 15 billion USA dollars per year worldwide in both agricultural and horticultural crops. Viruses penetrate into the plant only in a passive way, with or without the help of animal vectors. Horticultural grafting, plant wounding, vegetative (tubers, rhizomes, bulbs, roots) and generative organs (pollen, seed), lower fungi (*Olpidium, Polymyxa, Pythium* spp.), *Cuscuta* species and water are included in transmission without vectors. Nevertheless, insect vectors are most frequently involved in virus transmission under natural conditions (Fig 1), especially aphids (Harris, 1981; Horváth and Gáborjányi, 1999). In this chapter, we provide experiments for inoculation, isolation and identification of plant viruses as well as protocols and criteria for observation of the effects of viruses on plant growth and physiology.

Experiment 1: Mechanical inoculation of *Nicotiana tabacum* L. cv Xanthi with *Tobacco Mosaic Virus* (TMV)

Principle

Mechanical inoculation of plants is the most convenient and common method used for experimental virus transmission. It means bringing healthy plant parts, usually leaves, into contact with a virus containing suspension (inoculum). This type of virus transmission is a valuable method for disease diagnosis, local-lesion assay, propagation and maintenance of viruses, studying host–virus interactions and detecting viruses (Horváth, 1993). Most of the plant viruses, but not all, are mechanically transmissible. To achieve infection, small wounds must be made, as virus particles cannot penetrate in intact leaf surface, with its cuticle and cell walls acting as major barriers. The most effective way to wound the cell is by using abrasives, *e.g.*,

1. Georgikon Faculty of Agricultural Sciences Institute for Plant Protection, University of Pannonia, Keszthely, Hungary. E-mail: kg@georgikon.hu

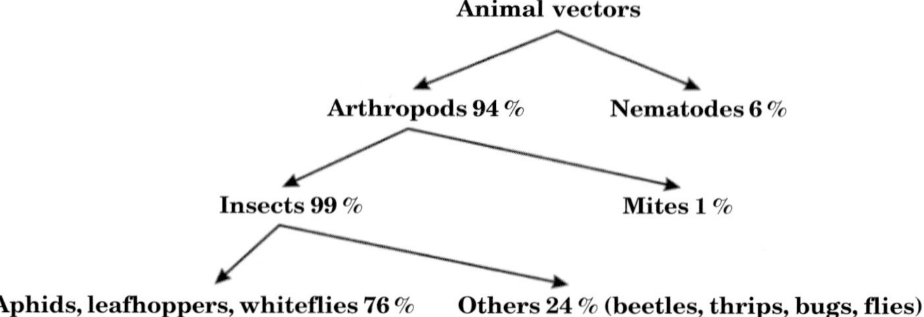

Fig 1. Transmission of plant viruses with animals

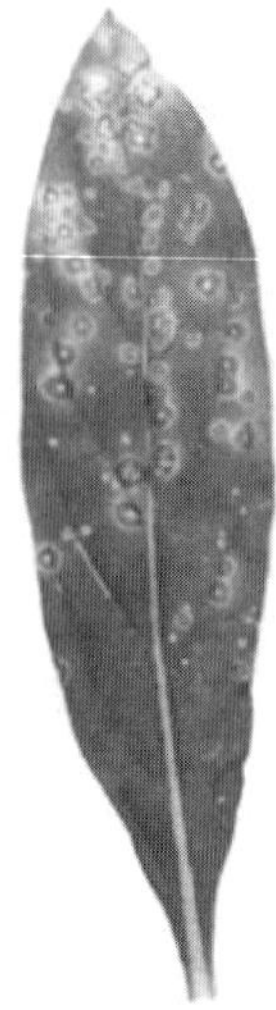

Fig 2. Local necrotic lesions on *Gomphrena globosa* leaves due to *Potato virus X* (PVX) infection

Fig 3. Systemic symptoms (blistering, leaf deformation, vein necrosis) on *Nicotiana tabacum* 'Xanthi' leaves due to *Potato virus Y* (PVY) infection

carborundum powder, dusted on the leaves before inoculation. The leaves are then inoculated by rubbing their surface with the inoculum. In compatible host–virus relations, appearance of local and/or systemic symptoms (Figs 2 and 3) can be expected generally 1–14 days after inoculations depending on susceptibility of test plants, characteristics of viruses (*i.e.*, stability, concentration and presence of other constituents in the inoculum) and environmental factors like temperature and light (Dijkstra and Jager, 1998).

Materials required

Inoculum source (TMV is maintained on *Nicotiana tabacum* 'Samsun' as a propagative host of this virus under greenhouse conditions), healthy *N. tabacum* 'Xanthi' test plants at 4–6 leaf stage, sterile mortar, glass spatula, carborundum (500 mesh with a particle size of 50 nm).

Reagents

Sörensen's phosphate buffer: Dissolve 3.628g of KH_2PO_4 and 7.122g of $Na_2HPO_4 \cdot 2H_2O$ in 1 L of distilled water.

Procedure

i. Take leaf sample from the upper parts of *N. tabacum* 'Samsun' showing strong mosaic symptoms due to TMV infection.

ii. Inoculum preparation: ground 1g of plant material with 2.5 mL of 0.1 M Sörensen phosphate buffer in a sterile mortar.

iii. Sprinkle uniformly and thinly carborundum powder on the leaves of the healthy test plants to be inoculated.

iv. The inoculum must be gently rubbed with a glass spatula over the leaf surface of the test plants from its base to the top. Generally, 3–4 leaves must be inoculated in each plant.

v. It is advisable to label the inoculated leaves with a small hole to distinguish the local and systemic symptoms later. After inoculation, rinse the infected plants with water and place them in a humid environment.

Observations

N. tabacum 'Xanthi' is a local host of TMV. Therefore, evaluate the number and size of local necrotic lesions on its leaves 2–3 days after inoculation.

2. VIRUS ISOLATION AND PURIFICATION

Our knowledge about the structure and function of viruses significantly increased after the successful isolation and purification of virions. The satisfactory procedure results in a relatively high amount and high purity of virions, which provide adequate facilities for electron microscopic studies and serological techniques, as well as for the determination of nucleic acid composition and sequence analysis. Generally, virus isolation means the separation of virions from the host plant tissues or the separation of relatively similar virus particles of multicomponent viruses. In a wider sense, it means the separation of viral nucleic acids or the coat protein from the nucleoprotein complex. Purification procedures are highly dependent on the chemical structure and the stability of viruses. However, general steps in all purification procedures are as under: (i) host selection, (ii) extraction methods, (iii) removal of contaminants by physical or chemical means, (iv) concentration of virions, (v) storage and detection of impurities and (vi) estimation of virus concentration. Isolation of TMV and extraction of its RNA are described below as a typical example of virus purification (Experiments 2 and 3). Nevertheless, various purification protocols (Noordam, 1973; Bos, 1983; Lane, 1986; van Regenmortel and Fraenkel-Conrat, 1986; Koenig, 1988; Milne, 1988; Foster and Taylor, 1998; van Regenmortel *et al.,* 2004) and molecular techniques are available in the literature (Crowther, 1955; Howell, 1985; Sambrook *et al.*, 1989; Stace-Smith and Martin, 1993; Rapley, 1996).

2.1. Host selection

The use of a pure virus culture is necessary to get satisfactory results. For this reason, it is advisable to choose a local lesion host and propagate a virus in susceptible plants. Holmes (1929) described that *Nicotiana glutinosa* produces tiny local necroses (called local lesions) instead systemic mosaic symptoms after TMV infection. In this resistant host, the infection remained localised in *N. glutinosa* leaves. The number of lesions could serve as a tool to calculate the infectivity of virus solutions. Virions in a single local lesion can represent more or less a genetically homogenous population. In mechanically non-transmitted viruses, only the insect transmission could be the first step to develop virus cultures on a host plant, adequate for the insect, and suitable for virus replication. Host plant selected for virus propagation must have the following characteristics: (i) it should produce the highest concentration of virus per gram tissue; (ii) it does not contain components, which interfere with extraction of virions; and (iii) it should grow well and should be easy to culture. For maximum recovery of virus material, one must harvest plants at the optimum time. Generally, the upper young, well-developed leaves contain the highest virus concentrations (Steere, 1964).

2.2. Extraction

Virus recovery from the host is highly dependent on the use of an appropriate buffer during maceration of plant tissues. The dilution, ionic strength, and pH of the buffer should also be the most favourable for the stability of virions, preventing aggregation or precipitation of virus particles. Attention should also be given to the presence of antioxidants such as ascorbic acid or enzyme inhibitors like sodium diethyl dithiocarbamate. Tannins or other phenols could be neutralized either by white egg or bone powder. Phosphate buffers are the most frequently used in the virus purifications, but before starting the extraction, it is necessary to examine the effect of buffers in a preliminary test.

The maceration of infected tissues depends on equipment availability. Tissues may be ground in a mortar and pestle, electric blender or a juice extractor. Maceration should be done at low temperature as quickly as possible, to avoid enzyme degradation. The use of liquid nitrogen and acetone can provide homogeneous tissue powders, which can be stored for longer times. Crude juice should be filtered through cheese cloth to remove the large debris, tissue fragments and pulps containing contaminants like chloroplasts, nuclei, and mitochondria.

2.3. Removal of contaminants

In the next step of purification, it is necessary to remove all non-virus particles and undesired salts from a suspension. This is essential, because viruses are often adsorbed to other components. First, the dark green or grey juice should be centrifuged at a low speed (3000g for 10–30 min) to separate the heavy tissue fragments. Addition of organic solvents, like chloroform, butanol or carbon tetrachloride to the crude juice extract will collect the majority of chlorophyll and lipoproteins by a rapid stirring to form an emulsion. One can use these solvents before and after the extraction too. During the low-speed centrifugation, the organic component forms a separate interphase, which will be easily separated both from the aqueous and from the solvent phase. The supernatant water phase contains the virions in a larger dilution. The use of organic solvents can lead to important loss of virus particles during clarification. Organic solvents cannot be used in viruses containing glycoprotein covering (Tospovirus, Rhabdovirus). The purification must be done at low temperature, using cold solvents.

2.4. Concentration of virions

After purification, the supernatant of the centrifuged juice will contain the virus particles, but in high dilution. Therefore, it is necessary to concentrate the virus content, remove the virus from the suspension and resuspend it in a small volume of the desired buffer. Several methods are available to concentrate the infected juice, like gel filtration, salting out, isoelectric precipitation or moving boundary electrophoresis. However, ultracentrifugation is the simplest way to concentrate the virus from the aqueous phase. During ultracentrifugation, using angle-head rotors

(180,000g for more than 2 h) virus particles and other larger particles move down and form a jelly-like pellet to the wall of the tube. The pellet is resuspended in the desired small amount of selected buffer. After resuspension, much of the non-virus proteins can be removed by another low-speed centrifugation. Usually, 2 or 3 cycles of high- and low-speed centrifugation are necessary to give a relatively pure virion suspension. Using sucrose density, gradient centrifugation in a swigged-bucket rotor, one can obtain more pure virus solution in a separation gradient. This method is also useful to separate the viral components of multicomponent viruses. During the centrifugation, sucrose gradient is formed in a centrifuge tube and separate the viral components in different layers. Heavy salts like caesium chloride can be used instead of sucrose in this method, since they also make a gradient (equilibrium centrifugation) and give the best results of separation. At the end of ultracentrifugation, one can get a relatively concentrated virus suspension in the proper solvent. Many viruses can be precipitated or crystallized out from virus suspension. Polyethylene glycol or high concentration of ammonium sulphate or other salts can precipitate virus proteins or virions. In these cases, irreversible aggregation of particles often occur and virus lose its infectivity.

2.5. Storage and detection of impurities

After purification, virus must be preserved in conditions, where it retains its integrity and infectivity. Viruses are less stable in pure form; therefore, a special care should be taken to choose the appropriate buffer. Sometimes proteins or glycerol is added to improve stability. Purified viruses can be stored in refrigerator at 4 °C, but most viruses retain their infectivity in a deep-freezer (at –80 °C) or in liquid nitrogen.

After melting the frozen virus suspension, one can remove the buffer salt components required for storage by dialysis. Contaminants such as proteins, nucleic acids or phenolic substances can be detected using electrophoresis, chromatography or density gradient centrifugation. Sedimentation coefficient (in Svedberg) estimated in analytical ultracentrifuge.

2.6. Estimation of virus concentration

Theoretically, virus solution contains only nucleoproteins. The purity and the virus concentration can be measured in ultraviolet (UV) light, at wavelengths between 220 and 300 nm, according to the optical density. Extinction coefficient represents a virus specific extinction of 1 mg mL^{-1} virus solution (measured in 1 cm light pathway at 260 nm). UV adsorption spectra of viruses are not virus-specific, but show the purity of the solution. Virions give adsorption maximum at 260 nm and minimum values at 240 nm. Proteins give a maximum adsorption at 280 nm, while nucleic acid's at

260 nm. Quotients of the two values (A 280/260) gave valuable information about the nucleic acid content (Francki and McLean, 1968).

Experiment 2: Extraction of *Tobacco mosaic virus* (TMV)

Tobacco mosaic virus is a representative member of Tobamoviruses and the most extensively studied phytopathogenic virus. It infects many cultivated crops, causing severe yield losses in tomato, pepper and tobacco. It forms 300 × 18 nm rigid rod-shaped particles. The single stranded RNA genome contains about 3600 nucleic bases, codes functional proteins for RNA replication (126 and 183 kDa), a protein responsible for intercellular movement (MP 30 kDa) and one for a protein coat (CP 17.6 kDa). The first two proteins are directly translated from the genome (the second with the read-through of a weak stop codon); the MP and CP are produced by translation of subgenomic RNAs. Extraction of TMV virions is achieved after precipitation with polyethylene glycol (PEG) as demonstrated by Gooding and Hebert (1967) and Leberman (1966). Extraction procedure was described in detail by Chapman (1998).

Principle

Hydrophilic polymer polyethylene glycol will destabilize the soluble form of virus emulsion and the virus particles will precipitate from the solvent. Virions could be simply collected by low-speed centrifugation, without using ultracentrifuge.

Materials required

All procedure should be done in cold chamber (about 4 °C), use ice bath and cold reagents, fume hood, warning blender, bench centrifuge with 20 mL polypropylene centrifuge tubes capable of 10,000g, Eppendorf micro centrifuge (13,000g*),* UV spectrophotometer capable to read absorbance in UV light (260–280 nm), cheese cloth (or miracloth).

Reagents

0.5 M phosphate buffer: prepare a 0.5 M solution of disodium hydrogen orthophosphate and adjust the pH to 7.2 with 0.5 M potassium dihydrogen orthophosphate.

Extraction buffer: Add 1 % (v/v) 2-mercapto ethanol to 0.5 M phosphate buffer just before use; Butan-1-ol; 20 % (w/v) PEG (about Mw: 8000) in distilled water; 5 M NaCl solution.

Procedure

i. Collect 20g of systemically infected young tobacco leaves. Grind these in a blender using 60 mL of extraction buffer.

ii. Filter the sap through two layers of cheese cloth, add butanol (0.8 mL/ 10 mL of filtrate), mix well the solvents, and incubate for 15 min at 4 °C. Remove the green upper, organic solvent.

iii. Centrifuge the aqueous phase at 10,000g for 30 min at 12 °C. Collect the upper light green water phase. Add 20 % PEG (Mw 8000) solution to 4 % final concentration. Mix centrifuge tubes several times and leave to stand them in an ice bath for 15 min.

iv. Centrifuge the tubes at 10,000g for 15 min at 4 °C. Remove all the supernatant carefully. Resuspend the pellet in 8 mL of 10 mM phosphate buffer (50-fold diluted from 0.5 M buffer), and centrifuge the virus suspension once again at 10,000g for 15 min at 4 °C.

v. Collect the supernatant into a fresh tube. Add 1.7 mL of 5 M NaCl solution and 2.42 mL of 20 % PEG. Mix it well and stand on ice to pellet the virions for 15 min.

vi. Collect pelleted virions by centrifugation (10,000g for 15 min at 4 °C). Discard the supernatant carefully and dissolve the pellet in 2 mL of 10 mM phosphate buffer. Divide the solution in microcentrifuge (Eppendorf) tubes and spin them at 13,000g for 30 min. Supernatant pure virion suspension will be collected in fresh tubes.

vii. Dilute small aliquots of virus suspension to determine the concentration measuring the absorbance at 260 and 280 nm. Absorption ratio 1:19 are expected. Yield should be about 20 mg viruses.

Precautions

Manipulations with 2-mercaptoethanol and butanol should be done in a fume hood.

Experiment 3: RNA isolation from *Tobacco mosaic virus* (TMV)

Principle

TMV virions contain single-stranded RNA (about 5 %) and coat protein (about 95 %). The mixture of phenol–chloroform separates the two components; RNA remains intact, but the protein precipitates. Overlying aqueous layer will contain the RNAs. Denatured proteins in the pellet can be removed by centrifugation. RNA will be further purified by ethanol precipitation as recommended by Chapman (1998).

Materials required

Ice bath and cold reagents, vortex mixer, Eppendorf micro centrifuge (capable 13,000g), deep-freezer (–20 °C), vacuum, UV spectrophotometer capable to read absorbance in UV light (260–280 nm).

Reagents

Reagents should be of molecular biology grade.

5× RNA extraction buffer: 5 extraction buffer contains 0.5 M sodium chloride, 5 mM ethylenediamine-tetra-acetic acid disodium salt, 5 % (w/v) sodium dodecyl sulphate, 0.1 M Tris (hydroximethyl-methylamine) adjusted to pH 8.0 with hydrochloric acid.

Phenol–chloroform 1:1 (v/v): Mix 250g phenol in 250 mL 0.1 M Tris-HCl, pH 8.0, and add 1.25g of 8-hydroxyquinoline. Equilibrate by extracting several times with 0.1 M Tris-HCl. Remove most of the overlying aqueous layer, add 240 mL chloroform and 10 mL of isoamyl alcohol. Mix and allow phases to separate. Store refrigerated and protect from light.

3 M sodium acetate, pH 5.0: Dissolve sodium acetate in water; adjust to pH 5.0 with glacial acetic acid. Adjust volume and treat with diethylpyrocarbonate (DEPC) to inactivate RNAse. To each liter of solution add 1 mL of DEPC, mix and incubate it overnight at room temperature in a loosely capped bottle. Autoclave it to destroy residual DEPC. Absolute ethanol, distilled water (treated with DEPC, autoclaved as above).

Procedure

i. Dilute an aliquot of purified virion suspension to 10 mg mL^{-1} with 10 mM phosphate buffer. Add 0.2 mL of 5× RNA extraction buffer to 0.8 mL of diluted virus suspension. Add 1 mL of phenol:chloroform (1:1; v/v). Vortex well, until an emulsion formation.

ii. Centrifuge emulsion in microcentrifuge tubes at 13,000g for 5 min at room temperature. Interface contains the denatured virus protein. Collect the upper, RNA containing aqueous phase and repeat the phenol extraction with phenol:chloroform twice. Collect aqueous phase of the third extraction. Add an equal volume of chloroform, vortex briefly to form an emulsion. Separate the water phase by microcentrifugation. The upper aqueous phase (about 0.7 mL) will be collected and separated into two 2 mL microcentrifuge tubes. Add to each tube 0.1 volume of 3 M sodium acetate and 2.5 volume of ethanol (about 0.9 mL). Mix it in a vortex and incubate the samples at –20 °C for 15 min to precipitate the RNA. Pellet the RNA by microcentrifugation at 13,000g for 15 min at 4 °C. Remove the supernatant. Repeat centrifugation once again and pipet off the residual liquid.

iii. Dry the pellet under vacuum for a few minutes. Add 0.2 mL of water and dissolve the pellet by gentle vortexing. Dilute some aliquots of purified RNA and measure the absorbance at 260–280. The preparation should have an $A_{260/280}$ ratio of 2.0. RNA yield can be estimated by assuming that 40 mg mL^{-1} solution of RNA has an absorbance of 1.

Precautions

Phenol and chloroform are toxic. Avoid contact with skin and inhalation. Both should be use in fume hood. Extraction of RNA needs a special precautions, all plastic or glassware should be treated by DEPC (see in reagents). Use plastic gloves to prevent contact with phenol and to avoid RNAse contaminations.

3. EFFECT OF VIRUS INFECTION ON PLANT PHYSIOLOGICAL PROCESSES

The plant physiological and biochemical aspects of virus infected crops have seen well studied. Hence, the effects of viruses on the photosynthesis (Funayama *et al.*, 1997; Almási *et al.*, 2000; Kazinczi *et al.*, 2006a), respiration (Gáborjányi and Fernandez, 1976; Gullner *et al.*, 1999; Király *et al.*, 2002), nucleic and protein synthesis (Salomon *et al.*, 1976), hypersensitive reaction (Király *et al.*, 1972), phytohormones (Pogány, 1999; Gáborjányi *et al.*, 1971), nutrient transport (Hall and Loomis, 1972), cell wall and intracellular membrane structures (Matthews, 1981) are well known. From a practical point of view, the viruses also kills the weeds, which adversely decreases the crops yields through direct competition for nutrients and water and also acts as alternative hosts to different pests. From virological point, weeds may play major role in the epidemiology of plant viruses, they are primary sources of infection in spreading the plant diseases, while their seeds and vegetative/reproductive organs are reservoirs of obligate parasites, such as plant viruses, by ensuring their overwintering (Kazinczi *et al.*, 2002a,b; 2005). Besides, weeds are also host plants of virus vectors. In sustainable agricultural practice and crop production, efforts are done to maintain biological diversity, hence, all the weeds are not killed and their population level is kept under at economic threshold. So far, little attention has seen paid to the reduced competitive ability of virus infected weeds. Biological decline of the diseased weeds mean advantages for the crop production. Competitive ability of diseased plants is reduced due to their decreased seed viability, germination, growth, water and nutrient uptake. Therefore, they do not mean competitive partners for the crops anymore, as compared to the healthy ones. It means that the viruses indirectly can contribute to weeds control (Kazinczi *et al.*, 2006b,c).

3.1. Germination and seed viability

Experiment 4: Germination tests

Principles

Some viruses are seed-transmitted and the extent of seed transmission varies (0 and 100 %), depending on host-virus relations, time of infection, virus strains and environmental factors (Bos, 1977; Mink, 1993; Johansen *et al.*, 1994; Kazinczi *et al.*, 2000). Even if a virus is not a seed-borne, it decreases the germination and viability of seeds from the virus–infected plants than healthy plants. The virus infection influences the germination and viability of seeds (Kazinczi *et al.*, 2000, 2002a; Kazinczi *et al.*, 2006c).

Materials required

Inoculum source [wheat plants infected with *Barley stripe mosaic virus* (BSMV)], wheat seeds, Petri dishes (12 cm diameter), filter paper, distilled water, incubator.

Procedure

i. Do mechanical infection to wheat seedlings with *Barley stripe mosaic virus* (BSMV) (see Experiment 1, *Procedure*). Uninfected plants serve as healthy controls. Collect and clean wheat seeds from the healthy and virus infected wheat plants at the end of vegetation period.

ii. Take four samples of 100 seeds from healthy and virus infected plants. Place each sample in a Petri dish on the top of filter paper. Add 10 mL of distilled water to each Petri dish. Transfer Petri dishes to an incubator at 20 °C.

iii. Observe the germination daily, until no germination occurs and remove the germinated seeds. Register germination as number of germinated seeds per 100 seeds at the end of the experiment.

Statisical analysis

Evaluate the results using analysis of variance (ANOVA). If germination percentage is under 100 %, and/or significant difference between the seed germination derived from healthy and virus infected plants occur, seed viability (TTC) test is required.

Experiment 5: Seed viability test

Principle

Reductive processes in the living cells can be detected by redox indicators. The viability of seeds is tested with 2,3,5-triphenyl-tetrazolium-chloride (TTC). When a seed is viable its dehydrogenases enzymes are active. TTC uptakes hydrogen ion (H^+) from the dehydrogenases and forms red-colored triphenyl-formazan. The dead cells do not reduce the TTC, therefore living (red) and dead (uncolored) parts of the seeds are well distinguished. Seed viability can be determined on the basis of the size and location of the red (living) and uncolored (dead) parts on the embryo and/or endospermium (Moore, 1985).

Materials required

Stereomicroscope, Petri dishes, filter paper, distilled water, incubator, folding knife, wheat seeds.

Reagents

Buffer mixture: dissolve 9.078g of KH_2PO_4 into 1000 mL distilled water (buffer I) and 11.876g of Na_2HPO_4 into 1000 mL distilled water (buffer II). Prepare 1000 mL of the mixture of buffer I and buffer II in a ratio of 2:3. Then, add 5g of TTC.

Procedure

i. Imbibe four samples of 100 wheat seeds (from healthy and BSMV infected plants; see Experiment 4, Procedure) in distilled water for 12 h.

ii. Make a longitudinal cut through the embryo and three-quarters part through the endosperm (Fig 4). Soak the seeds for 2 h in 0.5 % TTC solution.

iii. Evaluate the cut surface under a stereomicroscope. The criteria of seed viability: at least two root primordia and one third part of the scutellum are red-colored (Fig 5). Determine the proportion of the viable/non viable seeds.

Statistical analysis

Evaluate the results by analysis of variance (ANOVA).

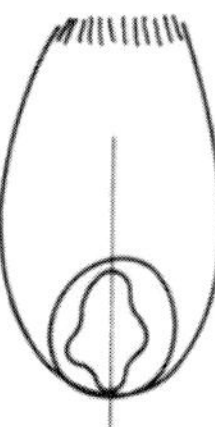

Fig 4. Longitudinal cut through the embryo and a part of the endospermium of wheat caryopses for preparing TTC test

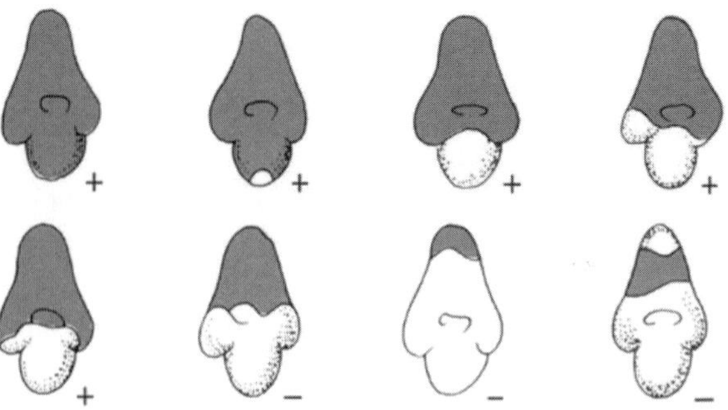

Fig 5. Evaluation of TTC test in wheat caryopses. Caryopses indicated with + are viable

3.2. Growth and nutrient uptake

The growth and development of the plants can be described by a logistic function, sigmoid (S) curve. The bell-shaped Gauss curve shows the increase (expressed either in dry matter or leaf area) within a time unit (absolute growth rate). The growth is slow at the beginning, then accelerates, reaches its maximum value (which coincides with the inflexion point of sigmoid curve and the peak point of Gauss curve) and finally slows down again (Fig 6). Growth analysis is a useful method to determine the dynamics of photosynthates produced (Blackman, 1919; Hunt, 1978). Different indices can be determined, if plant samples are taken regularly from a homogeneous stand weekly during the vegetation period and measure the dry weight and leaf area of the plant samples. From the values of growth indices and their changes in time, we can know the competitive ability of a plant species (Berzsenyi, 2000). So, under field conditions, we can follow the changes in the biomass production during the whole vegetation period. Under glasshouse conditions, in pot experiments, the effect of different treatments and environmental conditions on the early development of the plant can be observed (Experiment 6).

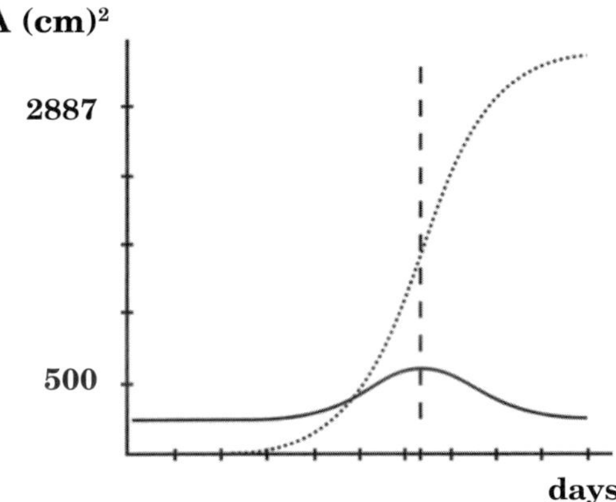

Fig 6. The changing of the leaf area of *Matricaria inodora* during the vegetation period (*Source*: Kazinczi, 1993)

The major plants nutriets are nitrogen (N), phosphorus (P), and potassium (K). Most of the nitrogen in plants is present as proteins. The N-content of plants dry matter is 0.1–6.0 %. Phosphorus is important for plant bioenergetics and its concentration in dry matter is much lower (0.01–0.6 %) than nitrogen. Potassium regulates the opening and closing of the stoma and plants contain 0.2–5 % K_2O. We can determine the nutrients content of the plant, on the basis of plant analysis. Results of plant analysis indicate the nutrient supply of the soil, and if nutrients content is insufficient we must apply fertilizers. The method for collection of plant samples' is determined by the aim of the analysis. Nutrient content of the plants mainly depends on plant species, plant parts, age of the plants, and nutrient supplying capacity of the soil. Nutrient content of plants indicate the nutrient level of the soil. We usually collect young plants or plant parts, as leaves, shoots and sometimes roots to estimate the nutrient supply of the plant (Debreczeni, 1991).

Experiment 6: The effect of TMV infection on the growth of *Solanum nigrum*

Principle

Reduction in plant growth and biomass production results from the plant–virus interactions, perhaps due to the reduced water and nutrients uptake owing to changes in the permeability of root cell membranes (Matthews, 1981), inhibited translocation of assimilates due to the phloem degeneration, changes in number and activity of enzymes responsible for transformation of carbohydrates (Faccioli *et al.*, 1977) and growth regulators (Kuriger and Agrios, 1977). In pot experiments, virus infections causes 29–82 % reduction in biomass than in healthy control plants (Kazinczi *et al.*, 2006b) (Fig 7). Little is known about the changes in the nutrient content of host plants due to various pathogens, including viruses. It seems that virus infections only reduces the concentration of micronutrients in the plant shoots, leadling to the reduced growth (Kazinczi *et al.*, 2006b).

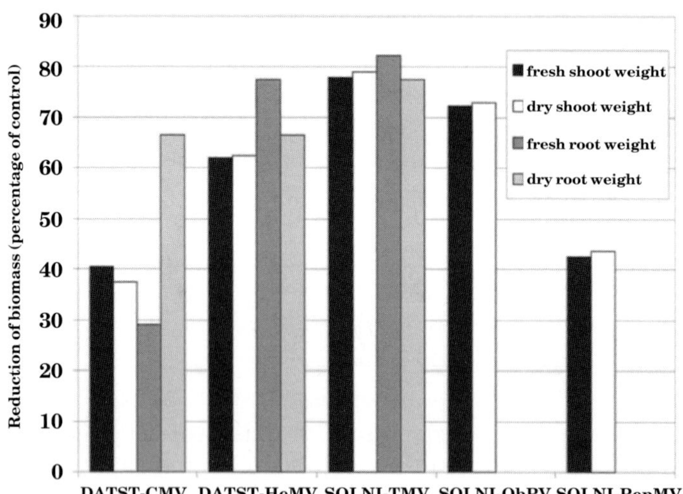

Fig 7. The effect of virus infections on the fresh and dry weight of test plants (root weight in *S. nigrum* – ObPV and *S. nigrum* – PepMV relations was not measured). Abbreviations in host-virus relations: DATST – *Datura stramonium*; SOLNI – *Solanum nigrum*; CMV – *Cucumber mosaic virus*; HeMV – *Henbane mosaic virus*; TMV – *Tobacco mosaic virus*; ObPV – *Óbuda pepper virus*; PepMV – *Pepino mosaic virus*

Procedure

i. Do mechanical infections to *S. nigrum* plants at 4–6 leaf stages with TMV (see Experiment 1). Uninfected plants serve as healthy controls.

ii. Two months after inoculation, cut the shoots at the soil surface and measure their fresh weight.

iii. The shoots must be immediately dried at 60 °C, until samples become crumbling by hand (usually 12–24 h). After drying, measure the dry weight of the plant samples.

iv. Take leaf samples from the upper parts of the shoots for the determination of NPK content.

v. Plant samples are prepared for NPK determination by mechanical grinding to particles of 0.5–1.0 mm size. The prepared samples can be stored in good closing glasses for 2 years.

Determination of NPK nutrients in plants:

Experiment 7. Wet digestion with sulfuric acid to determine NPK contents

Principle

The aim of digestion is to convert the nutrient elements of plant samples into ionic or inorganic compounds. The method is based on the wet oxidation of dried plant materials using sulfuric acid and digestion catalyst, which converts the organic nitrogen to the ammonium forms. The concentrated sulphuric acid effectively draws water from the organic material and liberates carbon, besides, the sulphuric acid heating at 338 °C breaks it down into water, sulphur dioxide and oxygen. Free oxygen combines with the elementary carbon and forms the carbon dioxide. While sulfur dioxide, reduces the nitrogen (of organic matter) to ammonia and forms the ammonium-sulphate.

Materials and equipments

Block digester (thermo block) with tubes (Fig 8), balance, chemicals spoon, 10 mL pipette, 50 mL volumetric flask, glass funnel.

Reagents

Concentrated sulphuric acid (H_2SO_4), catalyst mixture Se–K_2SO_4 (1:10; w/w), 30 % hydrogen peroxide (H_2O_2), pumice boiling granules.

Procedure

i. Weigh 0.2–1.0g plant sample and transfer into a digestion tube. Add few pumice boiling granules and about 0.5g catalyst mixture. Then, add 10

Fig 8. The block digester with tubes

mL concentrated sulfuric acid into the tube. Keep the prepared samples with sulfuric acid for a day or a night to precede reduction.

ii. Add 1–2 mL hydrogen peroxide before heating and mix carefully, because it will become hot. Close the tubes and keep a small glass funnel into the mouth of tubes to insure recirculation of digestion mixture and prevent the loss of sulfuric acid.

iii. Place tube into a thermo block digester and set at 180 °C for 30 min and then at 380 °C for 1–2 h. When samples will become dark brown again, remove the tube from the thermo block, and allow cooling to 30–40 °C. Then, add another 1–2 mL hydrogen peroxide into the tube. If any material is adhering to the neck of the tube, we can also wash down with this hydrogen peroxide. Put the tube into the digester block again and continue digestion at 380 °C, until sample becomes clear. Repeat this procedure till the solution becomes clear, then remove the tube, cool down and add 10–15 mL distilled water to the sample. Heat at 100 °C for 10 min to remove remainder of H_2O_2.

iv. When the digestion is complete remove the tubes from the block digester and cool down to room temperature. Pour the cooled sample into 50 mL volumetric flask with a funnel and fill up with distilled water. Sample will get warm again, so we must wait until the sample cools down again to the room temperature, and then make volume of the flask to 50 mL with distilled water. This solution is called Main Solution.

v. We also prepare a digested blank with 10 mL distilled water instead of plant sample with the same procedure discussed above.

Precautions

H_2O_2 and H_2SO_4 are corrosive materials. Avoid contact with skin and inhalation.

Experiment 8: Determination of total nitrogen by Kjeldahl method

Principle

After wet digestion, the nitrogen content of plants is often measured with titrimetry, as total Kjeldahl method, or spectrophotometry, potentiometry, and dead-stop indication methods (Fischl, 1987).

Total Kjeldahl method consists of three steps: (i) wet digestion of plant sample with sulphuric acid as described before in Experiment 7; (ii) distillation, when we distillate ammonia from main solution of plant sample with sodium hydroxide in a distillation apparatus and absorbs the nitrogen in boric acid solution; (iii) titration of ammonium borate with sulphuric acid.

Materials and equipments

Steam distillation unit (Fig 9), 100 mL titration flask, 10 and 20 mL glass pipettes, 50 mL burette.

Reagents

3 % boric acid solution (H_3BO_3); 33 % sodium hydroxide solution (NaOH); 1 % phenolphthalein indicator in alcoholic solution, mixed indicator: 0.2 % methyl red in alcoholic solution and 0.1 % methylene blue in alcoholic solution mixed in 1:1 ratio (color change at pH 5.4); 0.01 N standard sulphuric acid solution (H_2SO_4), distilled water.

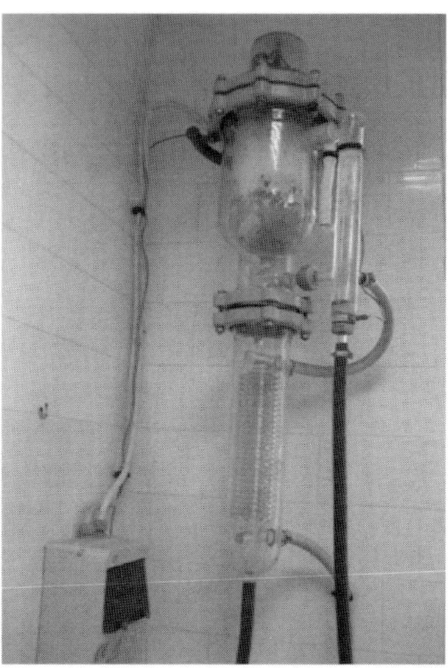

Fig 9. Steam distillation unit

Procedure

i. Proceed for wet digestion of plant sample with sulphuric acid as indicated in Experiment 7, *Procedure*.

ii. Set distillation unit and steam out the apparatus for at least 10 min.

iii. Pipette 20 mL of 3 % boric acid and take 2 drops mixed indicator into a titration flask. Take the flask under the cooling pipe in such a way that its mouth is dipped into the boric acid.

iv. Shake the main solution made from plant sample and pipette 10 mL into the distillation flask. Add 2 drops of phenolphthalein indicator to make alkaline pH and 20 mL of 33 % sodium hydroxide solution. Then, immediately close the flask and begin distillation. The finish of distillation is indicated by the change of indicator color in boric acid from lilac to green at pH 5.4. Continue distillation for 10 min and thereafter remove the titration flask from the distillation unit. Before connecting the next sample, rinse the apparatus with distilled water.

(v) Titrate the distilled liquid with 0.01 N H_2SO_4 until the color of mixed indicator becomes lilac again.

vi. **Blank test/digestion blank**: Distil 10 mL from the digested blank to check *N*-contaminations of reagents used.

Calculations

Nitrogen content of plant sample is calculated as under:

$$N\,(\%) = \frac{(V - B) \times F \times N \times R \times 14.01 \times 100}{W};$$

where N (%) is nitrogen concentration of examined plant sample, V is volume of 0.01 N H_2SO_4 needed for titration of the sample (mL), B is volume of 0.01 N H_2SO_4 needed for titration of the blank (mL), F is the correction factor of standardized 0.01 N H_2SO_4 solution, N is normality of H_2SO_4 solution, R is dilution of sample (ratio between total digested volume and distilled volume – this is 10 now), 14.01 is atomic weight of N, W is weight of plant sample measured to digestion (mg).

Precautions

NaOH is a corrosive material, hence, avoid the contact with skin and inhalation.

Experiment 9: Determination of phosphorus (P) content

Principle

There is a common method to determine P by spectrophotometer in the plant dry matter after digestion. It consists of wet digestion with sulphuric acid, which converts the P content of organic matter into (PO_4^{3-}) form and after that reaction with special reagent to develop color, which can be determined spectrophotometrically. We must prepare a standard solution series and after that a calibration curve, and read at the end the P concentration of sample from the standard curve. This method quantitatively determines the amount of *ortho*-phosphate (PO_4^{3-}) in plant materials (Thamm *et al.*, 1968).

Materials and equipments

Spectrophotometer, heater, oven, desiccator, beaker (1000 and 2000 mL), glass stick, glass funnel, measuring cylinder (500 mL), volumetric flask (50, 250, and 5000 mL), burette (50 mL), dark bottle (1000 mL), pipettes (5 mL, 20 mL, 100 mL, 150 mL).

Reagents

Ammonium heptamolybdate-ammonium vanadate solution in sulphuric acid: Prepare solution "A" slowly adding 285 mL of concentrated H_2SO_4 to 1500 mL distilled water and cool down the solution to room temperature. Prepare solution B, dissolving 4g ammonium metavanadate (NH_4VO_3) in 1500 mL boiling distilled water, and cool down to room temperature. Prepare solution C; dissolving 80 g ammonium heptamolybdate [$(NH_4)_6Mo_7O_{24}\cdot 4H_2O$] in 1500 mL distilled water at 50 °C and cool down to room temperature. Add all solutions A, B and C in a 5-L volumetric flask, mix carefully and make volume to 5 L with distilled water. This solution must be stored in dark bottles.

20 N Sulphuric acid (H_2SO_4): Carefully add 550 mL H_2SO_4 in 1000 mL beaker containing 400 mL of distilled water. Mix and leave to cool. Pour into 1000 mL volumetric flask with glass funnel and make-up volume with distilled water to 1000 mL.

Standard stock solution: We usually examine not only P but also K in the plant samples, hence, we practically make the standard stock solution containing these elements together. Dry about 2g potassium dihydrogen phosphates (KH_2PO_4) in an oven at 105 °C for 1 h and cool in a desiccator. Dissolve 1.917g dried potassium dihydrogen phosphate in distilled water in 1000 mL volumetric flask, in 500 mL distilled water. Dissolve 0.534g KCl heated at 500 °C in the same flask, after bring to 1000 mL volume with distilled water and homogenize. The standard stock solution prepared contains 1 mg P_2O_5 and 1 mg K_2O in 1 mL solution, respectively. Standard stock solution must be kept in a dark bottle, in a refrigerator.

Procedure

i. Proceed for wet digestion of plant sample with sulfuric acid as indicated in Experiment 7.

ii. **Prepare a series of standard solutions from standard stock solution**: Pipette 0, 2.5, 5, 10, 15, 20, 30, 40, 60, 80, 100, 125, 150 mL stock solution in 250 mL volumetric flasks. Add 45 mL of 20 N H_2SO_4 solution with a burette in every flask and make volume to 250 mL with distilled water.

iii. Pipette 5 mL from each solution into 50 mL volumetric flasks. Dilute to half with distilled water. Then, add 12.5 mL of ammonium heptamolybdate-ammonium vanadate reagent with burette and make volume to 50 mL with distilled water. These standard solutions contain 0 (blank), 0.05, 0.1, 0.2, 0.3, 0.4, 0.6, 0.8, 1.2, 1.6, 2, 2.5, and 3 mg P_2O_5 in 50 mL volume and used to measure the P content.

iv. Pipette 5 mL from digested main solution into 50 mL volumetric flask. Dilute to half with distilled water. Then, add 12.5 mL ammonium heptamolybdate-ammonium vanadate reagent with burette and bring to 50 mL with distilled water.

v. Calibrate the spectrophotometer at 410 nm with blank. After 30 min, record the absorbance of standard solution series and plant sample. Prepare a calibration curve for standards plotting absorbance against the respective P_2O_5 concentrations. Read the P_2O_5 concentration of unknown plant sample from the calibration curve.

Calculations

$$P\,(\%) = \frac{C \times 0.44 \times 100}{W};$$

where P (%) is phosphorus concentration of examined plant sample, C is P_2O_5 concentration read from the calibration curve, 0.44 is conversion rate from P_2O_5 to P, W is aliquot weight of plant sample in volume of main solution used for photometrical measurement.

Experiment 10: Determination of potassium (K)

Principle

Quantitative determination of K after sulphuric acid digestion is done from the main solution of plant sample by flame-photometer, measuring the intensity of light emission at a 766 nm wavelength with a red photocell. Emitted light is proportional to potassium concentration.

Materials required

Flame-photometer, 50 mL volumetric flask, 5 mL pipette.

Reagents

Distilled water, series of standard solution (see Experiment 9, *Procedure*).

Procedure

i. Proceed for wet digestion of plant sample with sulphuric acid as indicated in Experiment 7.

ii. **Standard solutions:** Make 10-folds dilution with distilled water from the series of standard solution prepared to determine phosphorus. Pipette 5 mL from each solution into 50 mL volumetric flasks and make volume to 50 mL with distilled water and shake thoroughly. These solutions contain 0 (blank), 0.05, 0.1, 0.2, 0.3, 0.4, 0.6, 0.8, 1.2, 1.6, 2, 2.5 and 3 mg K_2O in 50 mL.

iii. Pipette 5 mL from digested main solution of plant sample into 50 mL volumetric flask and make volume to 50 mL with distilled water and shake thoroughly.

iv. Calibrate the spectrophotometer at 766 nm with blank. Read the emission of series standard solutions and plant sample. Prepare a calibration curve for standards plotting light emission against the respective K_2O concentrations. Read K_2O concentration of unknown plant sample from the calibration curve.

Calculations

$$K\,(\%) = \frac{C \times 0.83 \times 100}{W}\ ;$$

where K (%) is potassium concentration of examined plant sample, C is K_2O concentration read from the calibration curve, 0.83 is converting rate from K_2O to K, W is aliquot weight of plant sample in volume of main solution using to measurement.

3.3. Effects of virus infection on the structure and function of photosynthetic apparatus

All studies on plants viruses reported the reduction in net photosynthetic rates (10–50 %). Inhibition in photosynthetic activity depends on the type of host parasite interaction and on the severity of symptoms. These reduction rates are seldom comparable, because the results has been expressed sometimes on the basis of leaf area, per gram of fresh or dry weight, or in other cases on per unit of chlorophyll content. Reduction in photosynthetic activity results in slow growth rate of leaves and stunting of diseased plants, often causing direct reduction in quality and yield.

4. EFFECT OF PLANT EXTRACTS ON VIRUS INFECTIONS AND VIRUS CONCENTRATION IN PLANTS

Experiment 11: Virus infected plants treated with a plant extract

Principle

Virus particles create extremely close biological units with the host cell. The virus biosynthesis is done by the organelles of the host cell. Therefore, chemical protection against plant viruses is unsuccessful *in vivo* and causes the death of the plant host cell and virus particles. Some natural substances inhibits the replication and cell–cell movement of viruses and reduce their concentration in host plant (Moraes *et al.,* 1974; Tóbiás and Gáborjányi, 1984; Baranwal and Verma, 1997; Manickam and Rajappan, 1998; Vivanco *et al.*, 1999). Natural virus inhibitors can be found (i) in the healthy plants or (ii) can be synthesized due to a stress effect, *e.g.,* virus infection (induced inhibitors). Mode of action of natural inhibitors is not known, perhaps these substances may modify specific receptor places on the plant cell surface. Hence, adhesion of virus particles do not occur (Gáborjányi and Tóbiás, 1986a,b). Some molecules are known to act as virus inhibitors, such as polysaccharides in *Abutilon striatum* (Moraes *et al.*, 1974), proteins in *Amaranthus caudatus, Dianthus caryophyllus, Chenopodium* spp. and *Phytolacca americana* (Ragetli and Weintraub, 1962; Wyatt and Shepherd, 1969; Smookler, 1971), glycoproteins in *Celosia cristata* (Baranwal and Varma, 1997) and glycoalkaloids in *S. nigrum* (Raychaudhury and Basu, 1983). Besides natural inhibitors (Kazinczi *et al.*, 2005), artificial substances inhibits the spread and multiplication of viruses and reduce the number of local lesions and virus concentration (Tallóczy and Gáborjányi, 1990; Kálmán and Gáborjányi, 1998; Horváth, 1999; Kazinczi *et al.*, 2002c).

Materials required

Inoculum source: *Nicotiana benthamiana* plants infected with *Tomato spotted wilt virus (TSWV), N. tabacum* 'Xanthi' test plants, *Abutilon theophrasti* shoot parts, grinder, filter paper.

Procedure

i. **Plant extract preparation:** Cut the collected fresh shoot parts of *A. theophrasti* into small pieces in a grinder. After grinding, stir 25g fresh biomass in 100 mL distilled water and left for a day. Next day, filter the mixture through a filter paper (MN 640w).

ii. Use the mixture for daily watering the test plants (*N. tabacum* 'Xanthi') from their 2–4-leaf stage until the end of experiments.

iii. Do mechanical inoculation of test plants with TSWV at 6–8-leaf stage (see Experiment 1).

iv. Four weeks after virus inoculation, apply DAS ELISA to detect TSWV and determine virus concentration (see Experiment 12). Uninfected healthy plants without extracts treatments serve as negative control, while TSWV-infected ones without any plant extracts treatments serve as positive control.

Experiment 12: Detection of TSWV by DAS ELISA test

Principle

The basic principle of this serological reaction is that alien materials with a large molecular weight (antigen) induce antibodies forming into the blood of an animal. When an animal is infected by a virus one response to the infection is the appearance in the blood stream of proteins which can combine specifically with the virus. Such proteins are known as antibodies. Antibodies are immunoglobulin (IgG) molecules produced by the B lymphocytes of animals in response to an antigenic stimulus. Serological tests are based on the specific antigen-antibody reactions. Among these, enzyme-linked immunosorbent assay (ELISA) is used most frequently for the detection and quantitative estimation of a large number of antigens from plants, including those of many plant pathogenic viruses (Clark and Adams, 1977; Hampton *et al.*, 1990). If an enzyme is linked covalently to the antibody the test is classified as a type of direct ELISA (Fig 10). ELISA registers the occurrence of antigen–antibody complexes by rapid enzymatic

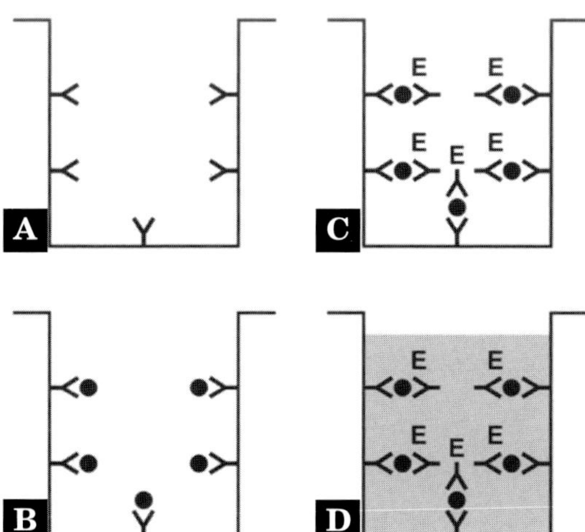

Fig 10. Main steps of DAS ELISA. A – incubation the microtiter plates with the antibody; B – formation of the antibody-antigen (virus) complex; C – formation of antibody-antigen (virus) + antibody-enzyme (conjugate) complex; D – enzymatic reaction (Source: Hampton *et al.* 1990)

development of a distinctly colored product. When the enzyme is alkaline phosphatase (AP), its substrate is the uncolored 4-nitrophenyl-phosphate. The presence of the specific virus antigen is indicated by the positive reaction of AP with its substrate yielding a yellow colored 4-nitrophenol.

Materials required

Microtiter plates, micropipettes, plate readers, ELISA kit (Boehringer, Sanofi, Loewe), containing antibody, antibody–enzyme conjugate, substrate, positive and negative control samples and different buffers.

Reagents

Coating buffer: dissolve 1.59g Na_2CO_3 and 2.93g $NaHCO_3$ in distilled water and dilute to 1 L. Adjust to pH 9.6; washing buffer: dissolve 8g NaCl, 2.9g $Na_2HPO_4{\cdot}H_2O$, 0.2g KCl and 0.5 mL Tween 20 in distilled water and dilute to 1 L. Adjust to pH 7.2–7.4.

Sample buffer: dissolve 20g polyvinyl pyrrolidone, 2g bovine serum albumin, 0.1g NaN_3 and dilute to 1 L. Adjust pH 7.4; Substrate buffer: 97 mL diethanolamine; 0.2g $MgCl_2{\cdot}6\,H_2O$; 0.2g NaN_3 in distilled water and dilute to 1 L. Adjust pH to 9.8 with 1 N HCl.

Procedure

i. Incubate the microtiter plates with the antibody. In this step, dilute IgG 1:500 in the coating buffer. Pipette 0.2 mL into each cell of microtiter plates and incubate at 37 °C for 4 h.

ii. Wash the cells four times with washing buffer. Let washing buffer stay in the wells for 3 min before removal.

iii. **Formation of the antibody–antigen complex**: Homogenize 0.4g fresh plant samples with 2 mL sample buffer. Pipette 0.2 mL for each well and incubate over night at 4 °C.

iv. Wash the wells five times with washing buffer. Let washing buffer stay in the wells for 3 min before removal.

v. **Formation of the antibody–AP conjugate**: Dilute antibody–AP conjugate 1:500 in sample buffer and pipette 0.2 mL in each well. Incubate at 37 °C for 4 h.

vi. Wash the wells five times with washing buffer. Let washing buffer stay in the cells for 3 min before removal.

vii. **Enzymatic assay**: Dissolve 4-nitrophenyl phosphate di-Na-salt in substrate buffer in a ratio of 1:1. Pipette 0.15 mL in each well and incubate at 25 °C for 1–2 h. Stop the reaction with 0.05 mL 3 N NaOH. Evaluate the reaction by an ELISA reader at 405 nm.

Calculations

Higher concentration of viruses in the plant samples will lead to higher extinction values. Therefore, virus concentration should be proportional to extinction values obtained. Test samples are considered resistant to virus infection, if their extinction values do not exceed two times than negative (uninfected) control.

Statistical analysis

Analysis of variance (ANOVA) must be done with susceptible, virus infected plant samples (where the extinction values are more than two times higher than the extinction values of the negative control samples), in order to determine the effect of plant extracts on the virus concentration in systemic host plants as compared to the positive control samples. Virus infected plants without plant extract treatments serve as positive control.

Precautions

Diethanolamine has an irritative effect, NaN_3 is a cell toxin. Avoid contact with skin.

Suggested Readings

Almási, A., Apatini, D., Bóka, K., Böddi, B. and Gáborjányi, R. (2000). BSMV infection inhibits chlorophyll biosynthesis in barley plants. *Physiological and Molecular Plant Pathology* **56**: 227–233.

Baranwal, V.K. and Verma, H.N. (1997). Characteristics of a virus inhibitor from the leaf extract of *Celosia cristata*. *Plant Pathology* **46**: 523–529.

Berzsenyi, Z. (2000). *Plant Growth Analysis in Crop Production*. PATE, Keszthely, Hungary.

Blackmann, V.H. (1919). The compound interest law and plant growth. *Annals of Botany* **33**: 353–360.

Bos, L. (1977). Seed-borne viruses. *In: Plant Health and Quarantine in International Transfer of Genetic Resources* (*Eds*., W.B. Hewitt and L. Chiarappa), pp. 39–69. CRC Press, Cleveland, USA.

Bos, L. (1983). *Introduction to Plant Virology*. Centre for Agricultural Publishing and Documentation, Wageningen.

Chapman, S. (1998). Tobamovirus isolation and RNA extraction. *In*: Methods in Molecular Biology Vol.81: *Plant Virology Protocols. Virus Isolation to Transgenic Plants*, (*Eds*., G.D. Foster and S.C. Taylor). pp. 123–129. Humana Press Inc., Totowa.

Clark, M.F. and Adams, A.N. (1977). Characteristics of the microplate method of enzyme-linked immunosorbent assay for the detection of plant viruses. *Journal of Genetical Virology* **34**: 475–483.

Crowther, J.R. (*ed*.) (1955). ELISA, *Theory and Practice. Methods in Molecular Biology*, Vol. 42. Humana Press. Inc., Totowa, NJ.

Debreczeni, B. (1991). *Agrochemical practices* (Agrokémiai gyakorlatok). Pannon Agricultural University, Keszthely, Hungary.

Dijksta, J. and de Jager, C.P. (1998). *Practical Plant Virology. Protocols and Exercises*. Springer, Berlin

Faccioli, G., Panopoulos, N.J. and Gold, A.H. (1977). Kinetics of the carbohydrate metabolism in potato leafroll infected *Physalis floridana*. *Phytopathologica Mediterranea* **10**: 1–25.

Fischl, G. (1987). *Comparative methodology for the determination of different N forms in plants.* (Összehasonlító módszertani vizsgálatok különböző N-formák meghatározására növényi mintában). Ph.D. Thesis, Keszthely, Hungary.

Foster, G.D. and Taylor, S.C. (1998). *Plant Virology Protocols: From Virus Isolation to Transgenic Resistance*. Humana Press, New Jersey.

Francki, R.I. and McLean, G.D. (1968). Purification of potato virus X and preparation of an infectious ribonucleic acid by degradation with lithium chloride. *Australian Journal of Biological Sciences* **21**: 1311–1318.

Funayama, S., Sonoike, K. and Terashima, I. (1997). Photosynthetic properties of leaves of *Eupatorium makinoi* infected by a geminivirus. *Photosynthesis Research* **53**: 253–261.

Gáborjányi, R. and Tóbiás, I. (1986a). Possibilities of chemical control of plant viruses. Induced antiviral substances and applied chemotherapeutics. *Növénytermelés* **35:** 341–350.

Gáborjányi, R. and Tóbiás, I. (1986b). Inhibitors of virus infection and virus replication in plants. *Növénytermelés* **35**: 139–146.

Gáborjányi, R. and Fernandez, T.F. (1976). Induced alteration of peroxidase activities and the growth of peppers inoculated with tobacco etch virus. *Acta Phytopathologica Hungarica* **11**: 277–281.

Gáborjányi, R., Balázs, E. and Király, Z. (1971). Ethylene production, tissue senescence and local virus infections. *Acta Phytopathologica Hungarica* **6:** 51–55.

Gooding, G.V. and Hebert, T.T. (1967). A simple technique for purification of Tobacco Mosaic Virus in large quantities. *Phytopathology* **57**: 1258.

Gullner, G., Tóbiás, I., Fodor, J. and Kõmíves, T. (1999). Elevation of glutathione level and activation of glutathione-related enzymes affect virus infection in tobacco. *Free Radical Research* **31**: 155–161.

Hall, A.E. and Loomis, R.S. (1972). An explanation of the difference in photsynthetic capabilities of healthy and beet yellows virus infected sugabeet (*Beta vulgaris* L.). *Plant Physiology* **50**: 576–580.

Holmes, F.O. (1929): Local lesions in tobacco mosaic. *Bot. Gaz.* Chicago **87**: 39–55.

Hampton, R., Ball, E. and De Boer, S. (1990). *Serological Methods for Detection and Identification of Viral and Bacterial Plant Pathogens.* A Laboratory Manual. APS Press, St Paul, USA.

Harris, K.F. (1981). Arthropod and nematode vectors of plant viruses. *Annual Review of Phytopthology* **19**: 391–426.

Howell, S.H. (1985): The molecular biology of plant DNA viruses. *CRC Critical Review in Plant Sciences* **2**: 287–316.

Horváth, J. (1993). Host plants in diagnosis. *In*: *Diagnosis of Plant Virus Diseases* (*Ed*. R.E.F. Matthews). pp. 15–48. CRC Press, Boca Raton.

Horváth, J. and Gáborjányi, R. (1999). *Plant Viruses and Virological Methods.* Mezõgazda Kiadó, Budapest.

Horváth, J. (1999). *Plant Virology*. Lecture notes. PATE GMK, Keszthely, pp. 237.

Hunt, R. (1978). *Plant Growth Analysis. Studies in Biology*. No. 96. Edward Arnold, London.

Johansen, E., Edwards, M.C. and Hampton, R.O. (1994). Seed transmission of viruses: Current perspectives. *Annual Review of Phytopathology* **32**: 363–386.

Kálmán, D. and Gáborjányi, R. (1998). Effect of salicylic acid and BTH (1,2,3-benzo-thiadiazole-7-carbothionic acis-S-methylester) on infection susceptibility and disease resistance to tobacco mosaic virus in tobacco plants. *Növényvédelem* **34**: 593–600.

Kazinczi, G. (1993). *Biology of weeds of winter wheat*. Ph.D. Thesis, Keszthely, Hungary.

Kazinczi, G., Gáspár, L., Nyitrai, P., Gáborjányi, R., Sárvári, É., Takács, A. and Horváth, J. (2006a). Herbicide-affected plant metabolism reduces virus propagation. *Zeitschrift für Naturforschung* **61C**: 692–698.

Kazinczi, G., Horváth, J. and Takács, A. (2006b). On the biological decline of weeds due to virus infection. *Acta Phytopathologica et Entomologica Hungarica* **41**: 213–221.

Kazinczi, G., Horváth, J., Takács, A.P., Béres, I., Gáborjányi, R. and Nádasy, M. (2005). The role of allelopathy in host–virus relations. *Cereal Research Communications* **33**: 105–108.

Kazinczi, G., Lukács, D., Takács, A., Horváth, J., Gáborjányi, R., Nádasy, M. and Nádasy, E. (2006c). Biological decline of *Solanum nigrum* due to virus infection. *Zeitschrift für Pflanzenkrankheiten und Pflanzenschutz Sonderheft* **20**: 781–786.

Kazinczi, G., Horváth, J. and Lukács, D. (2000). Germination characteristics of *Chenopodium* seeds derived from healthy and virus infected plants. *Zeitschrift für Pflanzenkrankheiten und Pflanzenschutz Sonderheft* **17:** 63–67.

Kazinczi, G., Horváth, J., Béres, I., Takács, A.P. and Lukács, D. (2002c). The effect of pendimethalin (STOMP 330) on some host–virus relations. *Zeitschrift für Pflanzenkrankheiten und Pflanzenschutz Sonderheft* **18**: 1093–1098.

Kazinczi, G., Horváth, J. and Lesemann, D.E. (2002b). Perennial plants as new natural hosts of three viruses. *Zeitschrift für Pflanzenkrankheiten und Pflanzenschutz* **109**: 301–310.

Kazinczi, G., Horváth, J., Takács, A.P. and Pribék, D. (2002a). Biological decline of *Solanum nigrum* L. due to tobacco mosaic tobamovirus (TMV) infection. II. Germination, seed transmission, seed viability and seed production. *Acta Phytopathologica et Entomolologica Hungarica* **37:** 329–333.

Király, Z., Barna, B. and Érsek, T. (1972). Hypersensitivity as a consequence, not the cause of plant resistance to infection. *Nature* **239**: 456–458.

Király, Z., Barna, B., Kecskés, A. and Fodor, J. (2002). Down-regulation of antioxidative capacity in a transgenic tobacco which fails to develop acquired resistance to necrotization caused by TMV. *Free Radical Research* **36**: 981–991.

Koenig, R. (ed.).(1988). *The Plant Viruses*. Vol. 3. Plenum Press, New York.

Kuriger, W.E. and Agrios, G.N. (1977). Cytokinin levels and kinetin–virus interactions in tobacco ringspot infected cowpea. *Phytopathology* **67**: 604–609.

Lane, L. (1986): Propagation and purification of plant RNA viruses. *Methods Enzymology* **118**: 687–696.

Leberman, R (1966). The isolation of plant viruses by means of "simple" coacervates. *Virology* **30**: 341–347.

Manickam, K. and Rajappan, K. (1998). Inhibition of antiviral activity of certain leaf extracts against tomato spotted wilt virus in cowpea. *Annals of Plant Protection Science* **6**: 127–130.

Matthews, R.E.F. (1981). *Plant Virology*. Academic Press, New York.

Milne, R.G. (*ed.*).(1988). *The Plant Viruses*. Vol. 4. Plenum Press, New York.

Mink, G.I. (1993). Pollen- and seed-transmitted viruses and viroids. *Annual Review of Phytopathology* **31**: 375–402.

Moore, R.P. (1985). *Handbook on Tetrazolium Testing*. International Seed Testing Association, Zurich.

Moraes, W.B., Luly, J.R., Alba, A.P. and Oliveira, A.R. (1974). The inhibitory activity of extracts of *Abutilon striatum* leaves on plant virus infection. II. Mechanism of inhibition. *Phytopathologische Zeitschrift* **81**: 240–253.

Noordam, D. (1973): *Identification of Plant Viruses*. Centre for Agricultural Publishing and Documentation, Wageningen.

Pogány, M. (1999). Plant viruses and plant regulators. *In*: *Plant Viruses and Virological Methods* (*Eds*., J. Horváth and R. Gáborjányi). pp. 284–340. Meõgazda Kiadó, Budapest.

Ragetli, H.W. and Weintraub, M. (1962). Purification and characterization of a virus inhibitor from *Dianthus caryophyllus*. I. Purification and activity. *Virology* **18**: 232–240.

Rapley, R. (*ed.*) (1996): *PCR Sequencing Protocols*. *Methods in Molecular Biology*, Vol. 65, Humana Press Inc., Totowa, NJ.

Raychaudhury, R. and Basu, P.K. (1983). Characterization of a plant virus inhibitor from two *Solanum* species. *Indian Journal of Experimental Biology* **21**: 212–215.

Salomon, R., Sela, I., Soreq, H., Giveon, D. and Littaur, U.Z. (1976). Enzymatic acylation of histidine to tobacco mosaic virus RNA. *Virology* **71**: 74–84.

Sambrook, J. Fritsch, E.F. and Maniatis, T. (1989). *Molecular Cloning: A Laboratory Manual*, 2nd ed. Cold Spring Harbor, New York.

Smookler, M.M. (1971). Properties of inhibitors of plant virus infection occurring in the leaves of species in the Chenopodiales. *Annals of Applied Biology* **69**: 157–168.

Stace-Smith, R. and Martin, R.R. (1993). Virus purification in relation to diagnosis. In: *Diagnosis of Plant Virus Diseases*, (*Ed*. R.E.F. Matthews). pp.129–158. CRC Press, Boca Raton.

Steere, R.I. (1964). Purification. *In*: *Plant Virology* (*Eds*., M.K. Corbett and H.D. Sisler) pp. 211–234. University of Florida Press, Gainesville.

Tallóczy, Z.S. and Gáborjányi, R. (1990). Inhibitory effect of L333, azadihydrouracyl (DHT) and cyanoguanidin (CG) on barley stripe mosaic virus in barley plants. *Növénytermelés* **39**: 245–253.

Thamm, F., Krámer, M. and Sarkadi, J. (1968). Determination of phosphorus content of plants and manures by the ammonium-vanadomolybdate methods. *Agrokémia és Talajtan* **17**: 145–156.

Tóbiás, I. and Gáborjányi, R. (1984). Virus research for the breeding of resistance. *Kertgazdaság* **17**: 51–53.

Vivanco, J.M., Querci, M. and Salazar, L.F. (1999). Antiviral and antiviroid activity of MAP-containing extracts from *Mirabilis jalapa* roots. *Plant Disease* **83**: 1116–1121.

van Regenmortel, M.H.V. and Fraenkel-Conrat, H. (*Eds.*) (1986). *The Plant Viruses*. Vol. 2. Plenum Press, New York.

van Regenmortel, M.H.V., Fauquet, C.M., Bishop, D.H.L., Carstens, E.B., Estes, M.K., Lemon, S.M., Maniloff, J., Mayo, M., McGeoch, D.J., Pringle, C.R. and Wickner, R.B. (*Eds.*) (2004). *Virus Taxonomy, Classification and Nomenclature of Viruses.* Seventh Report of the International Committee on Taxonomy of Viruses. Academic Press, San Diego, USA.

Wyatt, S.P. and Shepherd, R.J. (1969). Isolation and characterization of a virus inhibitor from *Phytolacca americana*. *Phytopathology* **59**: 1787–1794.

CHAPTER 12

Search for Antiphytopathogenic Compounds Involved in Plant Defence

Melina A. Sgariglia[1], Jose R. Soberón[1], Diego A. Sampietro[1], Emma N. Quiroga[1] and Marta A. Vattuone[1]

1. INTRODUCTION

Phytopathogenic organisms annually cause 20 % losses in crop yields globally in billion of dollars (Baker *et al*., 1997). Most agricultural products are subjected to diseases both in field conditions and after harvest, and bacteria and fungi are the major groups of pathogens involved (Staskawicz *et al.*, 1995). About 1600 bacterial species, including both Gram-negative and positive bacteria, mycoplasma-like organisms and spiroplasmas cause plant diseases. More than 8000 fungal species attack plants. Plant infections produced by yeasts are greater than those caused by filamentous fungi. However, filamentous fungi are less susceptible to commercially available antimicrobials (Taiz and Zeiger, 1998; Walton, 1997; Espinel-Ingroff *et al.*, 1997). Pathogens are controlled through intensive use of agrochemicals, leading to the development of resistant strains, especially among phytopathogenic fungi (Staub, 1991). This reality justifies the search for natural antimicrobial agents as an alternative to the synthetic ones (Wilson and Wisniewski, 1994) and promotes research of secondary compounds involved in plant–pathogen interactions. New antimicrobial leader structures with low toxicity for humans and animals, more selective against noxious organisms and relatively costless need to be discovered. Several workers have screened anti-phytopathogenic compounds from plant origin (Kurita *et al.,* 1981; Wilson *et al.,* 1997) and suggest that plants are an important source of these substances. This is expectable considering that plant antimicrobials are part of natural defensive responses against phytopathogens and may be biosynthesized in a constitutive (*i.e.,* phytoanticipins) or inducible (*i.e.,* phytoalexins) manner (Luque-Ortega *et al.,* 2004; Grayer and Kokubun, 2001; Satish *et al*., 1999; Osbourn, 1996).

1. Instituto de Estudios Vegetales "Dr. Antonio R. Sampietro". Facultad de Bioquímica, Química y Farmacia, Universidad Nacional de Tucumán. España 2903, 4000, San Miguel de Tucumán, Tucumán, Argentina. E-mail: instveg@unt.edu.ar

This chapter describes methods for *in vitro* evaluation of anti-bacterial and anti-fungal activities. Techniques and procedures described are useful tools to detect and quantify bioactivity of total extracts, purified extracts or pure compounds. The methodology employed includes diffusion methods, mainly oriented to determine the anti-microbial activity and dilution methods, which are appropriate for its quantitative analysis.

2. GENERAL ASPECTS INVOLVED IN PERFORMANCE OF ANTIMICROBIAL ASSAYS

The following aspects must be considered before conducting antimicrobial assays:

Microbial preservation: Short-term use of microorganisms implies their subculture at brief intervals. Bacterial and fungal cultures are grown by streaking on an enriched agar medium in Petri plates for bacteria and slant tubes (where agar was solidified while the tube is on a slanted surface) for fungi, respectively; incubate the plate only until formation of good cultures; store at 4 °C in refrigerator; subculture bacteria at 48 h and fungi at 10 day intervals.

Each streak plate should be carefully examined to ensure that only one colony type was formed. It is useful to compare colonies into old and new streak plates, observing if they have the same appearance. If more than a colony type is observed it is necessary to carry out a re-isolation of the original strains, under depending conditions of species (Nelson *et al.*, 1983; Dhingra and Sinclair, 1995; Riveros *et al.*, 2001).

Cryopreservation at –20 °C, –70 °C or in liquid nitrogen and drying by lyophilization are techniques often used for storage during large time intervals. For cryopreservation, bacteria are cultured in semi-solid nutritive media (*e.g.,* BHI) with a cryoprotectant (*e.g.*, 20 % glycerol or DMSO) while fungi are subcultured on rich agar slant tubes and stored at 4 °C. Bacteria and fungi should be subcultured in the preservation medium each 6 month or in a proper interval according to the species. Lyophilisation allows storing the dried cultures for 1 or 2 years.

Growth re-initiation: Since microorganisms purchased from commercial sources or preserved for long times are usually freeze-dried, their growth should be reactivated before use in antimicrobial assays. Re-initiation of microbial growth requires proper temperature and medium conditions. For example, bacterial strains preserved in ssBHI with 20 % glycerol, are incubated at 35–37 °C for 2 h. Lyophilized bacterial strains are usually re-hydrated in sterile nutritive broth (*e.g.,* Mueller Hinton broth, MHB) or sterile saline physiological solution (0.85 % NaCl) before incubation. Then, an isolated bacterial colony is streaked on Petri plates containing a non selective medium (*e.g.*, 10 ml, Mueller Hinton agar, MHA) to allow growth of individual colonies. After 24 h at 37 °C, bacterial colonies are isolated to prepare inoculum suspensions used for assays. Procedure for growth re-initiation of yeasts is similar to that described for bacteria, but the solid media are different and should be slanted in test tubes (5 mL per tube). For moulds, medium inoculation is achieved placing a 3 mm diameter plug of a growing mycelium of filamentous fungi or spore-forming fungi harvested from 8 to 10 day culture slant tubes (Fig 1). Yeasts and moulds growth is achieved after incubation at 30 °C for 48–120 h, according to the species. For some species temperature and incubation time are critical factors, hence, it is important to be sure about the optimal incubation conditions for each strain. For example fungi strains such as *Aspergillus flavus, Aspergillus fumigatus*, *Pseudallescheria boydii*, and *Rhizopus. arrhizus* usually need 35 °C for 7 days, and *Fusarium oxysporum* and *Fusarium solani* require 35 °C for 48–72 h and then at 25–28 °C for 7 days (Espinel-Ingroff *et al.,* 1997). Researcher should find out the accurate conditions for his test species.

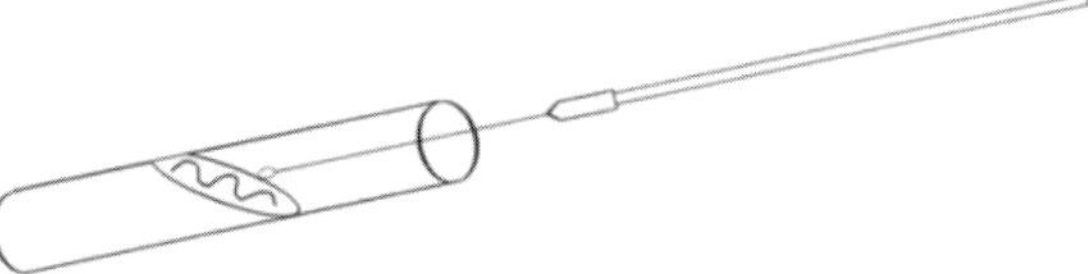

Fig 1. Streaking by a loop in slant tubes

Recommended culture media

For bacteria: Mueller Hinton agar (MHA) and Mueller Hinton broth (MHB) or Nutrient broth (NB) and Nutrient agar (NA) are used for continuous cultures and assays; semi-solid Brain–Heart Infusion (ssBHI) are used on bioautography assay; ssBHI with 20 % glycerol is employed for microbial preservation.

For fungi: Sabouraud Dextrose Agar (SDA) and Wort Agar (WA) are used for yeasts; Malt extract agar medium (MEA), Potato Dextrose Agar medium (PDA) and Sabouraud Maltosa Agar (SMA) for continuous cultures, preservation cultures and solid medium assays; Sabouraud Liquid Medium (SLM) or Malt Extract Broth (MEB) are used for liquid medium assays. For anaerobic microorganisms use the reducing media (The Oxoid Manual; Difco & BBL Manual; Fernandez de Caleya *et al.,* 1972; Hadacek and Greger, 2000; Sawaya *et al.,* 2002). Media composition are described in Table 1.

Sterilization methods

i. **Steam sterilization:** Autoclave culture media at 121 °C for 20 min fractionated in test tubes (mainly culture media, should be sterilized immediately after preparation); haemolysis tubes; 3 mL of distilled water, 0.85 % NaCl solution in haemolysis tubes; graduated pipettes (in port pipettes); open Eppendorf tubes (contained in semi-closed recipient); 0.22-µm Millipore filters (in Petri dish); port-filter (wrap in paper and in

Table 1. Composition of media commonly recommended for bacterial and fungal growth

BACTERIAL MEDIA	
	Solid Media
Mueller Hinton Agar (MHA)	*Composition*[a]: beef infusion, 300.0g; casein hydrolysate, 17.5g; starch, 1.5g; agar No 1, 10.0 g.
Nutrient Agar (NA)	*Composition*[a]: lab-Lemco powder, 1.0 g; yeast extract, 2.0g; peptone, 5.0g; sodium chloride, 5.0g; agar No 3, 15.0g.
	Semi-Solid Media
Brain Heart Infusion (BHIss)	*Composition*[a]: calf brain infusion solids, 12.5g; beef heart infusion solids, 5.0g; proteose peptone, 10.0g; dextrose, 2.0g; sodium chloride, 5.0g; disodium phosphate, 2.5g (add glycerol 20 % for preservation cultures). *Conditions*: pH 7.4 ± 0.2 at 25 °C
	Liquid Media
Mueller Hinton Broth (MHB)	*Composition*[a]: consists of the same ingredients than MHA but without the agar. *Conditions*: pH 7.4 ± 0.2 at 25 °C
Nutrient Broth (NB)	*Composition*[a]: consists of the same ingredients than NA but without the agar. *Conditions*: pH 7.4 ± 0.2 at 25 °C
FUNGAL MEDIA	
	Solid Media
Malt Extract Agar (MEA)	*Composition*[a]: maltose (technical), 12.75g; dextrin, 2.75g; glycerol, 2.35g; peptone, 0.78g; agar, 15.0g (The semi-solid medium for bioautography technique contains agar 7.5g).
Potato Dextrose Agar (PDA)	*Composition*[a]: potato Starch, 4.0g; dextrose, 20.0g; agar, 15.0g. *Conditions*: pH 5.6 ± 0.2 at 25 °C Adjust pH of approximately 3.5 with sterile tartaric acid (under sterility conditions).
Saborouraud Dextrose Agar (SDA)	*Composition*[a]: mycological peptone, 10.0g; dextrose, 40.0g; agar No 1, 15.0g. *Conditions*: pH 5.6 ± 0.2 at 25 °C
Saborouraud Maltose Agar (SMA)	*Composition*[a]: mycological peptone, 10.0g; maltose, 40.0g; agar No 1, 15.0g. *Conditions*: pH 5.6 ± 0.2 at 25 °C

Table 1. (Contd.)

FUNGAL MEDIA	
	Solid Media
Wort Agar (WA)	*Composition*[a]: malt extract, 15.0g; peptone, 0.78g; maltose, 12.75g; dextrin, 2.75g; glycerol, 2.35g; dipotassium phosphate, 1.0g; ammonium Chloride, 1.0g; agar No 1, 15.0g. *Conditions*: pH 4.8 ± 0.2 at 25 °C
	Liquid Media
Malt Extract Broth (MEB)	*Composition*[a]: malt extract, 6.0g; maltose (technical),1.8g; dextrose, 6.0g; yeast extract, 1.2g. *Conditions*: pH 4.7 ± 0.2 at 25 °C
Potato Dextrose Broth (PDB)	*Composition*[a]: consists of the same ingredients than PDA but without the agar. *Conditions*: pH 5.1 ± 0.2 at 25 °C
Saborouraud Liquid Medium (SLM)	*Composition*[a]: pancreatic digest of casein, 5.0g; peptic digest of fresh meat, 5.0g; dextrose, 20.0g. *Conditions:* pH 5.7 ± 0.2 at 25 °C

([a]): Approximate formula per litre

semi-closed recipient); paper discs (in Petri dish), tips for micro-pipettes; cork borer (Fig 7c); glass sprayers. Additives, such as antibiotics, hormones, vitamins and other compounds may be destroyed by heating. Therefore, they should be sterilized by filtration or other means and added after autoclaving the medium (Dhingra and Sinclair, 1995).

ii. **Dry sterilization:** Glass Petri plates are wrapped in brown paper and placed at 180 °C for 3 h in an oven.

iii. **Filtration:** Test and standard compounds can be sterilized by filtration with 0.22-µm filter membranes.

iv. Non-sterilizable materials (*e.g.,* stem of micropipette, external surface of containers, etc.) and laminar flow cabinet are sterilized with 70 % ethanol (v/v).

v. Sterile 96-well U-bottom microtitre plates are commercially available. However, no sterile well plates can be sterilized using UV light.

vi. All procedures that imply manipulation of axenic culture and/or sterile materials must be performed in laminar flow cabinet.

Strains:

They are selected as per the research purpose (James *et al.,* 1993; Schisler and Slininger, 1997). The inclusion of reference strains, such as those available in American Type Culture Collection (ATCC) or Spanish Type Culture Collection (CECT), is recommended because they allow comparative effects of a plant extract/bioactive metabolite on a specific microorganism respect to those observed on standard microorganisms.

3. DIFFUSION METHODS

In diffusion methods, an inert support containing the assayed compound/ plant extract is placed in contact with a medium (*e.g.*, agar) previously inoculated with the pathogen. After incubation, the diameter of the clear zone around the inert support (inhibition diameter) is measured.

Experiment 1: Disc diffusion test

Principle

In this method, filter paper discs containing different doses of the test compounds/plant extract, previously inoculated with the test organism, are placed onto the surface of an agar plate. The test compound/plant extract is left to diffuse into the medium, surrounding the disc (Fig 2). Susceptible microorganisms will not grow in the diffusion area. These areas are known as "inhibition zones" and their diameters are measured in mm. Degree of activity is evaluated comparing these areas with those obtained for known compounds. Although diffusion test is a useful tool to detect antimicrobial activity, it is not suitable for quantitative measurements because diffusion process is governed by several physical and biological factors.

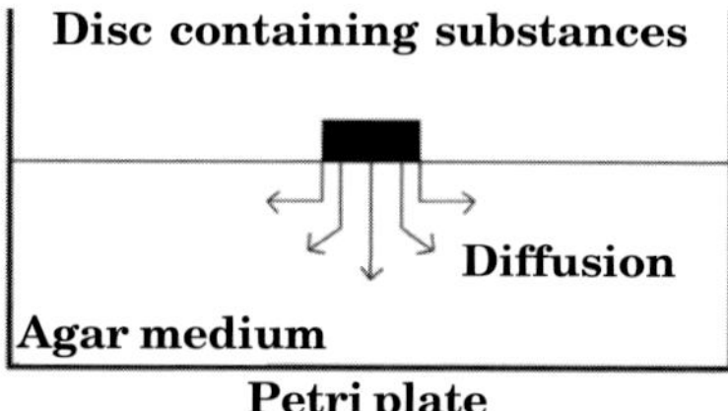

Fig 2. Diffusion of substances from disc to agar medium (*Source*: Dey and Harborne, 1991)

Materials and equipments

Agar culture media (see general aspects), Petri dishes (90 mm), test tubes, haemolysis tubes, general laboratory glassware (flasks, bottles, graduated cylinders), 1, 2, 5 and 10 mL sterile pipettes, 10–100 µL micropipettes, pipette tips, Whatman #4 paper discs (4, 6 or 9 mm in diameter), or commercial materials (*i.e.,* paper discs #2668/2 Schleicher and Schuell, Dassel, Germany; paper discs, Oxoid), loops, forceps, sterile applicator cotton swabs with wooden handles, cryogenic tubes for freezing down the working cultures, glass containers of several volumes, sterile Millipore filtration units of 0.22-µm-pore-size, elements for personal biosecurity (gloves, gowns, goggles or protective eye wear, chemical/biological safety hood), autoclave, boiling water bath for melting the solid media, analytical balance, incubator at 30 and 37 °C, horizontal laminar flow chamber or biological security chamber, refrigerator, freezer or liquid nitrogen tank, rotary evaporator, lyophilizator, spectrophotometer, vortex mixer, calliper or ruler.

Reagents

Test plant extracts; commercially available antibacterial or antifungal standard drugs; commercially available or isolated bacterial and fungal species (see strains in General Aspects); culture medium for assay: choose culture medium according to General Aspects, Recommended Culture Media. Prepare selected medium (Table 1) and fractionate 20 mL per tube. Mix thoroughly, heat with frequent agitation and boil for 1 min until total dissolution. Sterilize by autoclaving at 121 °C for 20 min. The medium may be fractionated in test tubes before sterilization. Culture medium for microbial suspensions can be 3 mL of sterile distilled water, 0.85 % NaCl or broth; Dimethyl sulphoxide (DMSO); Tween® 20 surfactant; 70 % ethanol.

Procedure

(A). For yeasts

1. **Sample preparation:** Extracts or purified compound solutions should be filtered and dried by lyophilization. The lyophilized residue should be weighed and dissolved in a known volume of solvent (the initial one or 100 % DMSO), later, should be filtered through a sterile 0.22-mm-size-pore membrane and received in sterile tube, into laminar flow chamber; the organics extracts, should also be filtered and dried under reduced pressure with a rotary evaporator, to be weighed and dissolved in a known volume of solvent (the initial one or 100 % DMSO) and can be used directly. It is suitable to concentrate the extracts as much as possible. The solutions should be stored at -20 or –80 °C.

2. **Discs Preparation**

   a. Load each sterile paper disc with known amount (between 50 and 200 µg) of test sample.

   b. Let it dry in laminar flow for 6–8 h. Rigorous drying of organic solvent is crucial and negative controls should be included to detect solvent effects.

   c. Discs impregnated with standard antimicrobial drugs and solvents used in test solutions should be included as positive controls (see *Observation* 2). Moreover, sterile discs without sample/drug addition should be included in the plates as controls.

3. **Preparation of Petri dishes:** Melt agar medium contained in tubes in the water bath and load the content of each tube in each Petri plate, in

a laminar flow. Dry the Petri plates in the laminar flow placing them with half-open cover. The surface of the dried medium should be smooth and should not show signs of desiccation, which can be observed as webbed ribbing patterns on the agar surface. Cover the plates until inoculation.

4. **Microbial suspension for inoculation:** Homogeneous inocula from bacterial and yeast cultures are taken during the logarithmic growth phase. In fungi, they are multicellular organisms and homogeneus inocula can be obtained at conidial stage, because conidia most closely represent unicellular life stage. However, homogeneity of fungal inoculum depends on conidia germination and full germination is not always possible. Then, spore quality will be not uniform, throughout the suspension and between suspensions (Hadacek and Greger, 2000).

For aerobic bacteria:

i. Pick up, with a sterile loop, well-isolated colonies from the bacterial subculture of up to 24 h, maintained in refrigerator and suspend them into 3 mL distilled water, 0.85 % NaCl or broth medium.

ii. Incubate aerobically at 37 °C for 4–5 h. Read absorbance at 625 nm against non inoculated medium.

iii. Repeat this operation until an absorbance of 0.08–0.1 is obtained at [625 nm (equivalent to N° 0.5 McFarland standard scale (Andrews, 2005)]. This suspension contains between 1×10^7 and 1×10^8 CFU mL^{-1}.

iv. Prepare 1/10 dilution with the suspension using the same medium before use, and inoculate 10 µL in each plate. This aliquot contains between 1×10^4 and 1×10^5 CFU per plate. These suspensions should be used up to 15 min after preparation.

For anaerobic bacteria:

i. Proceed as indicated above but suspend bacterium colonies from a subculture, previously maintained for up to 24 h in refrigerator, into broth (*i.e.,* MHB) and incubate at 37 °C for 4–6 h according with the species, in tubes with tight screwed cap until an absorbance of 0.08–0.1 is obtained at 625 nm (equivalent to N° 0.5 McFarland standard).

ii. Following incubation, keep the tube at room temperature in the dark colonies from a subculture (see general aspects, Growth re-initiation) with a sterile loop and suspend them in 3 mL of 0.85 % NaCl or broth (*e.g.,* MEB). To do it, vortex 20 s and add the medium for microbial suspension until a 0.08–0.11 absorbance is obtained at 625 nm.

iii. The resulting suspension contains $1–5 \times 10^6$ CFU mL^{-1}, which should be adjusted at $1–5 \times 10^5$ CFU mL^{-1} (1/10 dilution).

For moulds:

i. To obtain conidia, fungi should be cultured for 3–10 day on solid media.

ii. Spores are suspended in 3 mL broth or 0.85 % NaCl, vortexed by 20s and a small drop of Tween® 20 is added to obtain a homogeneous suspension.

iii. The optical densities of the conidial suspensions are read at 530 nm against the culture medium. Absorbance range required depends on conidial size of the tested moulds; *i.e.*, A_{530} = 0.09–0.11 is suitable for *Aspergillus* and *Sporothrix* species; A_{530} = 0.15–0.17 is used for *Fusarium, Pseudallescheria (Scedosporium),* and *Rhizopus* species; A_{530} = 0.2 for *Bipolaris* and *Histoplasma* species. These absorbance readings correspond to an inoculum size between 0.4–5 × 10^4 CFU mL^{-1} (NCCLS M38 – A standard for moulds, 2002).

5. Inoculation:

i. A sterile cotton swab is dipped into the inoculum suspension, within 15 min after adjusting its turbidity.

ii. The cotton swab should be rotated several times and pressed firmly on the inside wall of the tube above the fluid level. This will remove excess inoculum from the swab.

iii. The dried surface of an agar plate is inoculated by streaking the swab over the entire sterile agar surface. This procedure is repeated by streaking two more times, rotating the plate approximately 60° each time to ensure an even distribution of inoculum (Fig 3).

iv. The rim of the agar is swabbed. Use a distinct sterile applicator for each inoculated species. Sterility control should be included, this implies to leave plates without inoculating.

6. Disc application:

i. Sterile discs are dispensed onto the surface of the inoculated agar plate, including positive and solvent controls. Press down each disc, with a sterile loop or forceps, against the agar surface to ensure complete contact (Fig 4). All discs should be at the same distance from the edge of the plate and from each other. Preliminary tests may be needed to determine the best number of discs and distance among discs per plate, *e.g.*, a 90-mm plate will accommodate 6 discs without overlapping the inhibition zones (Andrews, 2005).

ii. Discs should be applied on agar surface within 15 min of inoculation and labelled as per their location on the plate.

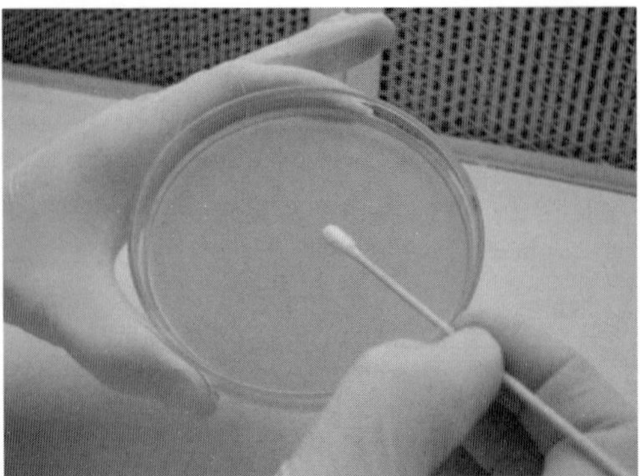

Fig 3. Inoculation on agar surface by streaking with a cotton swab

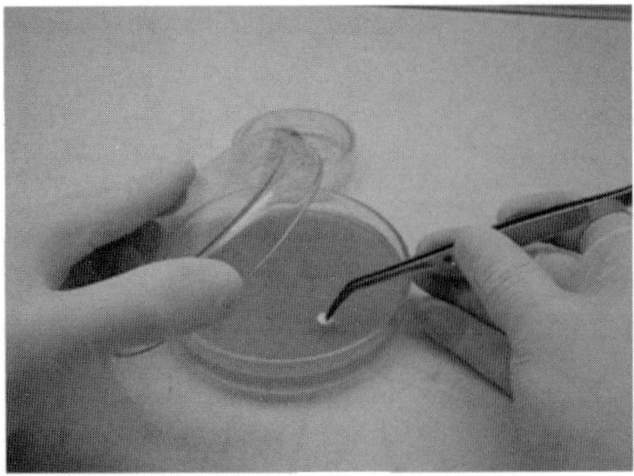

Fig 4. Disc application by a sterile forceps (*Source*: http://sps.k12.ar.us/massengale/aseptic_techniques)

7. Incubation conditions:

i. After disc application, cover and invert the plates.

ii. Plates should be placed for incubation within 15 min after disc application. Conditions for incubation vary according to microbe type:

Aerobic bacteria: At temperature of 35–37 °C for 16–24 h.

Anaerobic bacteria: At the same above conditions, but in 5 % CO_2 and humidified atmosphere.

For fungi: At temperature of 28–30 °C during 20–48 h for yeast and during 48–72–120 h for moulds.

8. Evaluation: Each plate is examined after a proper time of incubation. If the plate was satisfactorily streaked with an appropriate amount of inoculum, inhibition is observed as circular zones on uniform microbial lawn.

Calculations

Measure the diameter of inhibition zones (DZI) to the nearest millimeter including the disc diameter (Fig 5; see *Observations*). Zones are measured with sliding callipers or a ruler, which is held on the back of the inverted Petri dish (Fig 5b). Each diameter should be taken in six different directions. Calculate means for each diameter expressed in millimeter. The percentage of inhibition (% *I*) is determined as under:

$$\% I = [(\text{DZI in control} - \text{DZI in sample}) / \text{DZI in control}] \times 100$$

All the tests should be done in duplicate. Experiments should be repeated at least twice.

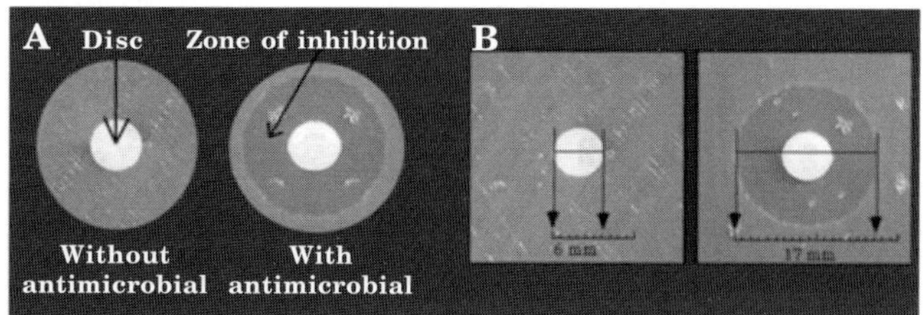

Fig 5. A) Plates containing disc without (left) and with (right) an antimicrobial compound. Note normal microbial growth in the left. B) Extended section of Zone of Inhibition

Observations

1. Clean micropipette stems with 70 % ethanol.
2. Discs containing antimicrobial agents (Difco®, Oxoid®) are placed directly into a freezer (–10 to –20 °C) until its utilization. All discs should be used before their labeled expiration date. A small working supply of discs may be kept in a refrigerator, in containers with a desiccant. To prevent condensation, the jars and disc dispensers should be allowed to reach room temperature before being opened.
3. To improve the visibility and measurement, plates can be kept above a black non-reflecting background to enhance the contrast.

Statistical analysis

The significance of differences between means of the variables examined is determined by the nonparametric Mann-Whitney U test ($p < 0.05$).

Experiment 2: The hole-plate method

Principle

In this method, holes are punched in the inoculated culture medium and the test compound/plant extract is pipetted into them, leaving to diffuse

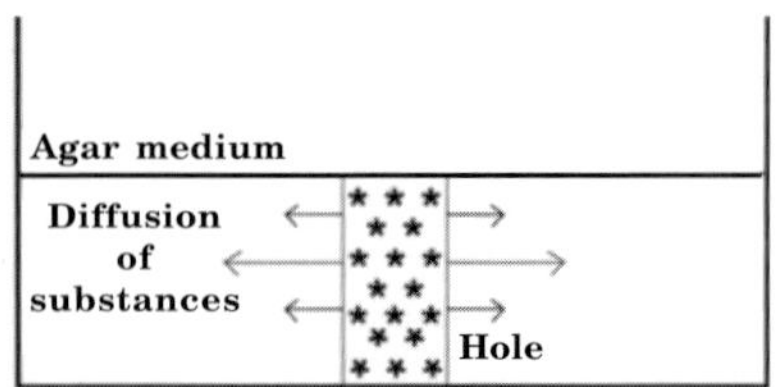

Fig 6. Diffusion of substances from the hole towards agar medium (*Source*: Dey and Harborne, 1997)

for a proper time through the solid agar layer within a Petri dish. The growth of the inoculated microorganisms is inhibited in a circular area or zone around the hole containing the test compound/plant extract (Fig 6). This procedure has the same principles as the disc diffusion test. Nevertheless, it is more suitable to test the aqueous solutions of plant extracts. In hole-plate method, suspended sample particles interfere less with diffusion of antimicrobial compounds into the agar (Dey and Harborne, 1991; Maleš, 2006).

Materials and equipments

Agar culture media (see general aspects), Petri dishes (90 mm), test tubes, haemolysis tubes, general laboratory glassware (flasks, bottles, graduated cylinders), 1, 2, 5 and 10 mL sterile pipettes, 10–100 µL micropipettes, pipette tips, a sterile cork borer for making the holes in agar, loops, sterile applicator cotton swabs with wooden handles, cryogenic tubes for freezing down the working cultures, glass containers of several volumes, sterile Millipore filtration units of 0.22-µm-pore-size, elements for personal biosecurity (gloves, gowns, goggles or protective eye wear, chemical/biological safety hood), autoclave, boiling water bath for melting the solid media, analytical balance, incubator at 30 and 37 °C, horizontal laminar flow chamber or biological security chamber, refrigerator, freezer or liquid nitrogen tank, rotary evaporator, lyophilizator, spectrophotometer, vortex mixer, calliper or ruler.

Reagents

Test plant extracts; commercially available antibacterial or antifungal standard drugs; commercially available or isolated bacterial and fungal species (see strains in general aspects); culture medium for assay: select the medium from general aspects, recommended culture media and prepare according to Table 1; culture medium for microbial suspensions: 3 mL of sterile distilled water, 0.85 % NaCl (dissolve 0.85g NaCl in distilled water and take up to 100 mL) or broth; dimethyl sulphoxide (DMSO); Tween® 20 surfactant; disinfectant solution for work space (70 % ethanol; v/v).

Procedure

i. See Experiment 1, *Procedure*, step 1.

ii. **Making holes:** 6-mm diameter holes are dug in agar plate using a sterile cork borer (see general aspect) and next these are removed from the seeded agar (Fig 7a,b) with the help of a sterile loop (Adelheid Brantner *et al.,* 1996; Mitra *et al.*, 2000). The criterion used for holes distribution in the agar plate is that used for the discs placing.

iii. Proceed as indicated in Experiment 1, *Procedure*, steps 3 to 5.

iv. **Load of holes:** The holes are aseptically filled with test solutions (100 µl/well) and controls, these involved the sterile solvents of solutions, any antimicrobial standard drug and drops of agar medium without inoculum as sterility control. To limit precipitation of substances as much as possible, pre-incubation should be done at room temperature (25 °C) rather than at 4 °C (Dey and Harborne, 1991).

v. See Experiment 1, *Procedure*, steps 7 to 8.

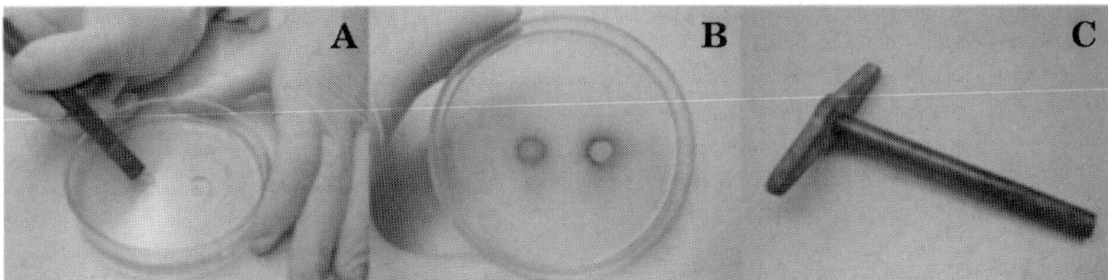

Fig 7. A) Making the holes in agar medium (*Source*: http://eve.kean.edu). B) Diffusion of a plant extract from holes towards agar medium. C) Cork borer

Calculations

Measure the diameters of inhibition zones (DZI) to the nearest millimeter including the diameter of the hole. Zones are measured sliding callipers or a ruler, which is held on the back of the inverted Petri dish (see Fig 5b). Each diameter should be taken in six different directions. Calculate means for each diameter expressed in millimeter. The percentage of inhibition (% I) is determined as follows:

$$\% I = [(\text{DZI in control} - \text{DZI in sample}) / \text{DZI in control}] \times 100$$

All the tests should be done in duplicate. Experiments should be repeated at least twice.

Observations

The incubated dishes are not inverted until all solution is absorbed into the agar medium.

Experiment 3: Inhibition of hyphal growth

Principle

A plug of a fungal mycelium is placed in the centre of an agar medium plate. After a first fungus growth, paper discs impregnated with different quantities of test compounds are placed at 0.5 cm from the mycelium border. Antifungal activity is observed as a crescent-shaped inhibitory zone at the mycelia front. This method allows determining the inhibitory activity of test compounds/plant extracts against filamentous fungi. It is suitable for fungi with inadequate sporulation, which can not be inoculated using conidial suspensions in disc diffusion test (Quiroga *et al.*, 2006).

Materials and equipments

Agar culture media for fungi (see culture media recommended in general aspects), Petri dishes (90 mm), test tubes, sterile haemolysis tubes, general laboratory glassware (flasks, bottles, graduated cylinders), 10 mL sterile pipettes, 10–100 µL micropipettes, sterile pipette tips, 4 mm diameter Whatman #4 paper discs, the discs must be prepared at 4, 6 or 9 mm in diameter with Whatman #4 or Whatman AA papers, or may be commercially acquired (paper discs #2668/2 Schleicher and Schuell, Dassel, Germany; paper discs, Oxoid), loops, forceps, glass containers of several volumes, sterile Millipore filtration units with 0.22-µm-pore-size, elements for personal biosecurity (gloves, gowns, goggles or protective eye wear, chemical/biological safety hood), autoclave, boiling water bath for melting the solid media, analytical balance, incubator at 30 and 37 °C, horizontal laminar flow chamber or biological security chamber, refrigerator, freezer or liquid nitrogen tank, rotary evaporator, lyophilizator, spectrophotometer, vortex mixer, calliper or ruler.

Reagents

Plant extracts; commercially available antifungal standard drugs; commercially available bacterial and fungal species that have been chosen (see strains in general aspects); culture medium for assay: select the medium from general aspects, recommended culture media and prepare according to Table 1; culture medium for microbial suspensions: 3 mL of sterile distilled water, 0.85 % NaCl (dissolve 0.85g NaCl in distilled water and take up to 100 mL) or broth; dimethyl sulphoxide (DMSO); Tween® 20 surfactant; 70 % ethanol.

Procedure

i. **Samples preparation:** see Experiment 1, *Procedure*.

ii. **Discs preparation:** see Experiment 1, *Procedure*.

iii. **Preparation of Petri dishes:** see Experiment 1, *Procedure*.

iv. **Inoculation:** (a) Take with a sterile loop a 3-mm-diameter plug of fungal mycelium cut from edge of active growing colony (see growth re-initiation in general aspects) and place it in the centre of agar plates. (b) Leave plates without inoculating as negative controls. Incubate the Petri dishes in the dark at 30 °C, until the mycelium attains 3 cm diameter.

v. **Disc application:** Place the disc on plates using a sterile forceps at 0.5 cm from the mycelium border, each disc must be pressed down to ensure complete contact with the agar surface. Besides discs containing test compounds/plant extracts, experiments should include positive controls consisting of an antifungal standard drug and solvent controls. For sterility control, it must place a sterile disc unloaded with substances should be placed on an agar plate without inoculating (see *Observations*). It is not recommendable to apply more than 4 discs per plate (see *Observations*).

vi. **Incubation:** Incubate the plates at 30 °C for 48–72 h in the dark.

vii. **Evaluation:** After appropriate time of incubation, each plate is examined. Antifungal activity is observed as a crescent-shaped inhibitory zone at the mycelia front as compared to the control plate of corresponding fungal species (Fig 8).

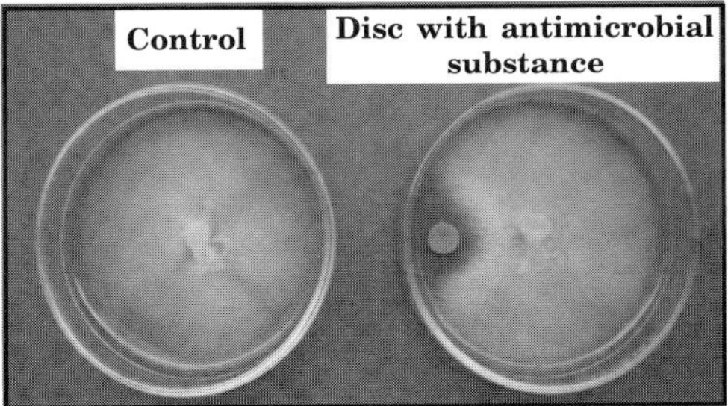

Fig 8. Petri plates showing normal fungus hyphal growth (left) and hyphal growth inhibition (right) around diffusion area of a disc impregnated with an antifungal substance

Observations

1. Experiments should be done in duplicate and repeated at least twice.
2. This method may be used to detect antifungal activity.

Calculations

Radial growth is measured when the hyphae control almost reach the edge of the discs (Fig 7b). Results may be expressed as hyphal growth inhibition (%) (Phongpaichit *et al.*, 2004).

Experiment 4: Direct bioautography

Principle

A suspension of microorganisms in a suitable melted semi-solid (0.6 % agar) culture medium is applied to a developed TLC plate. The substances diffuse from TLC plate towards culture medium. Incubation in a humid atmosphere allows microbial growth. Zones of inhibition are then determined by a dehydrogenase-activity-detecting reagent, *i.e.*, a tetrazolium salt. Metabolically active bacteria convert the tetrazolium salt into colored formazan. Thus, antimicrobial compounds appear as clear spots against a colored background.

Direct bioautography allows *in situ* localization of antimicrobial activity on a TLC plate; this method is widely used to screen antimicrobial activity on bacteria, yeast and spores forming fungi and bio-guided purification of plant extracts (Hamburg and Cordell, 1987; Rahalison *et al.*, 1994; Runyoro *et al.*, 2006).

Materials and equipments

Test tubes, sterile haemolysis tubes, general laboratory glassware (flasks, bottles, graduated cylinders), sterile Petri dishes (90 mm) with three divisions, Silica gel G60 F254 TLC plates, 0.2 mm thickness (20 × 20 cm, 250 µm thickness, Merck, Darmstadt, Germany) and 6 × 8 cm size (Quiroga *et al.*, 2004); UV lamps (366 and 254 nm), 10–100 µL micropipettes, sterile pipette tips, sterile Millipore filtration units of 0.22-µm-pore-size, sterile glass sprayers, loops, forceps (sterilized to flame), autoclave, laminar flow chamber, incubator chamber at 30 and 37 °C, boiling water bath for melting the solid media, water bath set at 45–48 °C, balances, vortex mixer.

Reagents

Microbial standard strains; solvents of analytical grade; sterile semi-solid culture media (see culture media recommended in general aspects and prepare according to Table 1, 5 mL per tube; pure or partially purified test compounds; 0.01 M, pH 7.0 sodium phosphate buffer/0.15 M sodium chloride [prepare Solution A: NaH_2PO_4 (PM: 102, 9), 13.9g/L (0.1M) and Solution B: Na_2HPO_4 (PM: 141, 96), 26, 82g/L ($Na_2HPO_4 \cdot 7H_2O$) or 35.85g/L ($Na_2HPO_4 \cdot 12H_2O$) (0.1 M); Mix: 39 mL A + 61 mL B, dilute with distilled water up to 1000 mL, measure pH; weigh 11.31g NaCl and dissolve into previously prepared buffer, measure pH, adjust it at 7.0 with solution A]; sterile 2.5 mg/mL MTT (3-[4,5-dimethylthyazol-2-yl]-2,5-diphenyl tetrazolium) solution: weigh MTT powder on analytical balance and dissolve in buffer 0.01 M previously prepared and filter this solution through a sterile 0.22-mm-size-pore membrane and received in sterile tube, store at 4 °C (Runyoro1 *et al.*, 2006); commercially available antimicrobial standard drugs; 70 % ethanol; sterile distilled water.

Procedure

i. **Sample preparation**: Stock solutions of pure compounds should be freshly prepared at 1 mg/mL in an appropriate solvent (volatile solvents are preferable to ensure a good evaporation). Aqueous extracts should be dried by lyophilization, weighed and dissolved in a known volume of solvent (methanol/water, 1:1, to enhance the steam pressure), then, filtered through a sterile 0.22-mm-size-pore membrane and received in a sterile tube. Organic extracts should also be filtered and dried under reduced pressure with a rotary evaporator, to be weighed and dissolved in a known volume of solvent. Extracts must be dissolved at a concentration of 10 mg/mL (Shittu *et al.*, 2006). The solutions should be stored at –20 or –80 °C.

ii. **Silica gel plates preparation:** Mark and cut 6 × 8 cm plates of silica gel, submerge them in 70 % (v/v) ethanol, remove them and leave them to dry in airflow under sterile conditions. Take plates with sterile forceps and introduce each silica plate in a sterile Petri dish, close them until the moment for being used.

iii. **TLC development:** Ten microliters of these solutions are seeded at the bottom of TLC plates in point or band formed by micro-pipettes with sterile tips (Shittu *et al.*, 2006; see *Observation* 1). They should be developed in duplicates in selected solvent systems (mobile phase) for each extract. After overnight air drying (in laminar flow cabinet) to remove the chromatographic developmental solvent completely, visualize the test compound under UV light (366 and 254 nm). Calculate and record the *Rf* of each spot. One of the duplicates is used as reference and the other is used for bioautography.

iv. **Inoculation**: Melt semi-solid culture media contained in tubes, maintaining them at 45 °C; prepare the microbial suspension according to Experiment 1, *Procedure*, step 4. Take up 500 µL of microbial suspension and load in a tube (1/10 dilution), vortex and distribute it over the chromatogram (see *Observation* 2), the medium solidifies as a thin layer (~1 mm thick) in airflow of laminar flow chamber after 15 min.

After solidification, the plates are placed in polyethylene container with three divisions; add on bottom of container a little amount of sterile distilled water to prevent the medium dryness.

v. **Incubation conditions**: As per Experiment 1, *Procedure*, step 7 (see *Observation* 3).

vi. **Evaluation**: Spray with MTT solution, incubate between 1 and 4 h at temperature corresponding to microbial culture type. The tetrazolium salt solution is changed to intense colored formazan by dehydrogenases of living microorganisms. Bacterial growth inhibition is observed as pale zones on a colored (blue) background of formazan (Runyoro *et al.*, 2006).

The areas of inhibition should be compared with the *Rf* of the related spots on the reference TLC plate (duplicate). Therefore, the spots associated with the antimicrobial activity are identified (Fig 9).

The test should be done in duplicate.

Observations

1. TLC spotting and development should be made in laminar flow too. Should be included a positive control (antimicrobial standard drug) per plate.
2. Is very important to do inoculation as fast as possible, to avoid temperature of the medium getting low and this one solidifies before being distributed on the chromatogram.
3. The bioautographic plates should be observed daily after its incubation before the addition of MTT.

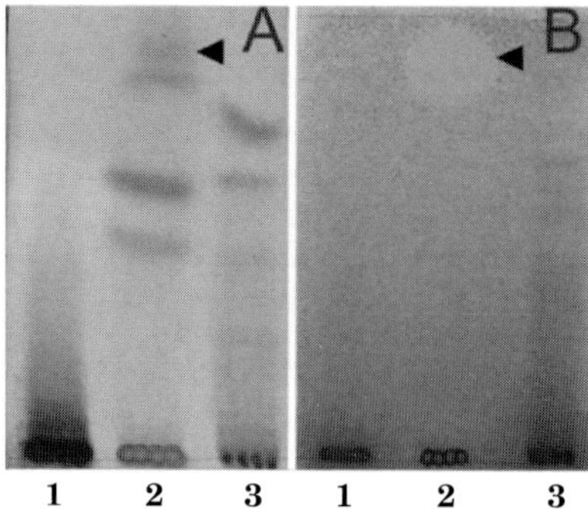

Fig 9. Bioautography technique: (A) Thin-layer chromatography performed in silica gel G60 F_{254} plates. (B) Alternatively, plates can be placed into Petri plates and covered with solid media. Growth inhibition can be seen in columns 1, 2, and 3 after spraying with methylthiazollyltetrazolium chloride

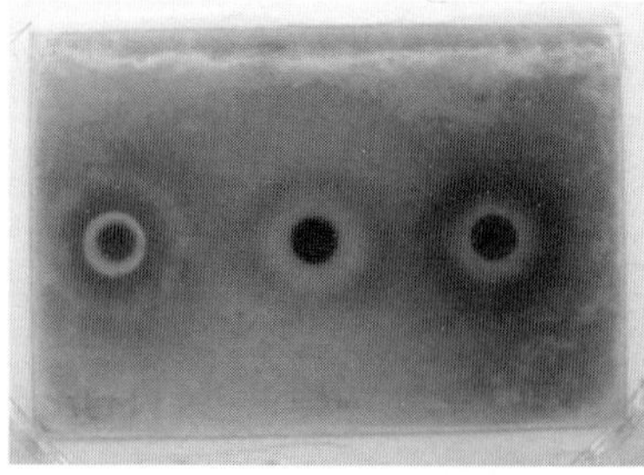

Fig 10. Bioautography of three purified extracts by "dot–blot" technique. Note inhibition zones surrounding the seeding spot

Optional: If identification of bio-active fraction/s is needed among several ones, do regular seeding of fractions on silica plates (without development in mobile phase), inoculate as indicated in *Procedure*, step 4. Incubate, spray with MTT solution and incubate. In the area of diffusion, around the sowing sites, the susceptible microorganisms will not grow and the inhibition halos will indicate which fractions are bio-active (Fig 10).

Bioautography method allows the use of very little amount of sample in bioassays. For this reason, this method is suitable for bioassay-guided isolation of compounds. Since crude extracts are resolved into their different components, bioautography method also simplifies the process of identification and isolation of active compounds.

4. DILUTION METHODS

Experiment 5: Agar dilution assay

Principle

This assay is often done as the first step to screen the antimicrobials. It allows simultaneous evaluation of several bacterial or yeast strains at a same dose of the test compound. This technique is also used as first approach to evaluate the MIC and the obtained results might be checked by more sensitive assays (as broth dilution assay). Serial dilutions of the evaluated substance are incorporated in to nutrient agar. Several bacterial or yeast strains are regularly placed in the solid medium surface at different places. The microbial growth is determined after the incubation (Cooper *et al.*, 2002). The absence of growth is interpreted as antimicrobial activity.

Materials and equipments

Autoclave, balances, biohazard bags, boiling water bath, cryogenic tubes for freezing down the working cultures, haemolysis tubes, forceps, general laboratory glassware (flasks, bottles, gradated cylinders), haemolysis tubes, horizontal laminar flow cabinet, incubation chamber, loops, micropipettes (adjustable volumes up to 10 and 100 µL); personal biosecurity elements (gloves, gowns, goggles or protective eye wear, chemical/biological safety hood), Petri dishes (90 mm), protective covers in work space, refrigerator/freezer or liquid nitrogen tank, spectrophotometer, sterile filter membranes supports, sterile filter membranes units with 0.22-µm-pore-size, sterile glass tubes, sterile pipettes (different volumes), sterile pipette tips, test tube racks, vortex mixer, water bath set at 43–48 °C (to keep temperature of melted agar).

Reagents

Pure or partially purified test compounds; commercially available bactericide and fungicide standards (see general aspects); nutrient agar media; media for the inoculum's suspensions: 3 mL sterile distilled water, 0.85 % NaCl or nutrient broth (the suitable media should be chosen taken into account the kind of microbial to be assayed); 70 % ethanol.

Procedure

The experiment is carried out on 3 consecutive days as under:

A. First day

i. Working materials should be sterilized as previously indicated (see Sterilization Methods in general aspects).

ii. Medium should be selected as per general aspects, recommended culture media. Prepare selected medium according to Table 1.

iii. Phytopathogenic strains should be activated as previously indicated (see Growth Re-iniciation in General aspects).

B. Second day

i. The test compound/plant extract to be evaluated should be sterilized through 0.22-µm membrane filter and collected into sterile glass tubes (see *Observation* 1). One milliliter of serial dilutions of the compound are prepared in sterile water or 96 % ethanol (see *Observation* 2).

ii. Incorporate each dilution into 9 mL of molten nutrient agar. After vigorous vortexing, the mixture is dispensed into sterile Petri dishes (90-mm diameter). Cover the plates half-open and leave till they solidify (10–15 min). Petri plates containing nutrient agar plus 1 mL of water or 96 % ethanol should also be prepared as the positive controls. Label each plate with the corresponding concentration level (see *Observation* 3).

iii. Prepare the microbial inoculum by picking with sterile wire at least three colonies from plates with activated strains, emulsifying them in 1 mL sterile 0.9 % NaCl (w/v) till reach 10^8 colony forming units (CFU) per mL (0.5 Mc Farland scale) (see Experiment 1, *Procedure*, step 4).

iv. Working microbial suspensions (WMS) are prepared diluting 50 µL of inoculum into 450 µL of sterile 0.9 % NaCl (w/v) in sterile Eppendorf tubes to reach 10^7 CFU mL^{-1} (see *Observation* 4).

v. Each strain is identified labelling the bottom of each Petri plate.

vi. Place punctually 2 µL of each WMS, using a micropipette with a sterile tip, on surface of the solid medium beginning with positive control plates, and following from the lowest to the highest sample concentration plates (see *Observation* 5).

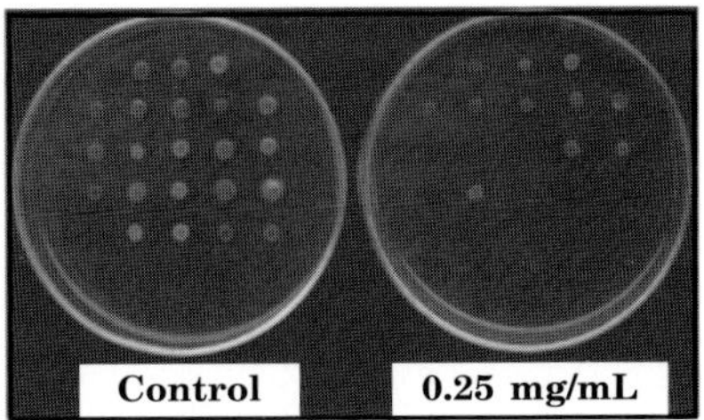

Fig 11. Agar dilution assay. Note normal growth of 17 bacterial strains (positive control in the right) and a plate containing a plant extract dilution with 4 inhibition growth zones (*Source*: Soberón *et al.*, 2006)

vii. Ghost controls (GC) should be kept by sowing 2 µL of inactivated WMS on Petri plates added with nutrient agar. The WMS inactivation is reached by heating the suspensions at 100 °C for 5 min (see *Observation* 6).

viii. Invert the inoculated plates and incubate them as per the microbial requirements (37 °C, 24 h for bacteria; 30 °C, 24 h for fungi).

C. Third day

i. Petri plates are inspected for microbial growth. Positive controls should exhibit punctual growth zones for all the strains. Microbial growth should be absent in GC (Fig 11).

ii. Microbial growth at each concentration level is compared with positive controls and GC.

iii. Minimal inhibitory concentration (MIC) is interpreted as the lowest concentration level with no growth zone for the strain.

Observations

1. The selection of reference antibiotic substances should be made as per the evaluated microbial species.
2. Dilutions should be prepared considering that they will be diluted 10 folds with nutrient agar.
3. At least three plates per concentration level should be prepared.
4. WMS should be employed up to 15 min after preparation.
5. This will reduce the carry over effect.
6. At least three independent experiments (N = 3) should be made (with three replicates for each concentration level of the assayed compound or extract, as previously stated).
7. The assay should be performed in horizontal laminar flow cabinet to avoid contamination.

Experiment 6: Broth dilution assay in 96-well plates

Principle

Liquid media posses higher homogeneity than solid ones and factors such as temperature, aeration, and pH are easily controlled (Jackson, 1997). This assay is often used in the final stages of antimicrobial research, to improve the results obtained from preliminary or exploratory assays (such as diffusion assays or agar dilution techniques). It offers the advantages: (a) fast, (b) inexpensive, (c) requirement of less sample quantity, (d) allows simultaneous testing of several combinations of samples or microorganisms, and (e) having accurate and reproducible results. In addition, broth dilution assay (called microdilution technique) opens the way to determine the minimal bactericidal (MBC) or minimal fungicidal concentrations (MFC). Serial dilutions of a test compound/plant extract are prepared in sterile polystyrene 96-well plates. Microorganisms are suspended in sterile broth and added into the plates. The covered plates are incubated at suitable conditions. The absence of microbial growth is interpreted as antimicrobial activity. The MBC (or MFC) could be determined in solid medium after incubating aliquots taken from the 96-well plates (see Experiment 8).

Materials and equipments

Sterile 96 multi-well microtitration plates (U-bottom). Materials and equipment are the same as those stated on agar dilution assay.

Reagents

Pure or partially purified test compounds; commercially available bactericide and fungicide standards (see general aspects); culture media (for choosing the medium see recommended culture media, in General aspects); 70 % ethanol.

Procedure

The experiment is carried out in at least 3 consecutive days as under:

A. First day

i. Working materials should be sterilized as previously indicated (see Sterilization methods in general aspects).

ii. Culture media should be prepared as previously indicated (see Recommended Culture Media in general aspects and Table 1).

iii. Phytopathogenic strains should be activated as previously indicated (see Growth Re-initiation in general aspects).

B. Second day

i. The test compound/plant extract (or standard antimicrobial compound) should be sterilized through 0.22 µm membrane filter and collected into sterile glass tubes (see *Observation* 1). One milliliter of serial dilutions of the compound is prepared in sterile water or 96 % ethanol (see *Observation* 2). In those experiments carried out with extracts containing essential oils, it is advisable to follow precautions of Mann and Markham (1998).

ii. Add 20 µL sample dilutions into the wells of the microtitre plate. A growth control should be prepared in a complete column of the plate, and should include the solvent used to prepare the sample. Sample color wells should be included, when sample dilutions are colored (to avoid the interference of color in the results), these should be prepared by adding a complete column (or file) of the plate with the employed sample dilutions (Fig 12).

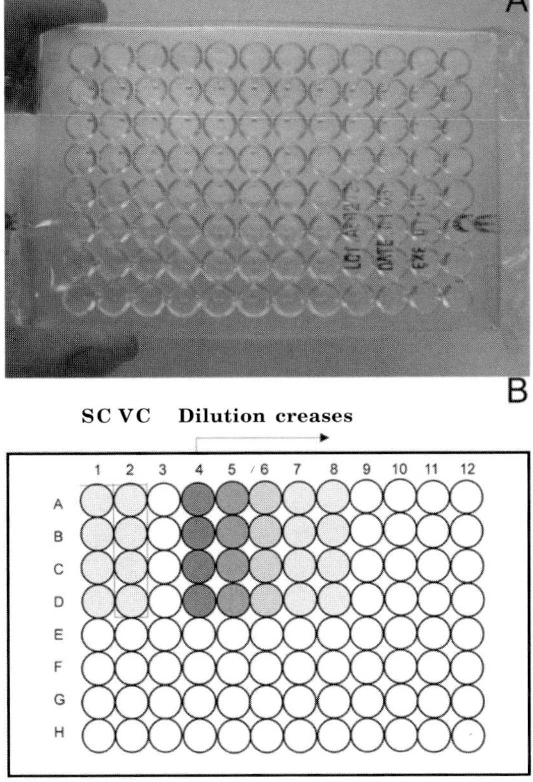

Fig 12. (A) A 96 well plate used for broth dilution assay (sterile enveloped) and (B) a suggested scheme of a 96 well plate showing the sterility control (SC) at column 1, the growth control (GC) at column 2 and dilutions of a tested compound from columns 4 to 12

iii. Wells of sterility controls are prepared adding only broth medium.

iv. Prepare the microbial inoculum by picking with sterile wire at least three colonies from plates with activated strains, emulsifying them in 1 mL sterile 0.9 % NaCl (w/v) till reach 10^8 CFU mL^{-1} which correspond to 0.5 Mc Farland scale (see Experiment 1, *Procedure*, step 4).

v. Working microbial suspensions (WMS) are prepared diluting 100 µL of inoculum into sterile tubes added with 9.90 mL of sterile broth (see *Observation* 3).

vi. Add 100 µL of WMS into the sample and growth control wells.

vii. Make up to 200 µL with sterile broth to reach 5×10^5 CFU mL^{-1}. Sterile broth should also be added into the sample color wells to reach 100 µL (see *Observation* 4).

viii. Mix thoroughly the microtitre plate.

ix. Seal the microtitre plate with plastic film to prevent dryness or contamination.

x. Incubate the sealed plate at suitable conditions (see *Observation* 5)

C. Third day

MICs are determined by visual examination of plate after incubation. Each well should be compared with the positive control and the sample color wells. The MIC value is taken as the lowest concentration of the test compound/plant extract that causes complete inhibition (100 %) of microbial growth (Shin *et al.,* 1998).

Observations

1. The selection of reference antibiotic substances should be made according to the evaluated species.
2. The dilutions should be prepared considering that they will be diluted 10-folds with the nutrient broth in each well of the microtitre plate.
3. WMS should be used up to 30 min after prepared.
4. It is a good practice to check the loaded inoculum concentration as follows: Add 10 µL from a growth control well, and dilute it 1/1000 with 10 mL of sterile nutrient broth, mix thoroughly and inoculate an agar nutrient plate with 100 µL from the above dilution. Incubate the plates in proper conditions (according to the strain). If 50 CFU per plate are observed after incubation the inoculum is considered well adjusted (in the microtitre plate).

5. Incubation conditions vary with the species. For most phytopathogenic bacteria, the incubation is usually performed between 27 and 29 °C for 16–24 h (Jabeen *et al.,* 2004; Vasinauskienë *et al.,* 2006). Fungi incubation at 35 °C, for 48–72 h (Espinel-Ingroff *et al.,* 1997). Sokoviæ and Van Griensven (2006) evaluated antimicrobial activity after 24 h at 28 °C for the bacterium *Pseudomonas tolaasii*, and 24 h at 25 °C for the fungi *Verticillium fungicola* and *Trichoderma harzianum* as incubation conditions. Lee *et al.* (2003) evaluated the antimicrobial activity of a phenazine-1-carboxylic acid against a broad group of pathogens, such as *Botrytis cinerea, Cladosporium cucumerinum* and *Sclerotinia sclerotiorum*, performing the incubation at 22 °C, and *Alternaria mali, Colletotrichum orbiculare, Cylindrocarpon destructans, Didymella bryoniae*, *Fusarium oxysporum*, *Magnaporthe grisea, Phytophthora capsici*, *Pythium ultimum*, *Rhizoctonia solani*, *Erwinia carotovora*, *Ralstonia solanacearum* and *Xanthomonas campestris* performing the incubation at 28 °C, for 2–5 days. However, as previously stated, researchers might find out the accurate conditions for the tested species.

6. The assay should be done in horizontal laminar flow cabinet to avoid contamination.

7. At least three independent experiments (N = 3) should be done with three replicates for each concentration of the assayed compound/plant extract (see Fig 12).

Experiment 7: Broth dilution assay in 96-well plates with endpoint improvement

Principle

The National Committee for Clinical Laboratory Standards (NCCLS) recommends the broth dilution assay previously described for MIC determination. Although this method demonstrates good intra- and inter-laboratory agreement, MICs of fungistatic agents (such as azole compounds) could be difficult to determine because of trailing endpoints caused by partial inhibition of fungal growth. Colorimetric methods often overcome this limitation. They generate clear-cut endpoints, which are visually detectable through colorimetric reactions. The most applied method employs alamar blue (AB), an oxidation–reduction colorimetric indicator that allows visualization of innate metabolic activity from microbial cells. When antimicrobial agent ceases this metabolic activity, AB remains blue (no growth is present). However, blue AB becomes red after a proper incubation period if microbial growth occurs. The AB method has an easy performance and allow obtaining reliable quantitative MIC values. It could be applied on both bacteria and fungi (Espinel-Ingroff *et al.,* 1995). An alternative colorimetric method employs the tetrazolium salt 2,3-bis(2-methoxy-4-nitro-5-sulfophenyl)-5-(phenylamino)carbonyl]-2*H*-tetrazolium hydroxide (XTT) and is closely similar to the AB method (Hawser *et al.,* 1998).

Materials and equipments

The materials and equipments are the same as those stated on broth dilution assay.

Reagents

AB sterile solution (commercially available). The reagents are the same as stated on broth dilution assay.

Procedure

The procedure is similar to that described in broth dilution assay, with the slight modifications indicated below:

Second day: Add 25 µL of AB solution into the wells containing WMS and the evaluated substance mixture (see Experiment 5, second day, step 7), make up to 200 µL with sterile broth, and continue as indicated in Experiment 5.

Third day: Plates are visually inspected for color changes. AB reduction (and therefore the end point reading) is assessed after 24 and 48 h. Growth control wells should exhibit a red color. Sample color wells should not shift the color to red. Cultures which remain blue are interpreted as negative growth and consequently as anntimicrobial activity. The MIC in the AB assay is determined visually as the lowest concentration of antimicrobial agent at which no color change occurred.

Experiment 8: Minimum bactericidal/fungicidal concentration

Principle

The minimum bactericidal concentration (MBC) or minimum fungicidal concentration (MFC) is the lowest concentration of a test compound needed to kill 99.9 % of the original microbial inoculum in a given time. MBC and/or MFC are determined after an incubation period (in solid medium) of a microbial inoculum pre-treated with an antimicrobial compound and represents the lowest concentration of the tested compound, in which no recovery of microorganism is observed (Vicedo *et al.,* 2006).

Materials and equipments

Test tubes, micropipettes (adjustable volumes up to 10 and 100 µL), sterile pipette tips, loops, forceps (sterilized to flame), sterile Petri dishes (90 mm in diameter), autoclave, horizontal laminar flow cabinet, oven incubator (adjustable temperature), boiling water bath for melting the solid media, balances.

Reagents

Nutrient agar media (see general aspects); 70 % ethanol.

Procedure

This experiment should be started the same day as broth dilution assay is finished (see Experiment 5, third day). The experiment takes at least 2 consecutive days for completion.

First day

i. Prepare the Petri dishes by adding 10 mL of molten nutrient agar media into sterile Petri dishes (90-mm diameter). Keep the plates half-open and leave until complete solidification (10–15 min). Then, close and label the plates indicating concentrations levels or growth controls (GC) (see *Observations* 1, 2 and 3)

ii. Take 50 µL of broth, using a sterile tip, from each MIC well and from the subsequent higher sample concentration wells and place it on the surface of nutrient agar plates. Take aliquots from growth control wells to include a positive control (GC) (see *Observation* 4).

iii. Spread inoculum with the aid of a sterile loop. Half-open the cover of dishes and leave till the agar absorbs the inoculum (about 15–20 min).

iv. Invert the inoculated dishes and incubate them as per the microbial requirements (see *Observation* 5).

Second day

Dishes are visually inspected to check the microbial growth. MBC (and so MFC) is defined as the lowest concentration of assayed samples, which produce 99.9 % reduction of CFU mL^{-1} respect to CFU observed in GC dishes (see *Observation* 6).

Observations

1. At least three petri dishes per concentration level should be made. Petri dishes could be prepared 1 day before performing the assay. Keep them at 5 °C until use.

2. Total Petri dishes needed could be estimated as under:

   No. of concentration levels × 3 + 3 (GC)

3. One Petri dish could be used for the inoculation of two or three aliquots from microtitre plate wells (make a suitable mark on the reverse side of the plate base).

4. Use a unique sterile tip for a single inoculation and then discard it.

5. The conditions usually employed are 48 h at 27–30 °C for plant pathogen bacteria, although some authors have employed higher temperatures (Moreno *et al.*, 2006), and 72 h at 26 °C for plant fungi (Lavermicocca *et al.*, 2003). The optimal conditions might be adjusted for the microorganism to be studied.

6. The presence of CFU indicates non-biocidal activity of the test compound/plant extract on microbial growth. The lack of CFU indicates biocidal activity.

Experiment 9: Inhibition of fungal radial growth

Principle

Fungal hyphal growth is inhibited in the presence of an antimicrobial substance/plant extract incorporated into a nutrient agar medium. The inhibition is a dose–response effect interpreted as antifungal activity (Quiroga *et al.*, 2006). This technique is usually used for mould species, filamentous and spore-forming fungi. The assay is often employed to find estimative MIC values.

Materials and equipments

Autoclave, balances, boiling water bath for melting the solid media, forceps (sterilized to flame), horizontal laminar flow cabinet, loops, micropipettes (adjustable volumes up to 10 and 100 µL), oven incubator, sterile filter membranes supports, sterile filter membranes units with 0.22-µm-pore-size, sterile Petri dishes (60 × 15 mm), sterile pipette tips, sterile test tubes, vortex mixer, water bath set at 45–48 °C to maintain temperature of melted agar.

Reagents

Fungal strains, nutrient agar medium, sterile test compounds, commercial antimicrobial chemicals, 70 % ethanol.

Procedure

The assay typically takes 3 days for completion as under:

First day

i. Working materials should be sterilized as previously indicated (see sterilization methods in general aspects).

ii. Culture media should be prepared as previously indicated (see recommended culture media in general aspects and Table 1).

iii. Phytopathogenic fungal strains should be activated as previously indicated (see growth re-initiation in general aspects).

Second day

i. The evaluated substance solution (plant extract or reference antibiotic) should be sterilized through 0.22-µm membrane filter and collected into sterile glass tubes (see *Observation* 1). One milliliter of serial dilutions of the compound are prepared in sterile water or 96 % ethanol (see *Observation* 2).

ii. Incorporate each dilution into 4 mL of molten nutrient agar. After vigorously vortex, the mixture is dispensed into sterile Petri dishes (90-mm diameter). Keep the plates half-open until complete solidification (10–15 min). Petri plates containing nutrient agar plus 1 mL of water or 96 % ethanol should be prepared as positive control. Label the plates indicating concentration levels (see *Observation* 3).

iii. Prepare the microbial inoculum, (see Experiment 1, *Procedure*, step 4). The suspension should contain between 1×10^7 and 1×10^8 spore mL^{-1}.

iv. When a filamentous fungus is used, dispose a 2.5-mm diameter plug of a growing mycelium of the fungus harvested from an 8- to 10-days-old culture over the centre of the prepared Petri dishes. When a spore-forming fungus is used, innoculate an aliquot of 10 µL of the prepared spore suspension (it will contain approximately 2.5×10^4 spore mL^{-1}).

v. Incubate the plates at 30 °C for 4–5 days in a moist chamber.

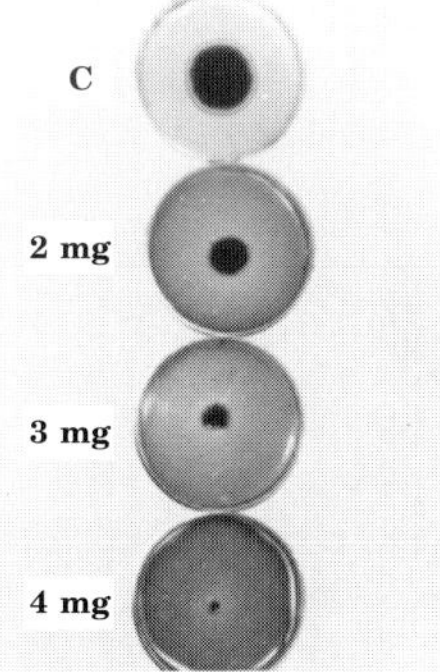

Fig 13. Hyphal Radial Growth Inhibition Test. The picture shows the growth control plate (C) and plates added with increasing test sample concentrations (*Source*: Quiroga *et al.*, 2001)

Third day

The plates are inspected for mycelial growth (Fig 13). Growing zones are measured (see *Observation* 4) and the growth inhibition (%) is then calculated according to Reyes Chilpa *et al.* (1997):

$$GI\ \% = MGc - MGs / MGc \times 100$$

where GI % is percentage of growth inhibition, MGc is mycelial growth in control (mm), and MGs is mycelial growth in the sample (mm).

Observations

1. The selection of reference antibiotic substances should be done as per test spp.
2. The dilutions should be prepared considering they will be 1/5 diluted with the nutrient agar.
3. At least three independent experiments ($N = 3$) should be made, with three replicates for each concentration level of the assayed compound or extract.
4. Use a suitable ruler to make measurements. Take at least three measurements in different directions and then calculate the mean.
5. The assay should be preformed in horizontal laminar flow cabinet to avoid contamination.

Suggested Readings

Agrios, G.N. (1997). Plant diseases caused by Mollicutes: Phytoplasmas and spiroplasmas. *In*: *Plant Pathology* (*Ed*., G.N. Agrios). pp. 457–470. Academic Press, New York.

Andrews, J.M. (2005). BSAC standardized disc susceptibility testing method (version 4). *Journal of Antimicrobial Chemotherapy* **56**: 60–76.

Baker, B., Zambryski, P., Staskawicz, B. and Dinesh-Kumar, S.P. (1997). Signaling in plant-microbe interactions. *Science* **276**: 726–733.

Bosio, K., Avanzini, C., D'Avolio, A., Ozino, O. and Savoia, D. (2000). *In vitro* activity of propolis against *Streptococcus pyogenes*. *Letters in Applied Microbiology* **31**: 174–177.

Brantner, A., Malesz, Z., Pepeljnjak, S. and Antolic, A. (1996). Antimicrobial activity of *Paliurus spina-christi* Mill (Christ's thorn). *Journal of Ethnopharmacology* **52**: 119–122.

Cooper, R.A., Molan, P.C. and Harding, K.G. (2002). The sensitivity to honey of Gram-positive cocci of clinical significance isolated from wounds. *Journal of Applied Microbiology* **93**: 857–863.

Dey, P.M. and Harborne, J.B. (1991). Assays for antifungal activity /Chapter 3: Screening methods for antibacterial and antiviral agents from higher plants. *In*: Methods in Plant Biochemistry. Vol. 6: *Assays for Bioactivity* (*Ed*., K. Hostettmann). pp. 70–96. Academic Press, London.

Dey, P.M. and Harborne, J.B. (1991). Screening methods for antibacterial and antiviral agents from higher plants. *In*: Methods in Plant Biochemistry. Vol. 6: *Assays for Bioactivity* (*Ed*., K. Hostettmann). pp. 47–69. Academic Press, London.

Dhingra, O.D. and Sinclair, J.B. (1995). *Basic Plant Pathology Methods*. Second Edition. CRC Press, Boca Raton, FL.

Espinel-Ingroff, A., Bartlett, M., Bowden, R., Chin, N.X., Cooper, J.R., Fothergill, A., Mcginnis, M.R., Menezes, P., Messer, S.A., Nelson, P.W., Odds, F.C., Pasarell,

L., Peter, J., Pfaller, M.A., Rex, J.H., Rinaldi, M.G., Shankland, G.S., Walsh, T.J. and Weitzman, I. (1997). Multicenter evaluation of proposed standardized procedure for antifungal susceptibility testing of filamentous fungi. *Journal of Clinical Microbiology* **35**: 139–143.

Espinel-Ingroff, A.M., Rodriquez-Tudela, J.L., and Martinez-Suares, J.V. (1995). Comparison of two alternative microdilution procedures with NCCLS reference macrodilution method M27P for *in vitro* testing of fluconazole-resistant and susceptible isolates of *Candida albicans*. *Journal of Clinical Microbiology* **33**: 3154–3158.

Fernandez de Caleya, R., Gonzalez-Pascual, B., Garcia-Olmedo, F. and Carbonero, P. (1972). Susceptibility of phytopathogenic bacteria to wheat purothionins *in vitro*. *Applied Microbiology* **23**: 998–1000.

Ferreira, J.C., Cardoso, M., de Souza, P.E., Miranda, J.C. and Barreto, S. (2005). Inhibitory effect of *Caesalpinia spinosa* leaflets crude extract on *Fusarium solani* and *Phoma tarda*. *Acta Scientiarum Biological Sciences Maringá* **27**: 185–188.

Hadacek, F. and Greger, H. (2000). Testing of antifungal natural products: Methodologies, comparability of results and assay choice. *Phytochemical Analysis* **11**: 137–147.

Grayer, R.J. and Kokubun, T. (2001). Plant–fungal interactions: The search for phytoalexins and other antifungal compounds from higher plants. *Phytochemistry* **56**: 253–263.

Hamburger, M.O. and Cordell, G.A. (1987). A direct bioautographic TLC assay for compounds possessing antibacterial activity. *Journal of Natural Products* **50**: 19–22.

Hawser, S.P., Norris, H., Jessup, C.J. and Ghannoum, M.A. (1998). Comparison of a 2,3-bis(2-methoxy-4-nitro-5-sulfophenyl)-5-[(phenylamino) carbonyl]-2H-tetrazolium hydroxide (XTT) colorimetric method with the Standardized National Committee for Clinical Laboratory Standards Method of Testing Clinical Yeast Isolates for Susceptibility to Antifungal Agents. *Journal of Clinical Microbiology* **36**: 1450–1452.

Herrera-Estrella, L. and Simpson, J. (1995). Genetically engineered resistance to bacterial and fungal pathogens. *World Journal of Microbiology & Biotechnology* **11**: 383–392.

Homans, A.L. and Fuchs, A. (1970). Direct bioautography on thin-layer chromatograms as a method for detecting fungitoxic substances. *Journal of Chromatography* **51**: 327–329.

Jabeen, N., Ajaz Rasool, S., Ahmad, S., Ajaz, M. and Saeed, S. (2004). Isolation, identification and bacteriocin production by indigenous diseased plant and soil associated bacteria. *Pakistan Journal of Biological Sciences* **7**: 1893–1897.

Jackson, M.A. (1997). Optimizing nutritional conditions for the liquid culture production of effective fungal biological control agents. *Journal of Industrial Microbiology & Biotechnology* **19**: 180–187.

James, J.R., Tweedy, B.G. and Newby, L.C. (1993). Efforts by industry to improve the environmental safety of pesticides. *Annual Review of Phytopathology* **31**: 423–439.

Jaufeerally-Fakim, Y., Autrey, J.C., Daniels, M.J. and Dookun, A. (2000). Genetic polymorphism in *Xanthomonas albilineans* strains originating from 11 geographical locations, revealed by two DNA probes. *Letters in Applied Microbiology* **30**: 287–293.

Kurita, N. (1981). Antifungal activity of components of essential oils. *Journal of Agricultural and Biological Chemistry* **45**: 945–952.

Lavermicocca, P., Valerio, F. and Visconti, A. (2003). Antifungal activity of phenyllactic acid against molds isolated from bakery products. *Applied & Environmental Microbiology* **69**: 634–640.

Lee, J.Y., Moon, S.S. and Hwang, B.K. (2003). Isolation and *in vitro* and *in vivo* activity against *Phytophthora capsici* and *Colletotrichum orbiculare* of phenazine-1-carboxylic acid from *Pseudomonas aeruginosa* strain GC-B26. *Pest Management Science* **59**: 872–882.

Leyns, F., Decleene, M., Swings, J.G. and Deley, J. (1984). The host range of the genus *Xanthomonas*. *The Botanical Review* **50**: 308–356.

Luque-Ortega, J.R., Martínez, S., Saugar, J.M., Izquierdo, L.R., Abad, T., Luis, J.G., Piñero, J., Valladares, B. and Rivas, L. (2004). Fungus-elicited metabolites from plants as an enriched source for new leishmanicidal agents: Antifungal phenyl-phenalenone phytoalexins from the banana plant (*Musa acuminata*) target mitochondria of *Leishmania donovani* promastigotes. *Antimicrobial Agents and Chemotherapy* **48**: 1534–1540.

Maleš, •., Brantner, A.H., Soviæ, K., Pilepiæ, K.H. and Plazibat, M. (2006). Comparative phytochemical and antimicrobial investigations of *Hypericum perforatum* L. subsp. perforatum and *H. perforatum* subsp. angustifolium (DC.) Gaudin. *Acta Pharmaceutica* **56**: 359–367.

Mann, C.M. and Markham, J.L. (1998). A new method for determining the minimum inhibitory concentration of essential oils. *Journal of Applied Microbiology* **84**: 538–544.

McCoy, R.E., Caudwell, A., Chang, C.J., Chen, T.A., Chiykowski, L.N., Cousin, M.T., Dale, J.L., DeLeeuw, G.T.N., Golino, D.A., Hackett, K.J., Kirkpatrick, B.C., Marwithz, R., Petzold, H., Sinha, R.C., Sugiura, M., Whitcomb, R.F., Yang, I.L., Zhu, B.M. and Seemuller, E. (1989). Plant diseases associated with mycoplasma-like organisms. *In*: *The Mycoplasmas* (*Eds*., R.F. Whitcomb and J.G. Tully). pp. 458–541. Academic Press, New York.

Mitra, S.K., Sundaram, R., Venkataranganna, M.V., Gopumadhavan, S., Prakash, N.S., Jayaram, H.D. and Sarma, D.N.K. (2000). Anti-inflammatory, antioxidant and antimicrobial activity of *Ophthacare brand*, an herbal eye drops. *Phytomedicine* **7**: 123–127.

Moreno, S., Scheyer, T., Romano, C.S. and Vojnov, A.A. (2006). Antioxidant and antimicrobial activities of rosemary extracts linked to their polyphenol composition. *Free Radical Research* **40**: 223–231.

NCCLS. 2003. *Method for Antifungal Disk Diffusion Susceptibility Testing of Yeasts, Proposed Guideline*. NCCLS document M44-P. NCCLS, Pennsylvania, USA.

NCCLS. 2002. *Reference Method for Broth Dilution Antifungal Susceptibility Testing of Filamentous Fungi; Approved Standard*. NCCLS document M38-A. NCCLS, Pennsylvania, USA.

Nelson, P.E., Toussoun, T.A. and Marasas, W.F.O. (1983). *Fusarium* species: *An illustration manual for identification*. p. 193. University Park, Pennsylvania State University Press, Pennsylvania, USA.

Osbourn, A.E. (1996). Preformed antimicrobial compounds and plant defence against fungal attack. *The Plant Cell* **8**: 1821–1831.

Padmanabhan, S.Y. (1973). The great Bengal famine. *Annual Review of Phytopathology* **11**: 11–26.

Phongpaichit, S., Pujenjob, N., Rukachaisirikul, V. and Ongsakul, M. (2004). Antifungal activity from leaf extracts of *Cassia alata* L., *Cassia fistula* L. and *Cassia tora* L. *Songklanakarin Journal of Science and Technology* **26**: 741–748.

Quiroga, E.N., Sampietro, D.A., Soberón, J.R., Sgariglia, M.A. and Vattuone, M.A. (2006). Propolis from the northwest of Argentina as a source of antifungal principles. *Journal of Applied Microbiology* **101**: 103–110.

Rahalison, L., Hamburger, M., Monod, M., Frenk, F. and Hostettinann, K. (1994). Antifungal tests in phytochemical investigations: Comparison of bioautographic methods using phytopathogenic and human pathogenic fungi. *Planta Medica* **60**: 41–44.

Riveros, F., Muño, G., González, L., Rojas, L., Alvarez, A.M. and Hinrichsen, R.P. (2001). Comparison between DNA and morphological analysis for identification of *Fusarium* species isolated from muskmelon (*Cucumis melo* L.). *Agricultura Técnica* **61**: 281–293.

Runyoro, D.K.B., Matee, M.I.N., Ngassapa, O.D., Joseph, C.C. and Mbwambo, Z.H. (2006). Screening of Tanzanian medicinal plants for anti-Candida activity. *BMC Complementary and Alternative Medicine* **6**: 11.

Satish, S., Raveesha, K.A. and Janardhana, G.R. (1999). Antibacterial activity of plant extracts on phytopathogenic *Xanthomonas campestris* pathovars. *Letters in Applied Microbiology* **28**: 145–147.

Sawaya, A.C.H.F., Palma, A.M., Caetano, F.M., Marcucci, M.C., Da Silva Cunhal, I.B., Araujo, C.E.P. and Shimizu, M.T. (2002). Comparative study of *in vitro* methods used to analyse the activity of propolis extracts with different compositions against species of *Candida*. *Letters in Applied Microbiology* **35**: 203–207.

Schisler, D.A. and Slininger, P.J. (1997). Microbial selection strategies that enhance the likelihood of developing commercial biological control products. *Journal of Industrial Microbiology & Biotechnology*, **19**: 172–179.

Shephard, M.C. (1987). Screening for fungicides. *Annual Review of Phytopathology* **25**: 189–206.

Shin, K., Yamauchi, K., Teraguchi, S., Hayasawa, H., Tomita, M., Otsuka, Y. and Yamazaki, S. (1998). Antibacterial activity of bovine lactoferrin and its peptides against enterohaemorrhagic *Escherichia coli* O157:H7. *Letters in Applied Microbiology* **26**: 407–411.

Shittu, O.B., Alofe, F.V., Onawunmi, G.O., Ogundaini, A.O. and Tiwalade, T.A. (2006). Bioautographic evaluation of antibacterial metabolite production by wild mushrooms. *African Journal of Biomedical Research* **9**: 57–62.

Soberón, J.R., Sgariglia, M.A., Sampietro, D.A., Quiroga, E.N. and Vattuone, M.A. (2006). Antibacterial activity of plant extracts from northwestern Argentina. *Journal of Applied Microbiology* **102**: 1450–1461.

Sokoviæ, M. and van Griensven, L.J.L.D. (2006). Antimicrobial activity of essential oils and their components against the three major pathogens of the cultivated button mushroom, *Agaricus bisporus*. *European Journal of Plant Pathology* **116**: 211–224.

Staskawicz, B.J., Ausubel, F.M., Barrer, B.J., Elis, J.G. and Jones, J.D.G. (1995). Molecular genetics of plant disease resistance. *Science* **268**: 661–667.

Staub, T. (1991). Fungicide resistance: Practical experience with antiresistance strategies and the role of integrated use. *Annual Review of Phytopathology* **29**: 421–442.

Strange, R.N. and Scott, P.R. (2005). Plant disease: A threat to global food security. *Annual Review of Phytopathology* **43**: 83–116.

Taiz, L. and Zeiger, E. (1998). *Plant Physiology*. Sinauer Associates, Inc., Publishers. Sunderland, Massachusetts.

Ullstrup, A.J. (1972). The impact of the southern corn leaf blight epidemics of 1970–71. *Annual Review of Phytopathology* **10**: 37–50.

Vasinauskienė, M., Raduðienė, J., Zitikaitė, I. and Survilienė, E. (2006). Antibacterial activities of essential oils from aromatic and medicinal plants against growth of phytopathogenic bacteria. *Agronomy Research* **4**: 437–440.

Vicedo, B., de la O Leyva, M., Flors, V., Finiti, I., Del Amo, G., Walters, D., Real, M.D., Garcia-Agustin, P. and Gonzalez-Bosch, C. (2006). Control of the phytopathogen *Botrytis cinerea* using adipic acid monoethyl ester. *Archives of Microbiology* **184**: 316–326.

Walton J.D. (1997). Biochemical Plant Pathology. *In*: *Plant Biochemistry* (*Eds.*, P.M. Dey and J.B. Harborne). pp. 487–502. Academic Press Limited, London.

Whitehead, N.A., Byers, J.T, Commander, P., Corbett, M.J., Coulthurst, S.J., Everson, L., Harris, A.K., Pemberton, C.L., Simpson, N.J., Slater, H., Smith, D.S., Welch, M., Williamson, N. and Salmond, G.P. (2002). The regulation of virulence in phytopathogenic *Erwinia* species: Quorum sensing, antibiotics and ecological considerations. *Antonie van Leeuwenhoek* **81**: 223–231.

Wilson, C.L., Solar, J.M., El Ghaouth, A. and Wisniewski, M.E. (1997). Rapid evaluation of plant extracts and essential oils for antifungal activity against *Botrytis cinerea*. *Plant Disease* **81**: 204–210.

Wilson, C.L. and Wisniewski, M.E. (1994). *Biological control of postharvest plant disease of fruits and vegetables: theory and practice*. CRC Press, Boca Raton.

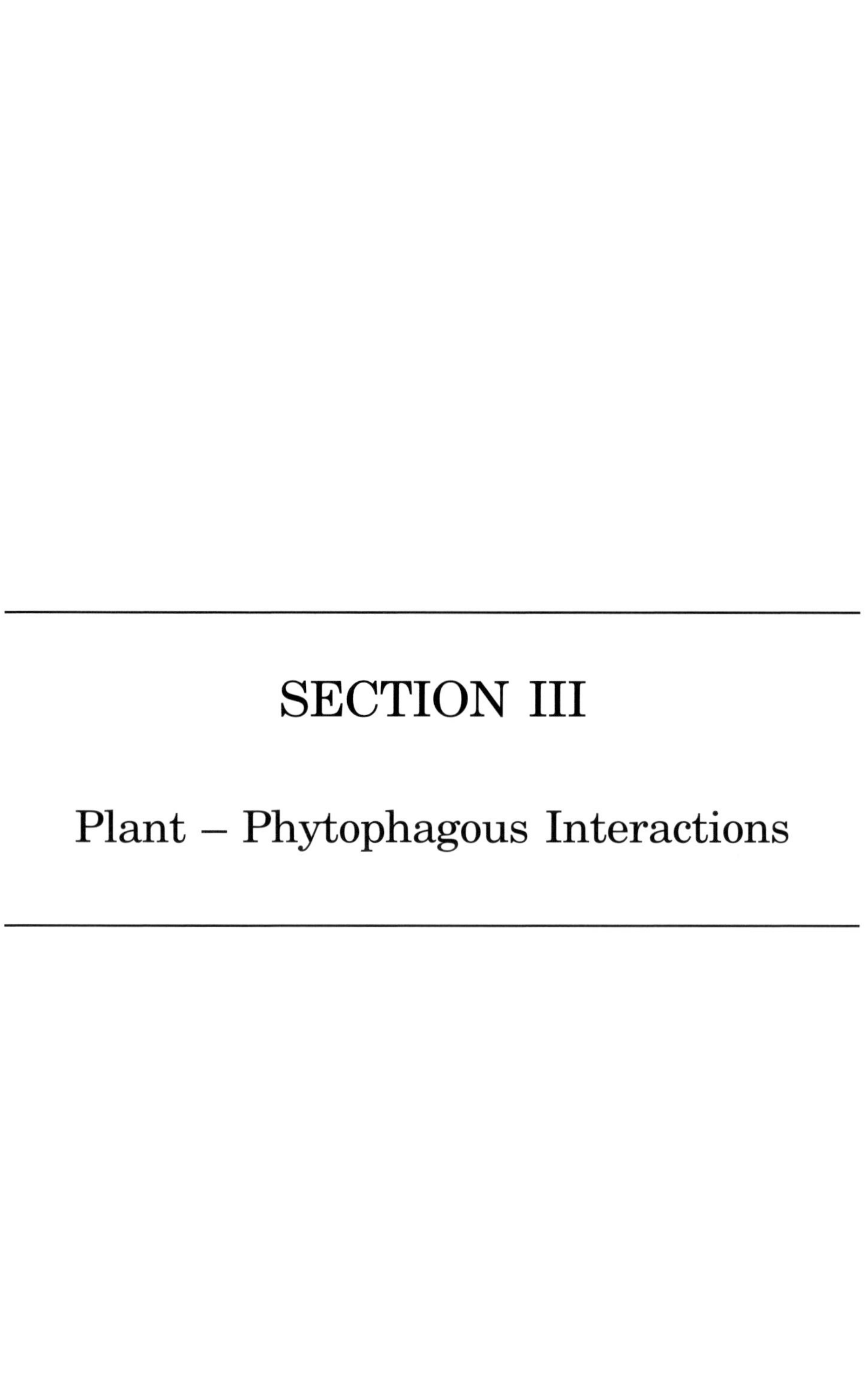

SECTION III

Plant – Phytophagous Interactions

CHAPTER 13

Bioassays of Phytochemicals to Control Plant – Insect Interactions

SUSANA B. POPICH[1]

1. INTRODUCTION

Agricultural production in the world is orientated towards the use of good agricultural practices (GAP), a management of agrochemicals that ensures the integrity, transparency, and harmonization of the global agricultural rules, providing harmlessness on food production, health conditions, safety, and well being of workers, animals and the environment. The 30 % of pesticides in developing countries are not up to international standards and these deficiencies include active ingredients beyond permissible thresholds. Current development of agrochemicals requires compounds with permissible thresholds below 0.05–5 µg/kg (FAO/OMS, 2003), and quick biodegradation. In this context, plants may provide such type of compounds because they produce allelochemicals which are, in fact, natural pesticides and herbicides. The chemistry of plant defense involves the production of compounds with a great chemical heterogeneity (Einhellig and Barkosky, 2003), including substances with repellent, deterrents, antifeedant or toxic activities (Primo Yufera, 1991). Repellents generally prevent landing of adult insects or movement of larvae onto plant surfaces. Antifeedants inhibit or deter feeding by those insects that venture onto the plant, whereas, toxins either kill or immobilize them after feeding. First insect–plant interactions are often mediated by volatile phytochemicals which attract or repel phytophagous arthropods. Then, non-volatile allelochemicals influence, by gustatory stimuli, insect acceptance or rejection of a plant for feeding. Therefore, the sense of taste plays a major role in determining host ranges or dietary discrimination of insects. This chapter provides several assays involving target insects with different feeding habits allowing to know how these organisms respond to plant extracts/ allelochemicals.

1. Departamento de Ciencias Básicas y Aplicadas, Universidad Nacional de Chilecito, Ruta Los Peregrinos S/N. CP: 5360 Dpto., Chilecito. La Rioja, Argentina. E-mail: spopich@undec.edu.ar

2. BIOASSAYS FOR STORED GRAIN PESTS

Experiment 1: Toxicity bioassay on *Sitotroga cerealella* (Olivier)

Principle

The grain moths includes only lepidopterous, such as *Sirotroga cerealella*, capable of destroying unbroken grain kernels, when stored at inadequate moisture levels. *Sitotroga cerealella* is a serious pest of stored grains worldwide (Fig 1). The insect lays its eggs on the grain surface (Fig 2). After hatching, the larvae bore into the grain and complete their development

Fig 1. Normal adult of *Sitotroga cerealella* on wheat grains (*Source*: http://www.sgrl.csiro.au)

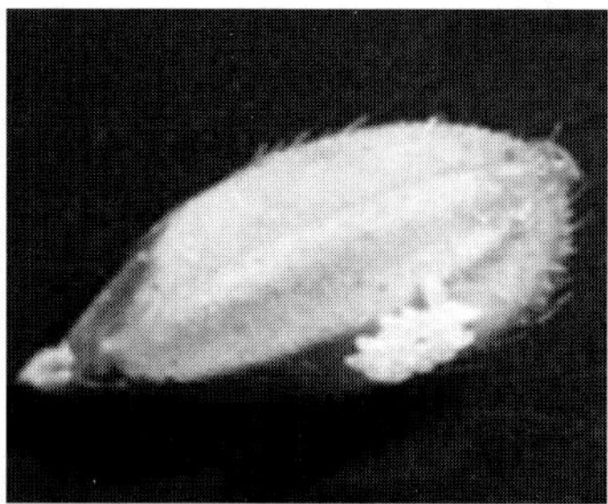

Fig 2. Normal eggs on wheat grain of *Sitotroga cerealella* (*Source*: http://www.nfri.affrc.go.jp)

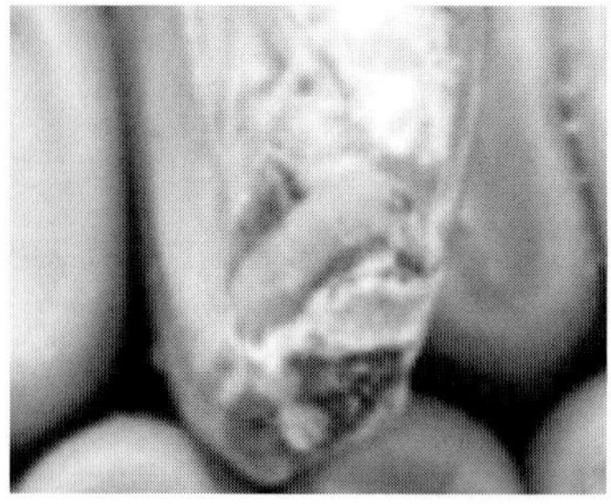

Fig 3. Last larval stage of *Sitotroga cerealella* on maize (*Source*: www.viarural.com.ar/viarural.com.ar/agricultura/granos)

within the endosperm. The larvae are always inside the grain kernel except in the first instar (Fig 3). Each female lays about of 40 eggs, deposited either singly or in clusters. The following toxicity bioassay involves eggs of *S. cerealella* which are placed in contact with a substrate exposed to different concentrations of a plant allelochemical/extract (treated diet), including a control without it (control diet). The larvae and pupal development periods are measured and the proportion of normal and deformed adults is recorded.

Materials and equipments

Eggs from *S. cerealella* obtained from females maintained in laboratory colonies and not exposed to insecticides, grains of wheat (*Triticum aestivum* L.), 500 mL Erlenmeyer flasks, culture tubes, beaker, cotton croks, cotton cloth, forceps, analytical balance, pyrex brand vacuum desiccator, freezer, thermometer, small paint brush, rearing cabinet, a plant allelochemical/ extract, stove capable of 180 °C, containers.

Procedure

i. **Diet preparation:** Grains of wheat (*T. aestivum*) are wrapped in a clean and wet piece of cotton cloth and placed into a beaker. Then, they are kept in a stove at 180 °C for 6 h. The grains are removed at intervals of 2 h and dampened with distilled water. After 6 h, the grains are cooled, placed in close containers and maintained in a freezer until use.

ii. A portion of the diet is impregnated with the solvent used for dilution of the plant allelochemical/extract (control diet). Another portion is prepared with a solution of the plant allelochemical/extract under study (treated diet).

iii. After *in vacuo* evaporation of solvent in vacuum desiccator, 100g of each diet is placed in a 500-mL Erlenmeyer flasks followed by 0.5g of *S. cerealella* eggs. Each treatment has three replications.

iv. The Erlenmeyer flasks are maintained at 27 ± 2 °C and 50–60 % RH in a rearing cabinet until the emergency of the larvae.

v. The larvae are maintained on the wheat grains for about 15–20 days. Then, larvae is molted to pupae. This moment is characterized by an increase in temperature of wheat grains, that is easily determined from on exterior of Erlenmeyer flasks.

vi. Pupae are easily observed between wheat grains. This stage finishes after 7–10 days, when adults emerge.

vii. Proportion of normal and malformed adults or absence of adult emergency is recorded.

Observations

Bioactive plant compounds/allelochemicals can produce deformations on body, legs and wings of adult insects. This information is very important to evaluate allelochemical effects (Fig 4).

Statistical analysis

The results can be expressed as mean ± SE for number of adults. The differences among means are detected through analysis of variance (ANOVA). Tukey's or LSD tests are used for pairwise multiple comparisons of groups at level of $p < 0.05$.

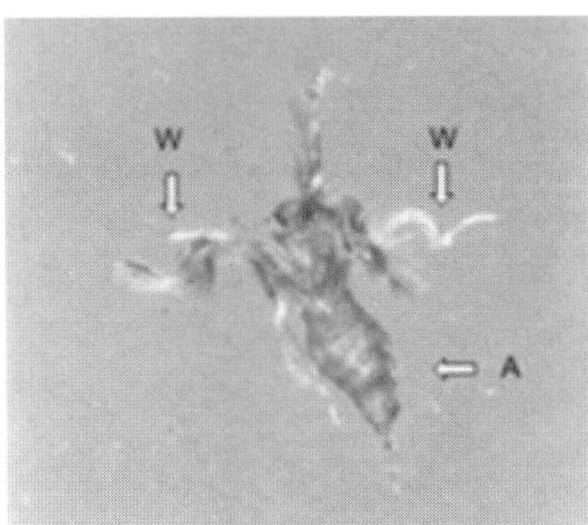

Fig 4. Deformed adult of *Sitotroga cerealella*. Note malformations of wings (W) and abdomen (A) (*Source*: Bardón *et al*., 1999)

Experiment 2: Effects of allelochemicals on fecundity and eggs fertility of *S. cerealella* from filial 1.

Principle

Oviposition capacity can be significantly affected by substances added to insect diets. Females at larvae stage eating a diet treated with plant allelochemicals/extracts could oviposite number of eggs than those fed with control diets. On the other hand, these females could produce infertile eggs. This phenomenon, known as filial 1 sterility, is an effective and environmentally safe strategy to control the lepidopteran pests.

Materials and equipments

Assay tubes one for each couple (OD 16× L 125 mm), 250g of wheat, 50 mL flask vials, light microscope (40× objective), slides, microtome (Standard Microtome Blade Holder), normal adult couples from filial 1, at least 6 adult couples of 2-day-old for each treatment, small paint brush for to seize the larvae, culture stove at 25 °C, forceps, cotton croks to cover Erlenmeyer's flask, spatulas, thermometer.

Reagents

Karnovsky solution: Add 20 mL of 16 % *p*-formaldehyde, 8 mL of 50 % glutaraldehyde, and 25 mL of 0.2 M sodium phosphate buffer to 25 mL of distilled water, hematoxilyn-eosin, test compounds.

Procedure

i. Twelve normal adults are selected from controls of Experiment 1. The same amount is chosen from treated diet. Each pair is placed in a separate glass tube to lay eggs. The tubes are placed in a culture stove at 25 °C. After 48 h, the females lay eggs on the grains. Eggs are visible as small and oval bodies. The chorion exhibits a regular sculptured pattern with prominent ribs. The eggs must be carefully separated from wheat grains with a small spatula. For microscopic observations, 30 % of the eggs are fixed in Karnovsky solution, dyed, transversally sectioned by microtome and mounted on microscope slides.

ii. Strong mitotic activity is observed as abundant stained nuclei, which is expected in unaffected embryo into normal eggs (Fig 5). In contrast, few dyed nuclei indicate low mitotic activity and therefore, a possible chemical sterilization effect (Fig 6).

iii. The rest of eggs (70 %) are placed in flask vials (30 eggs each) with 10g of untreated diet (wheat grains) and incubated at 25 °C for 4 days.

iv. The number of eggs is recorded and percentage of hatching is determined.

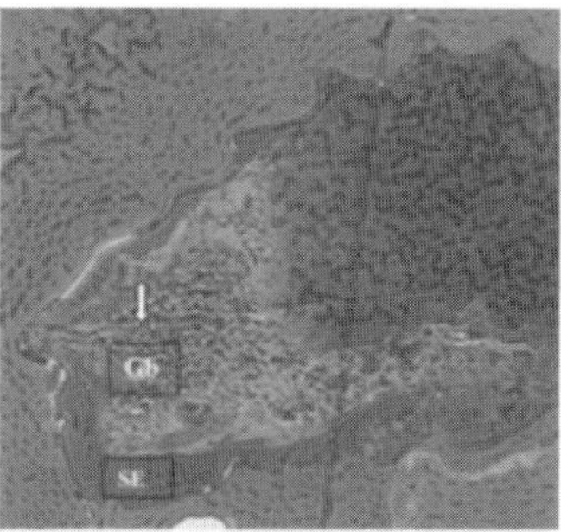

Fig 5. Longitudinal section of control eggs. Serosa (SE) is clearly observed. Cell of the germinal band (Gb) are in active mitotic division which is evident by the stained nuclei (*Source*: Bardón *et al.*, 1999)

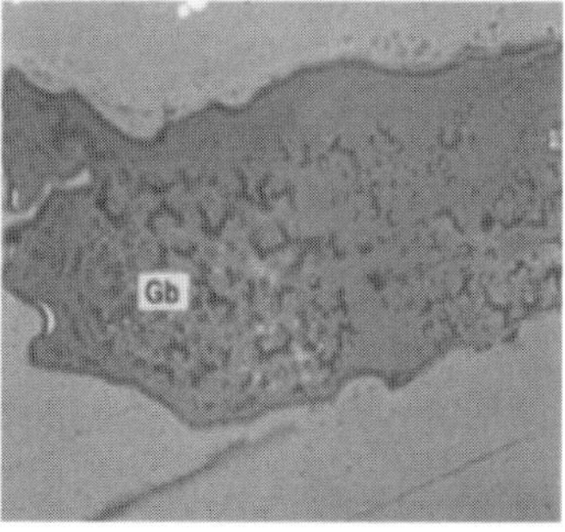

Fig 6. Longitudinal section of eggs laid by females from test. Serosa is not observed. Cell of the germinal band (Gb) show a decrease in mitotic activity (*Source*: Bardón *et al.*, 1999)

Observations

Any substance applied on the natural diet of *S. cerealella* can produce a remarkable reduction in egg production. For example, females fed with sesquiterpene lactones from *Cyrtocymura cincta* (Asteraceae) can suffer a reduction of 60 % in egg production compared with control females. These eggs are often infertile.

Experiment 3: Toxicity bioassay on *S. granarius*

Principle

Sitophilus granarius is a common pest of millets and other grains (Fig 7). It is a small coleopteran whose life cycle is developed only inside the grain (Stejskal and Kucerova, 1996). A simple and no expensive laboratory assay can be performed using this insect as target. Artificial or natural insect diet is impregnated with the test solution/plant extract. Then, the insect is forced to ingest the treated diet, observing mortality produced by the assayed compounds/extracts.

Fig 7. Adults of *Sitophilus granarius* on corn (*Source*: http://www.cmis.csiro.au)

Materials and equipments

Artificial diet: Bean flour (200g); agar-agar (10g); yeast extract (3g); mixture of vitamins (0.4g); and distilled water (250 mL); mixture of vitamins contains: calcium pantothanate (9.6g), nicotine acid amide (4.8g), riboflavin (2.4g), folic acid (2.5g), thymine hydrochloride (1.2g), pyridoxine hydrochloride (1.2g), biotin (0.096g), vitamin B_1 (0.048g); flask vials of wide mouth (60 mL); cotton corks; test solutions; culture stove at 23 °C; forceps; spatulas; thermometer analytical balance.

Procedure

i. **Diet preparation:** Ten gram of agar are dissolved in 125 mL hot distilled water (90 °C). The bean flour, yeast extract and the mixture of vitamins are added after cooling to 40 °C. Then, the volume is taken up to 250 mL. After that, diets are dehydrated for 96 h at room temperature.

ii. A known portion of artificial diet is mixed with the solvent and employed as control. Another portion is impregnated with the test solution (treated diet). After solvent evaporation, the treated and control diets are divided into equal portions, weighed and placed in vials. Each treatment is performed with three replications.

iii. Ten adult insects of similar weight and age are placed in each vial and kept at 23 °C in the culture stove for 96 h. Then, mortality and weight of uneaten diet are determined.

Observations

Sitophilus granarius is sensitive to several deterrent substances, which significantly decrease its growth (Bernays *et al.*, 2000). Mortality in short time (*i.e.*, 96 h) allows a clear evaluation of such effect (Table 1).

Table 1. Effects of methanolic extract of *H. argentea* and *M. acuminata* on the growth and mortality of *S. granarius*

Evaluated parameters	**Control**	***M. acuminata* EM**	***H. argentea* EM**
Antifeedant index (%)	0	72.99	66.17
*Growth modification (mg)	2.06 ± 0.14^{a}	1.39 ± 0.18^{b}	1.32 ± 0.31^{b}
Mortality (%)	0	15	40

*Means within a column followed by the same letter not significantly different ($p > 0.05$) according to Tukey's multiple range test

Calculations

The following indices are calculated:

$$\text{Antifeedant Index: } [1 - W_d / W_{td}] \times 100$$

where W_d is weight of eaten treated diet and W_{td} is weight of eaten control diet.

$$\text{Mortality (\%): } [1 - N_e \times N_i / N_{ti} \times N_{tc}] \times 100$$

where N_e is the number of insects treated till the end of assay, N_i is the number of control insects at the beginning of the assay, N_{ti} is the number of insects fed with the treated diet at the beginning of the assay and N_{tc} is the number of insects fed with the control diet at the end of the assay.

Statistical analysis

Weight data can be expressed as mean ± S.D. The differences are evaluated by analysis of variance (ANOVA). For the comparison of groups the Tukey's and LSD tests can be used.

3. BIOASSAYS ON LEPIDOPTERAN PEST: *SPODOPTERA FRUGIPERDA* AND *SPODOPTERA LATIFASCIA* (NOCTUIDAE)

3.1. Choice and no choice tests for Lepidopteran species

Principle

Choice and no choice tests allow determining the antifeedant activity of either extracts or pure compounds in a short time. In a paired choice test, an insect is forced to choose between a treated and an untreated diet. In a no-choice test, half of the insects are forced to feed on a treated diet, while, the other half is fed with the untreated diet (Fig 8). If the substance acts as an antifeedant, then insects will eat more of the untreated than the treated diet (Cole, 1994). The difference between the amount of consumed untreated and treated diets is used to calculate the feeding deterrence. If the substance acts as an antifeedant, then insects will eat more of the untreated than the treated diet. The experiment is stopped when half larvae in the control eats 50 % of the control diet. No choice test is used to check the strength of antifeedant compounds, to confirm primary or secondary antifeedant effects and to detect the growth regulator compounds or toxicants.

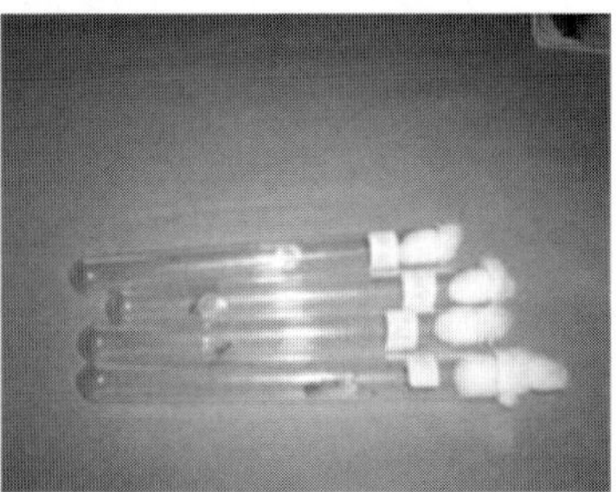

Fig 8. Example of no choice test

These assays can be done using artificial diets. The test substance (extracts or pure compounds) is dissolved in water or a proper organic solvent at the desired concentrations and incorporated to artificial diet. For choice test, a known portion of control and the same amount of treated diet is placed in a glass tube and offered to insect. Between the two diet portions, a larva is introduced in the tube and allows choosing the diets. After a time, the remaining diets (control and treated) are weighed. Weight of consumed diet is easily calculated by difference. For No choice test, the same amount of diets (control and treated) are placed in separated tubes with a larva in every tube and forced to feed the diet. Losses of weigh by dehydration of the diet are avoided.

Materials and equipments

Treated and untreated diet; glass tubes (OD 13 × L 100 mm), cotton corks, analytical balance, forceps, small paint brush, rotary evaporator, cotton corks; larvae of seconds instar of *Spodoptera frugiperda*. Artificial diet

Preparation: 12.5g of agar is dissolved in 125 mL of hot distilled water. Bean boiled and milled (250g), wheat germ (12.5g) and yeast (3g) are mixed in 125 mL distilled water and added to the agar solution. After cooling until 40 °C, ascorbic acid (1.5g), methyl *p*-hydroxybenzoate (1.5g), formaldehyde (4 mL of a 38 % water solution), distilled water (500 mL), and antibobies (1.5g) are incorporated. The diet must be stored in refrigerator at 4 °C.

Experiment 4: Choice test

Procedure

i. The test substance is dissolved in a proper solvent according to its solubility.

ii. A portion of artificial diet is mixed with the solvent. This portion is employed as control diet. Another portion is mixed with test solution. Then, the solvent is removed *in vacuo*. A known portion of dried treated diet and the same amount of dried control diet are placed inside each glass tube. Prepare at least 10 glass tubes.

iii. A larva of second instar is introduced into each tube, between the two diet portions. Then, tubes are closed with cotton corks.

iv. The larva is allowed to choose the diet. The experiment concludes after that 50 % of control larvae eat 50 % of control diet.

v. The remaining control and treated diets are weighed and feeding deterrence (FR_{50}) or feeding deterrence index (FDI) are calculated.

Experiment 5: No choice test

Procedure

i. The test substance/plant material is dissolved in a solvent according to its solubility.

ii. A portion of artificial diet is mixed with the solvent. This portion is employed as control diet. Another portion is mixed with solution test. Then, solvent is removed *in vacuo*.

iii. Prepare a set of tubes, placing a known weight (*i.e.*, 70 mg) of treated diet inside each glass tube. A set of tubes containing control diets is also prepared as previously indicated for treated diets.

iv. A larva of second instar is introduced in each tube. Then, close the tubes with cotton corks.

v. The larva is allowed to eat the diets. The experiment concludes after that 50 % of control larvae eat 50 % of control diet.

vi. The remaining control and treated diets are weighed and feeding deterrence (FR_{50}) is calculated.

Calculations

i. Amount of consumed control (C) and treated (T) diets is calculated as under:

$$C = W_{ic} - W_{ec} \text{ and } T = W_{it} - W_{et}$$

where W_{ic} and W_{et} are weights of control and treated diets at the beginning of the experiment, respectively. W_{it} and W_{et} are weights of control and treated diets at the end of the experiment, respectively.

ii. Feeding deterrence (FR_{50}) is calculated as under:

$$FR_{50} = T / C$$

The value of FR varies between 0 and 1. Antifeedant effect is more important as well as a small value is recorded.

Feeding deterrence index (FDI %) can also be calculated as under:

$$\text{FDI } (\%) = (C - T) / (C + T) \times 100$$

The value of FDI varies between 0 and 100. Higher FDI (%) value indicates higher antifeedant effect.

Statistical analysis

Results are reported as mean ± SEM. Differences in the mean values are evaluated by analysis of variance (ANOVA). The Tukey's test is used for all pairwise multiple comparison of groups. In all statistical analysis $p > 0.05$ are considered not significant.

3.2. NUTRITIONAL INDICES AND TOXICITY BIOASSAY

Experiment 6: Nutritional indices

Principle

Insect requirements for dietary constituents are determined by intake measurement. The efficiency of digested food for growth depends not only on energy requirements for maintenance but also on balance of nutrients. The nutritional indices determines the effect of different doses of a test compound on the growth, weight and the metabolism of an insect forced to ingest a diet treated with the test substance (Fig 9). The nutritional indices are as under:

i. **Conversion of ingested food (ECI):** It is the measure of insect ability to use ingested food for growth.

ii. **Relative growth rate (RGR):** It is the mean of larval weigh increment per day corrected for initial body weight.

iii. **Food utilization rate (CI):** It is the consumption rate corrected for final body weight (Alvarez Colom *et al.*, 2007).

iv. **Approximate digestibility (AD):** It is the difference between the weight of ingested food and the weight of the feces.

Calculation of these indexes requires measurement of the following parameters: the weight to the food ingested, the weight of the insect feces and the weight gained by the insect during the experiment.

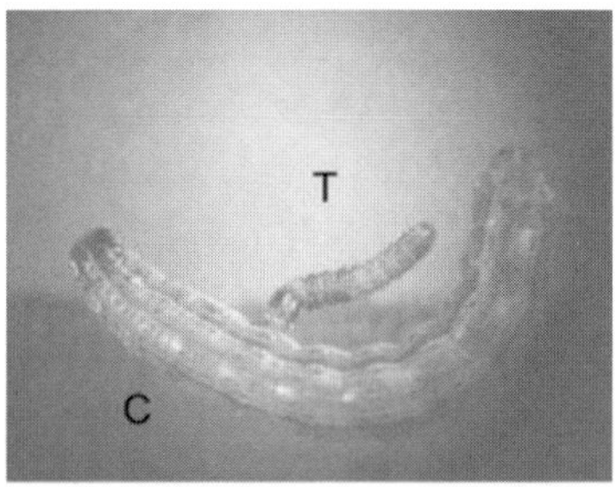

Fig 9. Effects of allelochemicals substances on *Spodoptera frugiperda* development. Both larvae have the same age. C, control larva and T, larva affected by allelochemicals substances

Materials and equipments

Control and treated diets: the same ones used in the choice experiment, analytical balance, insects rearing cabinets adjusted to 25 °C, photoperiod of 8/12 h (light/dark) and relative humidity of 60 %, culture tubes (OD 13 × L 125 mm), cotton corks, treated and control diets, forceps, small paint brush, ethanol 70 %, refrigerator (10 °C), 1500 mL plastic bottle.

Procedure

i. Known weights of control and treated diets are placed in separate test tubes. Each treatment comprises 20 tubes.

ii. One second instar larva of known weight is introduced in each tube.

iii. All the tubes are covered with cotton corks and placed into the rearing cabinet.

iv. Diets are replaced every 2 days, to ensure fresh material for insect feeding. These amounts of diet must be weighed before introducing them into the tubes.

v. After 10 days larvae, diet and feces are weighed. This time may vary as per the insect species under study. Then, nutritional indices are calculated.

vi. The same larvae are placed in the same tubes to continue with the experiments of toxicity. Then, the possible toxicity of treated diet is determined.

Calculations

The Nutritional Indices are calculated as follow:

$$ECI = RGR / CI$$

$$RGR = (A - B) / t$$

$$CI = D / t$$

$$DA = D - F / t$$

where D is the food eaten during the experiment, B is the initial larval weight, A is the final larval weight, t is the time interval from the beginning to the end of the experiment and F is the weight of feces.

Observations

Second instar larvae are very small and fragile. Hence, they are entered very carefully into the tubes. The larvae should neither disturbed nor be removed from the tubes throughout the test. In the test of nutritional indices, it is common to observe effects on the larvae, *i.e.*, retention of exuviae (old larval tegument that cannot be removed correctly).

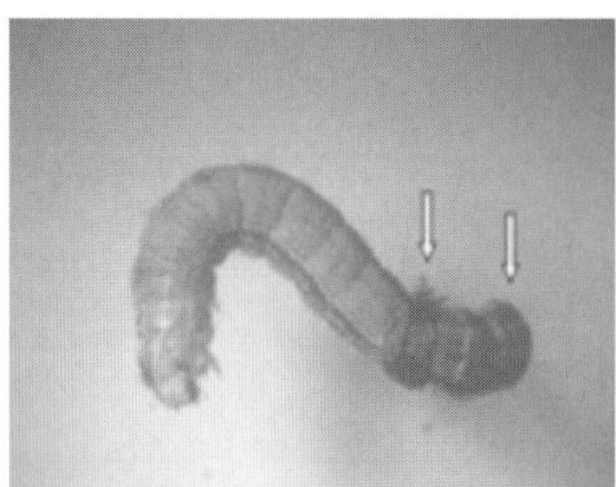

Fig 10. Fourth instar larva of *Spodoptera frugiperda.* Arrows indicate retention of successive exuvies

Statistical analysis

All results are expressed as mean ± SEM. Differences in the mean values by analysis of variance (ANOVA) and. Tukey's and LSD tests are used for all pair-wise multiple comparison of groups.

Experiment 7: Toxicity bioassay on Lepidopteran species

Principle

In toxicity bioassay, a target insect is forced to ingest a diet previously treated with a plant extract or a pure substance. Diets, tubes and larvae are prepared as indicated for nutritional indices (Experiment 6). A new treatment consisting in artificial diet treated with a commercial insecticide is also used as positive control. The bioassay is maintained in a rearing cabinet under controlled conditions of temperature, light, and relative humidity. Mortality is determined after 48 h by counting the number of living larvae. If pupation occurs, the pupae should be removed and their malformations observed. The toxicity is determined by LD_{50} is the dose that kills half (50 %) of the animals tested (LD is "lethal dose"). Malformations in larvae (Fig 11A, B), pupae and adults must be registered in the surviving insects.

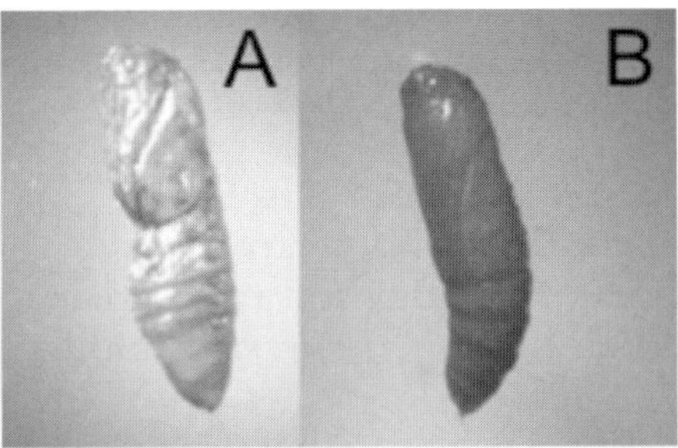

Fig 11. Pupae of *Spodoptera frugiperda*: (A) normal pupe and (B) pupe affected by an allelochemicals

Materials and equipments

Commercial insecticide at recommended doses, treated and control diets, analytical balance, insects rearing cabinets adjusted to 25 °C, photoperiod of 8/12 h (light/dark) and relative humidity of 60 %, assays tubes (OD 13 × L 125 mm), cotton corks, forceps, small paint brush, ethanol 70 %, refrigerator (10 °C), 1500 ml plastic bottle. Insect diet: 6g of sucrose and 2 mL of bee honey are dissolved in 1 L of hot distilled water (90 °C). 1.5g of methyl *p*-hydroxybenzoate is added after cooling until 40 °C. This solution must be conserved in refrigerator at 4 °C.

Procedure

i. A portion of artificial diet is impregnated with the solvent, which is further evaporated under vacuum. This diet is the control. Another portion is mixed with commercial insecticide at recommended doses and is used as positive control. The treated diet is prepared by impregnation with the test compound or plant extract.

ii. Known amounts of each diet are placed in separated test tubes. Each treatment should comprise 20 replications.

iii. It is very important to use a homogenous population of second instar, which are obtained within 3 days of hatching.

iv. One second instar larva of well-known weight is introduced in each tube. All the tubes are closed with cotton corks and placed in rearing cabinet.

v. Diets are replaced every 2 days, to ensure fresh material for insect feeding.

vi. The pupal stage does not feed or drink, but periodic light misting to prevent desiccation is important. The pupae are extracted from the containers sexed and placed in new containers for adult emergency and oviposition. During all experiment, special attention must be paid to larval behavior, molt process, pupation process, malformations, and mortality in all stages, adult emergency and number of adults. All these parameters must be recorded.

vii. Additionally, it is convenient a Filial 1 fertility study. Twenty adults, are placed in 1 L plastic bottles, ten per bottle, with a small amount of cotton impregnated with the diet and an absorbent paper like oviposition substrate. After 4 days, females laid eggs in the paper.

ix. Carefully, the paper is removed from the bottle and the area containing eggs is cut.

xi. Paper and eggs are washed with a solution of sodium hypochlorite at 5 %.

xii. The eggs are counted and placed in plastic containers (60 mL capacity) until larval emergency.

xiii. Emergence percentage is determined.

Observations

1. Second instar larvae are very small and fragile. Hence, they are very carefully entered into the tubes.

2. Amount of diet provided in each tube should be low enough to prevent strong bacterial growth that can kill larvae. Replacement of diets every 2 days ensures fresh material for insect feeding and prevents significant bacterial activity.

3. It is important neither to touch nor to move the assay tube or larvae when they are molting. They are very susceptible to damage during this vulnerable transition. The same applies to the larva when it is becoming a pupa.

Calculations

Mortality can be determined as under:

$$\text{Mortality (\%)} = X - Y / (X \times 100)$$

where *X* is survival (%) in control test and *Y* is survival (%) in treated test.

Statistical analysis

It is preferable to use a statistical program, such as SAS, which incorporates a Log Probit Analysis. The toxicity of an unknown product is determined by estimation and comparison of the LC_{50}s of the tested product and control calculating percentage of mortality at different assayed concentrations.

Suggested Readings

Alali, F.Q., Zhang, Y., Rogers, L. and McLaughlin, J.L. (1997). (2,4-*cis* and *trans*)-gigantecinone and 4-deoxygigantecin, bioactive nonadjacent bis-tetrahydrofuran annonaceous acetogenins, from *Goniothalamus giganteus*. *Journal of Natural Products* **60**: 929–933.

Alvarez Valdes, D., Elias, A. and Bardon, A. (2001). Allelopathic agents from *Cyrtocymura cincta*. *Natural Products Letters* **15**: 445–450.

Alvarez Colom, O., Neske, A., Popich, S. and Bardón, A. (2007). Toxic effects of annonaceous acetogenins from *Annona cherimolia* (Magnoliales: Annonaceae) on *Spodoptera frugiperda* (Lepidoptera: Noctuidae). *Journal of Pest Science* **80**: 63–67.

Anaya, A.L., Macias-Rubalcava, M., Cruz-Ortega, R., Garcia-Santana, C., Sanchez-Monterrubio, P.N., Hernandez-Bautista, B.E. and Mata, R. (2005). Allelochemicals from *Stauranthus perforatus*, a Rutaceous tree of the Yucatan Peninsula, Mexico. *Phytochemistry* **66**: 487–494.

Asakawa, Y., Toyota, M., Von Konrat, M. and Braggins, J.E. (2003). Volatile components of selected species of the liverwort genera Frullania and Schusterella (Frullaniaceae) from New Zealand, Australia and South America: a chemosystematic approach. *Phytochemistry* **62**: 439–452.

Bernays, E.A., Oppenheim, S., Chapman, R.F., Kwon, H. and Gould, F. (2000). Taste sensitivity of insect herbivores to deterrents is greater in specialists than in generalists: A behavioral test of the hypothesis with two closely related caterpillars. *Journal of Chemical Ecology* **26**: 547–563.

Borkosky, S.A., Alvarez Valdés, D., Bardón, A., Diaz, J.G. and Herz, W. (1996). Sesquiterpene lactones and other constituents of *Eirmocephala megaphylla* and *Cyrtocymura cincta*. *Phytochemistry* **42**: 1637–1639.

Burnett, W.C., Jones, S.B., Mabry, T.J. and Padolina, W. (1974). Sesquiterpenes lactones-insect feeding deterrents in Vernonia. *Biochemical Systematics and Ecoogy* **2**: 25–29.

Carpinella, M.C., Miranda, M., Almiron, W.R., Ferrayoli, C.G., Almeida, F.L. and Palacios, S.M. (2007). *In vitro* pediculicidal and ovicidal activity of an extract and oil from fruits of *Melia azedarach* L. *Journal of the American Academy of Dermatology* **56**: 250–256.

Cavé, A., Figadére, B., Laurens, A. and Cortes, D. (1997). Acetogenins from Annonaceae. *In*: Progress in the Chemistry of Organic Natural Products (*Eds*., W. Herz, G.W. Kirby, R.E. Moore, W. Steglich and Ch. Tamm). pp. 457–496. Springer-Verlag, New York.

Cole, M.D. (1994). Key antifungal, antibacterial and anti-insect assays – A critical review. *Biochemical Systematics and Ecology* **22**: 837–856.

Cortes, D., Myint, S., Dupont, B. and Davoust, D. (1993). Bioactives acetogenins from seeds of *Annona cherimolia*. *Phytochemistry* **32**: 1475–1482.

Degli Esposti, M., Ghelli, A., Ratta, M., Cortes, D. and Estornell, E. (1994). Natural substance (acetogenins) from the family Annonaceae are powerful inhibitors of mitochondrial-NADH dehydrogenase (Complex I). *Biochemical Journal* **301**: 161–167.

Einhellig, F.A. and Barkosky, R.R. (2003). Allelopathic interference of plant–water relationships by *p*-hydroxybenzoic acid. *Botanical Bulletin of Academia Sinica* **44**: 53–58.

FAO/OMS. 2003. Comisión del Codex Alimentarius. Programa Conjunto sobre las Normas Alimentarias. 25° Período de sesiones. Incluye Informe sobre la 34° Reunión del Comité del Codex sobre Residuos de Plaguicidas 2002, 2003.

Florentino, A., Abrosca, B.D., Izzo, A., Pacifico, S. and Monaco, P. (2006). Structural elucidation and bioactivity of novel secondary metabolites from *Carex distachya*. *Tetrahedron* **42**: 3259–3265.

Harborne, J.B. (1992). Plant toxins and their effects on animals. *In*: *Introduction to Ecological Biochemistry*. Fourth Edition. pp. 316–340. Academic Press, Great Britain.

Hernandez, L.R., Riscala, E., Catalán, C., Diaz, J.G. and Herz, W. (1996). Sesquiterpene lactones and other constituents of *Stevia maimarensis* and *Synedrellopsis grisebachii*. *Phytochemistry* **42**: 681–684.

Kasten, J.R.P., Preceti, A. and Parra, J.R.P. (1978). Comparative biological data of *Spodoptera frugiperda* (Smith) in two artificial diets and a natural substrates. *Revista de Agricultura* **53**: 68–78 (In Portugues).

Khalid, S.A., Farouk, A., Geary, T.G. and Jensen, J.B. (1986). Potential antimalarial candidates from African plants: And *in vitro* approach using *Plasmodium falciparum*. *Journal of Ethnopharmacology* **15**: 201–209.

Koul, O., Koul, S., Sing, C., Taneja, S.C. Shammu-Gavel, M., Kampasi, H., Saxena, A.K. and Qazi, G.N. (2003). *In vitro* cytotoxic elemanoides form *Vernonia laioipus*. *Planta Medica* **69**: 164–166.

Koul, O., Singh, G., Singh, R., Daniewski, W.M. and Berlozecki, S. (2004). Bioefficacy and mode-of-action of some limonoids of salannin group from *Azadirachta indica* A. Juss and their role in a multicomponent system against lepidopteran larvae. *Journal of Bioscience* **29**: 409–416.

Linch, M.J., Rápale, L.D., Mellor, D., Spare, D.P. and Inwood, M.J.H. (1972). Staining of plant cuttings. *In*: *Laboratory Methods*. Nueva Editorial Interamericana. Mexico (In Spanish).

Macías, F.A., Molinillo, J.M.G., Galindo, J.C.G., Varela, R.M., Torres, A. and Simonet, A.M. (1999). Terpenoids with potencial use as natural herbicide templates. *In*: *Biologically Active Natural Products*: *Agrochemicals* (*Eds*., H.G. Cutler and S.J. Cutler). pp. 234–241. CRC Press, Boca Raton, USA.

Marston, A. and Hostettmann, K. (1991). Assays for molluscicidal, cercaricidal, schistosomicidal and piscicidal activities. *In*: *Methods in Plant Biochemistry* (*Eds.,* K. Hostettman and P.M. Dey). pp. 153–178. Academic Press, London.

Ministerio de Salud Pública. (1998). *Normativa*. Res.:280/98 Sustancias Químicas Prohibidas o Restringidas para la República Argentina.

Morton, J.F. (1987). *Fruits of warm climates*. Media Inc., Greensboro, NC.

Muhammad, I., Li, X.C., Dumbar, D.C., Elsohy, M.A. and Khan, I.A. (2001). Antimalarial (+)-*trans*-hexa-dihydrobenzopyran derivatives from *Machareium multiflorum*. *Journal of Natural Products* **64**: 1322–1325.

Nawrot, J., Blosszyk, E., Harmatha, J., Novotna, L. and Drozdz, B. (1985). Action of antifeedant of plant origin on beetles infesting stored products. *Acta Entomologica Bohenoslovaca* **83**: 327–335.

Picman, A.K. (1986). Biological activities of sesquiterpene lactones. *Biochemical Systematics and Ecology* **14**: 255–281.

Popich, S.B., Alvarez Valdés, D. and Bardón, A. (2005). Toxicity og terpenoids from *Cyrtocymura cincta* (Asteraceae) on life cycle of *Spodoptera latifascia* (Lepidoptera: Noctuidae). *Boletín de Sanidad Vegetal. Plagas* **31**: 179–185 (In Spanish).

Primo Yufera, E. (1991). *Chemical Ecology: New Methods to Fight Against Insects.* Mundi-Prensa, España (In Spanish).

Rodríguez, E., Towers, G.H.N. and Mitchel, J. (1976). Biological activities of sesquiterpene lactones. *Phytochemistry* **15**: 1573–1580.

Rosselli, S., Maggio, A., Bellone, G., Formisano, C., Basile, A., Cicala, C., Alfieri, A., Mascolo, N. and Bruno, M. (2007). Antibacterial and anticoagulant activities of coumarins isolated from the flowers of *Magydaris tomentosa. Planta Medica* **73**: 103–116.

Seaman, F.C. (1982). Sesquiterpene lactones as taxonomic characters in the Asteraceae. *Botanical Review* **48**: 121–325.

Sharma, V.P., Ansari, M.A. and Razdan, R.K. (1993). Mosquito repellent action of neem (*Azadirachta indica*) oil. *Journal of the American Mosquito Control Association* **9**: 359–360.

Stejskal, V. and Kucerova, Z. (1996). The effect of grain size on the biology of *Sitophilus granarius* L. (Col., Curculionidae). I. Oviposition, distribution of eggs and adult emergence. *Journal of Applied Entomology* **120**: 143–146.

Stored Grain Pest Management. (2002).CD–ROM University of Illinois. Extension. College of Agricultural Consumer and Environmental Science.

CHAPTER 14

Bioassays on Plant – Nematode Interaction

SERGIO MOLINARI[1]

1. INTRODUCTION

Plant parasitic nematodes are microscopic soil-borne animals that attack and damage almost all crops. Worldwide plant parasitic nematodes annually cause 12.7 % losses in crops. Almost all nematode families are obligate root parasites, which develop from eggs to adults through four molts. Nematode families differ in morphology, life styles, parasitism, and host ranges. However, all known plant parasitic nematodes have a stylet in their mouth to pierce and suck the nutrients from plant cells. According to the degree of penetration of the nematode body inside the root, nematodes are divided into ecto- and endo-parasites. Nematodes that insert all or most of the body inside the root are called endoparasites, whilst ecto-parasites insert only their stylet inside the root, and live and reproduce in soil. If after egg hatching, all life-stages maintain the ability to move inside or outside the roots, nematodes are called migratory; conversely, if a mobile infective stage is followed by developmental immobile stages, nematodes are called sedentary.

Plant parasitic nematodes are divided into two orders: Tylenchida and Dorylaimida. All the genera belonging to Dorylaimida (*Xiphinema*, *Longidorus*, *Trichodorus,* and *Paratrichodorus*) are constituted by migratory ecto-parasites, which attack trees and herbaceous crops, and can severely damage the infested plants by transmitting plant pathogenic nepo- and tobra-viruses. Many economically important genera of migratory endoparasites (*Pratylenchus, Radophulus, Anguina,* and *Ditylenchus*) belong to Tylenchida: they damage plants because their movement and feeding inside the roots cause cell death and tissue necroses. However, the most important nematode families of this Order, either as models of plant–pathogen interaction or as crop pests, are constituted by sedentary endoparasitic nematodes (Heteroderidae, Nacobbidae). The family Heteroderidae includes the most diffused genera, such as *Globodera*, *Heterodera* and *Meloidogyne*. *Globodera* and *Heterodera* are cyst-forming nematodes; cyst-forming adult females, which protrude from the roots with the majority of their body, lay their

1. Institute of Plant Protection, National Council of Research, C.N.R., Via G. Amendola 122/D, 70126 Bari, Italy. E-mail: s.molinari@ba.ipp.cnr.it

eggs inside the body, and at the end of their life-cycle, have their external cuticle brown colored (from a whitish or yellowish color when they are alive) and hardened. These hardened, brown, and dead females are called cysts, which protect the eggs inside from the environmental challenges and are the means for spreading the infestation. These cysts can be collected either attached to infested roots or, more easily, dispersed in the soil. Conversely, females of *Meloidogyne* spp., generally known as root-knot nematodes (RKNs), result, at the end of their life-cycle, completely inserted into the roots and lay eggs in an external gelatinous matrix, which is clearly visible outside the roots. Moreover, nematode action induces hypertrophy and hyperplasia of the surrounding tissues, thus causing the formation of the familiar galls on roots of susceptible hosts. Sedentary endoparasitic nematodes do not kill parasitized cells and have evolved very specialized and complex relationships with their hosts. Cyst-forming nematodes induce the formation of feeding sites, called syncitia, in which few cells merge by dissolving their cell walls, whilst RKNs induce the formation of few discrete giant or nurse cells; both types of feeding site have the role of actively transferring solutes and nutrients toward the developing nematode. *Meloidogyne* species differ from most other sedentary endoparasites by having extensive host ranges, parasitizing more than 2000 plant species, although relatively few host–parasite relationships have been investigated. This chapter will focus mainly on bioassays which characterize the most studied interactions between such sedentary endoparasites and model hosts.

2. ISOLATION AND REPRODUCTION OF NEMATODES

2.1 Preparation of inoculum of root-knot nematodes (RKNs)

The infective stage of RKNs is the vermiform and motile second-stage juvenile (J2), which emerges from the egg after the first molt. The J2 penetrates roots and establishes a feeding site. After two molts they develop into adult females which lay hundreds of eggs in a gelatinous egg sac protruding outside the roots (Fig 1). The collection of living J2 from the infested plants is carried out by two main procedures: (i) isolation of eggs from plant roots and (ii) isolation of intact egg masses from roots.

Experiment 1: Isolation of nematode eggs from plant roots

Materials and equipments

Fresh nematode-infested roots, plastic buckets, plastic bags, plastic bottles, glass beakers, graduated cylinders, scissors, commercial bleach (4–5 % sodium hypochloride), wash room, timer, electric shaker, 500-, 200-, 45-mesh sieves, stereoscope, counting slide, stirrer, Eppendorf pipettes.

Reagents

Egg-extracting solution: mix 140 mL commercial bleach with 860 mL tap water (~ 0.56 % of sodium hypochloride) in a 1 L graduated cylinder.

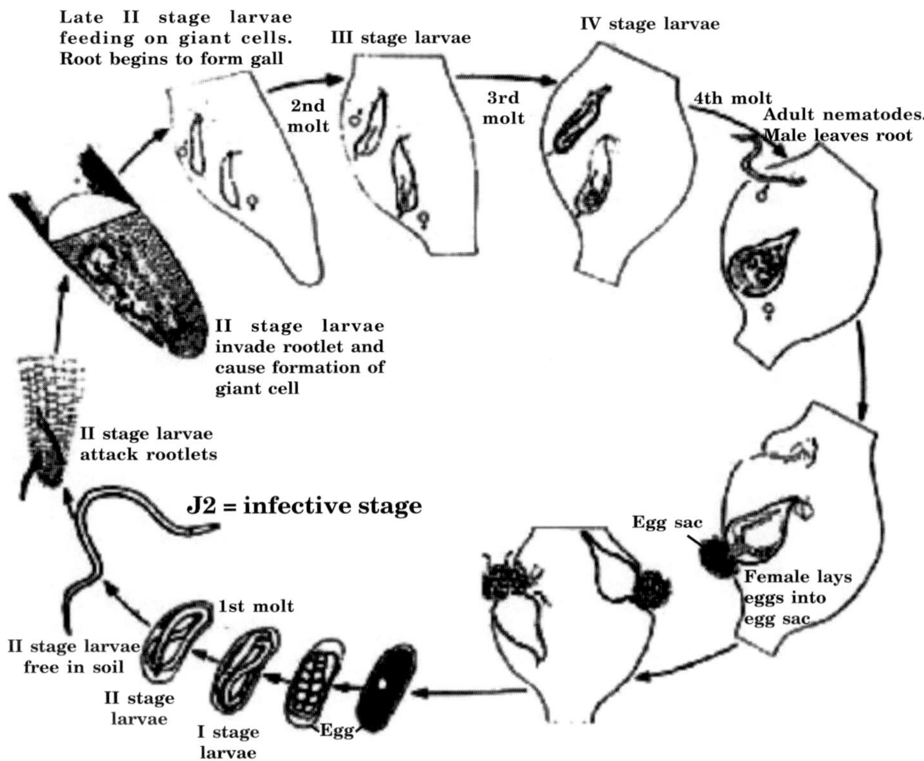

Fig 1. Life cycle of *Meloidogyne* spp.

Procedure

i. The infested roots must be set free from the soil. If the plants were grown in pots, the root system attached to the soil should be extracted from the medium as intact as possible. The root system should be immersed in water and gently moved allowing the soil to drop off. Dripping roots can be temporarily put into plastic bags and stored in the refrigerator to be processed later on.

ii. Roots must be dried and cut into small pieces by scissors.

iii. Dried roots should be weighed, in case the number of eggs per gram root fresh weight is to be calculated (see below).

iv. The minced roots must be put into plastic bottles of increasing volume according to the weight of roots.

v. The extraction of eggs from the gelatinous matrix of the egg masses is done by sodium hypochlorite. Hypochlorite is contained at 4–5 % in the easily available commercial bleach. According to the time interval after which the eggs will be used for inoculum, the percentage of bleach may vary. Ten percent bleach neither harm the amount of eggs extracted nor significantly decrease egg hatch over a period of 2 weeks. For a longer period of egg storing (in refrigerator), a solution of 140 mL bleach in 1 L of tap water may be used.

vi. Set a timer onto 4 min, pour into the bottles with the minced roots enough bleach solution to make them freely float in the bottles, which should be filled at half of their capacity and shake vigorously for 4 min. The weak bleach solution breaks down the gelatinous matrix surrounding the eggs thus releasing them from the root.

vii. According to the size of the root samples, the procedure may be more or less simple. The objective is to have suspensions of eggs with the lowest contamination of root debris as well as the shortest time of manual labor:

a. If samples are constituted by single root systems grown in small pots (~ 80 cm^3), roots (1–4g) are put into 100-mL plastic bottles. As a sieve apparatus, any broad screen over a small 500-mesh sieve can be used.

b. If samples are constituted by large root systems or groups of root systems (up to 40–50g), a pyramid of sieves should be set up as follows: at the top a 35–45-mesh sieve, in the middle a 170–200-mesh sieve and at the bottom a 500-mesh sieve. A spray nozzle is used to thoroughly rinse the top sieve on which bleached root suspension is poured, leaving root material to be discarded. The same procedure is done with the middle sieve until only the ground up root material appears to be left on the screen which is then discarded. The eggs are collected on the bottom sieve and poured into a glass beaker. If a large amount of water is used to wash the first two sieves, a bucket can be used to collect such water, which is then filtered through the 500-mesh sieve. The egg suspensions can be replaced into the plastic bottles which should be marked with the identity of the nematode population (species, geographical origin, isolate, date of harvest, date of egg extraction, and other relevant information).

The eggs will not be killed. They are not exposed to the bleach solution for more than 5 min.

viii. Egg suspensions should be brought to the same volume with tap water. Then, the number of eggs per milliliter is counted using a counting slide, which consists of a glass or plastic cell of 1-mL volume over which a grid is marked (a small plastic Petri dish with lines engraved on the base can be used). The suspension containing eggs is thoroughly mixed by stirring and then 1 mL of solution is pipetted onto the counting slide, and following the grid, the total number of eggs in 1 mL is counted under a stereoscope (25× magnification). This should be repeated three times and the average of three counts should be obtained. The system of egg counting and egg appearance under the stereoscope is shown in Fig 2.

ix. Egg suspensions must be diluted to have no more than 200 eggs in 1 mL.

x. If a degree of infestation is requested, the index of reproduction must be determined as number of eggs per gram root fresh weight or per root system.

Calculations

Index of reproduction may be expressed as:

$$(A \times B) / n \qquad \text{or} \qquad (A \times B) / g$$

where A is number of egg/mL (obtained from stereoscope analysis), B is total volume of suspension in mL (the volume may be detected directly on the egg-containing beaker or suspension may be poured onto a graduated cylinder), n is number of root systems from which eggs have been extracted, g is grams of fresh weight of roots from which eggs have been extracted.

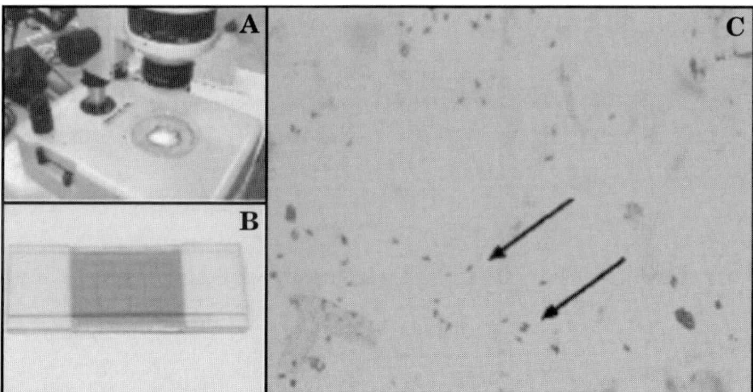

Fig 2. Counting of nematode eggs. (A) Stereoscope apparatus, (B) counting slide, and (C) arrows indicate eggs observed at 12× magnification.

Experiment 2: Isolation of intact egg masses from the roots

Materials and equipments

Fresh nematode-infested roots, scissors, forceps, Eosin Yellow (Sigma Co.), stereoscope.

Reagents

Solution of colorant: dissolve 0.1g Eosin Y in 1 L tap water, store at room temperature; physiological solution: dissolve 9g NaCl in 1 L of distilled water.

Procedure

i. Roots are cut by scissors into very small pieces, dried, and weighed.

ii. Samples of 1g can be prepared from the root systems and immersed in the colorant solution. The roots must be stored in the refrigerator for at least 1 h. The gelatinous egg masses are stained red and are easily recognizable outside the roots under a stereoscope (6× magnification).

iii. Egg masses can be separated from the infested roots by forceps, counted, collected and placed in physiological solution (0.9 % NaCl) for storage.

iv. Also in this case, the number of egg masses per gram or per root system can be determined.

Experiment 3: Hatching of infective juveniles (J2) from free eggs or egg masses

Materials and equipments

Glass beakers, small 500-mesh sieves, glass Petri dishes, growth chamber.

Procedure

i. Egg or egg mass suspensions are poured onto small 500 mesh sieves placed into glass Petri dishes (Fig 3). A layer of common lab paper (not filter paper) can be arranged on the sieve to retain soil and plant debris and allow only living juveniles to pass through the screen into the water reservoir. For the system considered, Petri dishes should be filled with about 35–40 mL of tap water, to establish a continuous exchange of water between the surface of the sieve and the water reservoir. A smaller system should be adopted, if a single egg mass is used for hatching (Fig 3).

ii. Petri dishes are placed into a growth chamber at 27–28 °C in darkness and eggs are incubated for 48 h. According to the time and quality of storage of eggs and egg masses, the speed and magnitude of hatching will greatly vary. Normally, two collections of juveniles carried out in 2

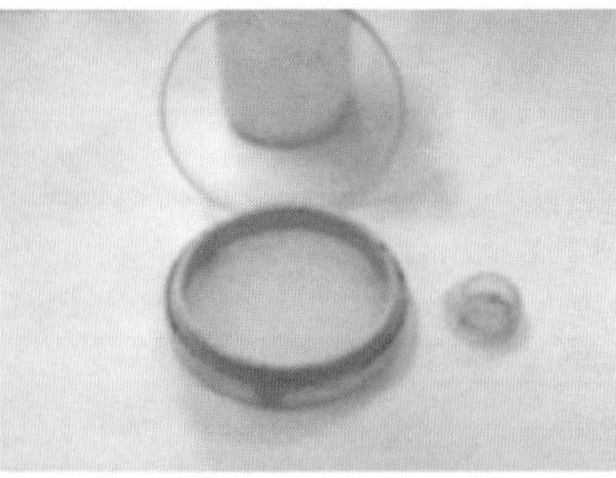

Fig 3. Filters utilized for J2 hatching. The small filter is for single egg mass hatching.

subsequent days will be enough to accumulate juveniles for suitable culturing of nematodes.

iii. Suspensions of freshly hatched living juveniles can be saved in glass beakers for little time in the refrigerator.

Calculations

The number of juveniles per milliliter is counted by the same procedure used for eggs (see *Calculations*, Experiment 1). Under the microscope, living juveniles appear vermiform, characterized by slow sinuous movements (Fig 4). Vacuolated immobile rod-like or C-shaped individuals are most probably dead specimens; fast and very fast movements are characteristic of saprophytes, which, obviously, must not be counted. Juvenile suspensions must be diluted to have not more than 200 individuals in 1 mL.

2.2. Culturing root-knot nematodes

For many non host-specific species, root-knot nematodes may be cultured on a susceptible variety of tomato (*e.g.,* "Moneymaker" or "Rutgers"). Host-specific species must be cultured on their particular host.

Experiment 4: Inoculation of tomato with root-knot nematodes (*Meloidogyne* spp.)

Materials and equipments

Seeds of tomato, temperature-controlled greenhouse, pots, sterilized soil, stirrer.

Reagents

Preparation of reagents for Hoagland's solution: dissolve 34g KH_2PO_4 in 250 mL distilled water (1 M), 50.5g KNO_3 in 500 mL d.w. (1 M), 118.1g $Ca(NO_3)_2$ in 500 mL d.w. (1 M), 61.6g $MgSO_4$ in 250 mL d.w. (1 M). Salt solution: per 100 mL distilled water dissolve 0.29g H_3BO_3, 0.18g $MnCl_2 \cdot 4H_2O$, 0.02g $ZnSO_4 \cdot 7H_2O$, 0.01g $CuSO_4 \cdot 5H_2O$, 2 mg $NH_4MoO_4 \cdot 2H_2O$. Preparation of 5 L Hoagland's solution: 5 mL KH_2PO_4 1 M, 25 mL KNO_3 1 M, 25 mL $Ca(NO_3)_2$ 1 M, 10 mL $MgSO_4$ 1 M, 5 mL salt solution, 5 mL $FeSO_4 \cdot 7H_2O$ 0.1 M. Use distilled water and set pH at 6.5 with KOH.

Procedure

i. Seeds of tomato are placed in tap water for some days to be imbibed. Then, they are sown in a sterilized medium. About 15 days after germination, seedling are transplanted in pots of suitable size filled with sterilized soil and are left to grow in pots until first leaf stage, or until their shoots are about 10 cm high. This allows growth of some roots prior to nematode invasion, to prevent rapid development of the parasite overwhelming the host. At this early growth stage, watering with Hoagland's solution is recommended.

ii. Inoculation of seedlings can be made with free eggs or living juveniles. Egg or J2 suspensions are poured onto the soil close to the plant stem by making two holes in the soil and then very lightly watering. Additionally, inoculum can be added to small wet pots just before the transplanting of well-grown seedlings. When a comparison of infestation degrees among different pots is required, inoculation should be done by a continuous stirring of inoculum suspension from which determined aliquots of volume are taken using pipettes. For 2–3 days following inoculation, plants should be only very lightly watered and subsequently waterlogging should be avoided.

iii. Use the free egg inoculum (1000–5000 eggs) only when the knowledge of the level of initial population is not necessary (*i.e.,* to reproduce populations). When a comparison of reproduction indexes of different nematode species, populations and isolates is required, inoculation should be carried out by the same number of freshly hatched living J2. Normally, 200–300 J2 are enough to have a marked and homogeneous infestation of susceptible tomato root systems without causing a premature death of the plants.

iv. Environmental conditions are very important to let the infested plants grow up to the end of root-knot nematode life cycle (fully developed egg masses). Temperature of soil in pots should be maintained at 24–26 °C. In these conditions, 40 days after inoculation are sufficient to produce a fully developed infestation of the roots.

Observations

Greenhouses provided with controlled environmental conditions are the most suitable. Nematode infested tomato plants are more sensitive to water stress than healthy plants. Therefore, watering should be consequently programmed. When several species or different populations and isolates are cultured together, contamination is prevented through splash barriers between pots. Avoid contact between the bottom of the pot and the bench by standing pots on supports.

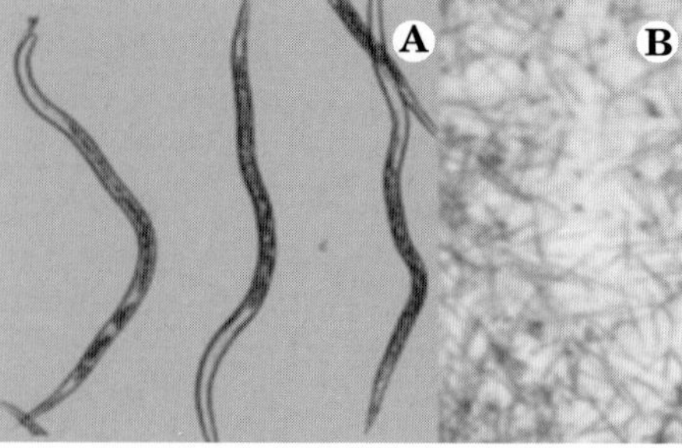

Fig 4. Infective juveniles of *Meloidogyne* spp. (A) at high magnification and (B) under a stereoscope.

2.3. Reproduction of root-knot nematodes in axenic cultures

This technique allows the reproduction and the maintenance of a high number of RKN populations and isolates without the necessity of having greenhouse facilities and all the devices required for growing plants. Moreover, studies on root–nematode interactions can be made without the interference of additional or contaminating microorganisms. All the steps are performed in strictly sterile conditions. The basic strategy is to build up an *in vitro* culture of roots, which are infested with sterilized nematode inoculum. Also in this case, susceptible tomato is commonly used.

Experiment 5: Inoculation of cultured tomato roots by root-knot nematodes

Materials and equipments

Seeds of tomato, commercial bleach, Tween 20 (Sigma Co.), Agar or Gelrite (Sigma Co.), Sigma Gamborg's B5 with minimal organics, Sigma Nitsch and Nitsch vitamin powder, Amphotericin B, Streptomycin sulphate, Clorexidine diacetate, deionized and sterile water, filtration apparatus, 0.22 µm filters, autoclave, plastic Petri dishes (9 cm diameter), laminar flow hood, stereoscope.

Reagents

Preparation of 1 L medium for root culture (B5 medium): in 800 mL distilled or deionized water add 3.3g Gamborg's B5 and 20g sucrose, set pH at 5.7, add 20g Gelrite and bring volume to 1000 mL, autoclave for 40 min, after autoclaving add 1 mL vitamin solution passed through 0.22 µm filter for sterilization.

Procedure

i. Sterilizing seeds for tissue culture:

   a. Wrap groups of 25 seeds in cheesecloth pieces of 10 cm^2.

   b. Place in a beaker.

   c. Prepare sterilizing solution (20 % bleach with 2 drops Tween 20 per 100 mL bleach solution).

   d. Immerse seeds and place in vacuum for 10 min.

   e. Rinse with autoclaved water three times.

   f. Prepare 1 % agar or Gelrite (Sigma, Co.): Slowly pour the powder in distilled or deionized water, while, stirring, fill the half of a flask with the solution, autoclave (121 °C, 1.05 kg cm^{-2}) for 40 min.

g. Proceed in a laminar flow at sterile conditions. Let the medium cool down at 40–45 °C and pour in Petri dishes (9 cm diameter).

h. Place the cheesecloth containing sterilized seeds on sterile Petri dish and unwrap seeds.

i. Place seeds (25–30) on the gels in Petri dishes and store in darkness at 27 °C.

j. Check daily for germination and contamination: If seeds are contaminated throw away entire plate.

ii. **Starting root culture:** Many chemicals, salts, and microelements are needed for *in vitro* growth of excised roots. Generally, pre-mixed preparations of these products are commercially available. For the present procedure, Sigma Gamborg's B5 with minimal organics can be used as basal medium associated with Sigma Nitsch and Nitsch vitamin powder as a vitamin support.

a. Take germinated seeds with emerging roots about 2-cm long and place on sterile Petri dishes.

b. Cut off and place roots on B5 media (4–5 per dish).

c. Let grow for 3–5 days at 27 °C in darkness.

d. Check all dishes for contamination.

iii. **Inoculation of roots with nematodes:** Nematode egg masses is the inoculum for *in vitro* infestation of cultured roots as their sterilization is easier and the toxicity of disinfectants is better supported. For sterilization, following methods may be used:

a. Clean carefully the egg masses excised from roots by stirring in sterile water changing several times the cleaning water.

b. Stir the egg masses in a solution of 0.05 mg mL^{-1} Amphotericin B (mycostatic) and 1 mg mL^{-1} streptomycin sulphate (antibiotic) for 1 h.

c. Rinse in sterile water.

d. Wash the egg masses with 1 % clorexidine diacetate for 15 min.

e. Rinse in sterile water.

Sterilized egg masses must be placed in a sterile Petri dish. Egg masses are placed close to cultured roots in sterile conditions (4–5 masses per plate) under a laminar flow hood. Incubate at 27 °C in darkness and check under a stereoscope for hatching of nematodes. Contaminated dishes must be discarded. Nematode reproduction *in vitro* is shown in Fig 5.

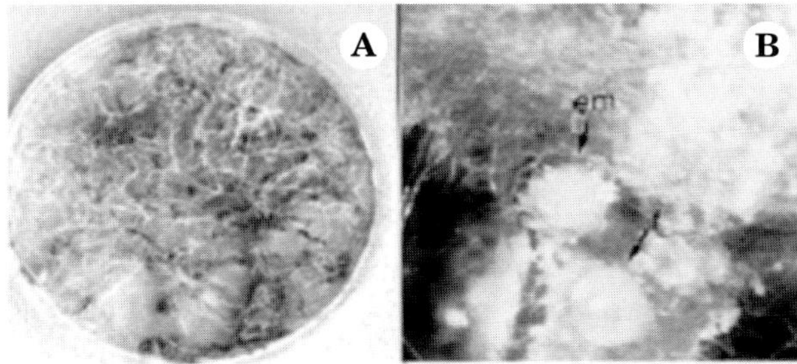

Fig 5. *Meloidogyne* spp. growing on roots cultured *in vitro*. (A) cultured roots heavily infested by RKNs, (B) a female (f) separated by its own egg sac (em).

2.4. Preparations of samples for biochemical experiments

Samples for biochemical experiments can be arranged from different life stages according to the families of nematodes which we are interested to investigate. For instance, when species identification of RKNs is to be carried out, the first step of the experiment consists in the isolation of females from infested roots. If cyst-forming nematodes are being studied, cysts must be isolated from soil. Also ectoparasites, as the virus vector family of Longidoridae, must be isolated as motile individuals from collected soil samples.

Experiment 6: Isolation of root-knot nematode females from roots

Heavily infested tomato roots are the most suitable starting material to have consistent amount of isolated females to be used for molecular identification experiments.

Materials and equipments

Plastic bottles, Pectinex® Ultra SP-L/Celluclast® (can be requested or purchased from Novo Nordisk, Denmark), NaCl, kaolin powder, sucrose, 25- and 60-mesh sieves, Eppendorf tubes, glass beaker, thermostatic water bath, electric blender, centrifuge, 500-mL centrifuge tubes, Vibromixer.

Reagents

Preparation of root-degrading solution: Pour 150 mL Pectinex® Ultra SP-L (15 %) and 350 mL Celluclast® in a 1 L cylinder, bring to volume with distilled water; preparation of the sucrose solution: dissolve 385g sucrose in 1 L distilled water [38.5 % (w/w) or 1.18 specific gravity], or use $ZnSO_4$/ $MgSO_4$ solutions of the same specific gravity.

Procedure

There are different methods to extract living females from the roots; the one suggested here is as under:

i. Chop up rinsed roots and place in a 1 L plastic bottle.

ii. To macerate roots, pour 500 mL of root-degrading solution in the bottle containing minced roots: note that roots should “float” in the medium. Use more bottles for more consistent root samples:

  a. Put the bottles in a water bath at 30–35 °C and let the sample under a gentle agitation for at least 2 h up to overnight.

  b. Pour the suspension on to a 25-mesh sieve nested with a 60-mesh sieve and wash. The resulting root debris can be collected from the top sieve and chopped in an electric blender. The chopping time required depends on the type of mixer used and needs to be continued long enough to allow the remaining females, which had not been removed by maceration, an easy egress from the roots but not to damage them. Cover the macerated roots with 0.9 % NaCl solution and grind for about 15s. Recover the females and place in a glass beaker.

i. If a limited number of females are to be used, females can be separated from root debris under a stereoscope (6× magnification) and saved in 0.9 % NaCl solution. Samples of 10–20 females can be distributed in Eppendorf tubes, solution drained off and females stored dry in the freezer or at –80 °C.

ii. If bulk isolation is needed, female suspension must be entirely set free from root debris. In this case, a centrifugal flotation is required:

  a. Pour the washings from the sieve procedure (point 4) into 500-mL centrifuge tubes and add 5 mL of kaolin powder. Thoroughly mix with a Vibromixer.

  b. Place the tubes in the centrifuge. Be sure to balance the tubes.

  c. Centrifuge at 1500g for 4 min.

  d. Decant water from the tubes: Nematodes are in the pellets together with root debris.

  e. Re-suspend the pellets in the sucrose solution with a Vibromixer for at least 30s.

  f. Balance and centrifuge at 1500g for 4 min.

  g. Supernatants are poured onto a 500-mesh sieve and purified females are collected in a beaker. Bulk samples of females can be stored as described in point (iii).

Precautions

Wear clothes when handling root-degrading products because they may be allergenic.

Experiment 7: Concentration of J2 to be used as biochemical samples

Samples of hundreds of thousands infective J2 should be used for biochemical applications. Such high number of J2 can be obtained by hatching, in tap water at 27 °C from egg masses manually collected on infested tomato roots, as described in Experiment 3.

Materials and equipments

Amphotericin B, streptomycin sulphate, filtration apparatus, 1.0–5.0 µm cellulose nitrate membrane filters (Whatman Ltd., Maidstone, UK), Eppendorf tubes, 1 mL syringes.

Procedure

i. It is necessary to thoroughly rinse the egg masses and set them free from any soil or root debris, to have J2 suspensions sufficiently clean to be used as biochemical samples. If the J2 suspension appears contaminated by a lot of junks, apply centrifugation flotation described in Experiment 6, point (iv). The drawback of this procedure is that the high osmotic pressure of sucrose solution can deform and even kill the living juvenile.

ii. After recovering from sucrose solution, J2 must be washed with distilled water many times. After the purification step, J2 must be concentrated. Use a filtration apparatus provided with 1.0–5.0 µm cellulose nitrate membrane filters (Whatman Ltd., Maidstone, UK). Such filters retain nematodes and allow their collection in minimal volumes of distilled water (0.5–1 mL) or directly of grinding medium. If homogenization can be done in distilled water (*i.e.*, DNA/RNA isolation) and smaller volumes are needed, nematodes placed into Eppendorf tubes can be centrifuged at maximum speed in a bench centrifuge. Excess of water can be carefully drained off by a syringe provided with a very thin needle. If nematodes are recovered from the filter with sucrose-containing media, they are not pelletable and the restriction of the extraction volumes not possible any more. In the case that contamination by microorganisms must be avoided, it is better to thoroughly wash nematodes with distilled water, surface sterilize by incubation for 1 h in a chilled solution of 5 µg mL^{-1} amphotericin B and 1 mg mL^{-1} streptomycin sulphate and finally rinse with sterilized water before placing them in the grinding buffer.

Experiment 8: Isolation of cysts from soil samples

Materials and equipments

Cyst-containing soil, Fenwick can, 60-mesh sieve, filter paper (18.5-cm diameter), stereoscope, glass Petri dishes, camel's hair brush.

Procedure

One of the most used methods require the equipment Fenwick, which proceeds as follows:

i. Mix the soil thoroughly.

ii. Fill the modified Fenwick can with water. Place the sample of 100 cm^3 of well-mixed soil in the top sieve (18- or 24-mesh sieve).

iii. Wash the sample into the apparatus via the funnel. The coarse material is retained on the top sieve, heavy particles sink to the bottom of the apparatus and the floating cysts are carried off over the overflow collar.

iv. Cysts are collected on a 60-mesh sieve together with soil and organic debris.

v. Wash and collect the debris on a filter paper (18.5-cm diameter) and let it dry at room temperature.

vi. Place the filter in a Petri dish and view it through a stereoscope with overhead light. Cysts appear as shown in Fig 6 (right). Pick up the cysts with a camel's hair brush (No. 00, 0, or 1).

vii. Collect 100–200 cysts per sample for biochemical application and put them in a plastic tube in water. Rinse thoroughly the samples with distilled water.

viii. Water can be drained off and cysts handled with a spatula for their transfer to grinding devices.

Fig 6. (A) Fenwick apparatus and (B) cysts as they can be observed under a stereoscope.

Experiment 9: Isolation of ectoparasites from soil samples

The economically most important ectoparasites belong to the virus-vector family of Longidoridae (Dorylaimida). The procedure will describe the isolation of the two main genera: *Longidorus* and *Xiphinema* spp. (Fig 7, left). To extract these long vermiform nematodes from soil, the Cobb wet-sieving technique is used. Most adults of large dorylaims are caught on a

60-mesh sieve, adults of average-size nematode on a 167-mesh sieve and many juveniles on a 325-mesh sieve.

Materials and equipments

Plastic buckets; 25-, 60-, 167-, 325-mesh sieves; plastic beakers.

Procedure

i. The soil sample (200 mL) is placed in a bucket and covered with water; any soil chunk should be gently broken.

ii. The mixture is vigorously stirred to suspend particles and after about 25s, the supernatant is poured through a 25-mesh sieve into another bucket; the sediment on the sieve is thoroughly rinsed: this operation is repeated twice.

iii. This procedure of filtration may be repeated with 60-, 167-, and 325-mesh sieves; use a gentle stream of water to wash the residue accumulated on the sieves into a glass beaker. Most individuals of longidorids are recovered on 167-mesh (90 μm) sieve.

iv. If the contents of the beakers appear cloudy, they should be poured back onto the corresponding sieve and rinsed again before proceeding.

v. To avoid clogging of the finest sieves with soil particles, pour the suspension on the sieve inclined at an angle of about 30° and, when the suspension has entirely been poured in, pat the underside of the sieve and left it in and out.

vi. Leave the suspensions in the beakers settle for 2 h. Gently pour off the supernatant leaving about 40 mL of residue. Mix the remaining suspension and pour onto a small 500-mesh placed in a Petri dish, as described in Experiment 3, point (i).

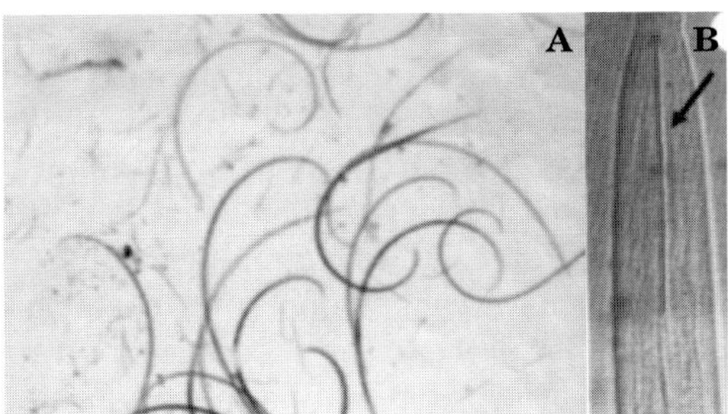

Fig 7. (A) Specimen of *Xiphinema index* observed under a stereoscope; in contrast, the much smaller size of *Meloidogyne* J2 is shown. (B) Arrow shows the stylet of a *Xiphinema* individual.

vii. Most nematodes pass through the filters in 24 h and can be collected, under a stereoscope, almost free from debris.

viii. An easy way to discriminate plant ectoparasites from the many contaminating saprophytes is to individualize the stylet characterizing the mouth apparatus of any plant parasitic nematode (Fig 7, right).

Observations

Proportion of nematodes is not retained when a suspension is poured even onto the finest sieve; therefore, it is advisable to pour the suspension many times and to collect the residue off the sieve at each time.

3. MOLECULAR IDENTIFICATION OF NEMATODES OF MAJOR INTEREST

The most common technique used for molecular identification of nematodes is protein electrophoresis. However, the major problem dealing with biochemical application of nematology is the scarce amount of proteins that can be recovered from the various nematode life stages. Furthermore, in some cases, such as the identification of the species or pathotypes forming a mixed population, there is need for identifying single individuals. Only adult living females of sedentary endoparasitic nematodes, where the economically important families *Meloidogyne* spp. (root-knot nematodes), and *Globodera* / *Heterodera* spp. (cyst nematodes) are found, are relatively rich in proteins and more specifically, enzymes that can be used for biochemical identification. Actually, the protein content of single RKN adult females has been measured in my laboratory as about 1 µg of *total proteins* (see below for this procedure). If we consider that most of the proteins of an endoparasitic female are structural and contribute to assemble the cuticle of the nematode, it is reasonable that the relative amount of the specific enzymes that we use for identification tests is very low. However, innovative miniaturized electrophoresis equipments that are commercially available since 10 years, allow the analysis of very tiny volumes of samples (1–4 µL), which can save at the most concentrated level the proteins extracted from nematode individuals. Single endoparasitic females can be tested by means of this technique. Few large adult females of the ectoparasite *Xiphinema index* can be ground to have proteins sufficient for obtaining visible band patterns on minigels. Conversely, eggs and endoparasitic vermiform invasive juveniles contain negligible amount of proteins; thus, the corresponding samples must be formed by a conspicuous number of individuals. In the procedures used in my laboratory, 100 cysts represent a suitable sample for miniaturized electrophoresis; conversely, only hundreds of thousands of invasive juveniles allow collecting enough proteins for any biochemical assays. On the other hand, when we deal with sedentary females and we are confident in their genetic homology (females belonging to a standard population or coming from individuals of a single egg mass), samples can be more easily formed by 10–20 components. In case, the identification is requested of a single cyst, or groups of single cysts, those single cysts must be used for single cyst inoculation of suitable hosts and the hatched juveniles should be allowed to reproduce forming a genetically homologous population of cysts.

Generally, the obtaining of full visible bands and the staining of all the bands available for the enzyme activities commonly used is guaranteed if the loaded samples contain an amount of total proteins as low as 8–10 µg. Thus, the request to operate with equipments at their lowest threshold of sensitivity for proteins leads to the problem of homogenizing such parasites in very small volumes of medium, or alternatively, to concentrate proteins coming from larger samples. In the next experiment, preparation of samples constituted by minimal volumes will be described.

Experiment 10: Preparation of nematode samples for miniaturized electrophoresis

Homogenization of nematodes should be done in appropriate media provided with pH buffers, osmotic regulators and possibly, proteinase inhibitors. For obtaining acceptable patterns of enzyme bands on the gels, some authors have proposed standard electrophoresis systems with very simple extraction media, such as a sucrose–Triton one constituted by 20 % (w/v) sucrose or glycerol, 2 % Triton X–100 and distilled water. The medium suggested in this experiment for operating with miniaturized electrophoresis procedure is a modification of the standard Tris–borate buffer: it is buffered at an alkaline pH to aid the dissolution of the protein mixture in the medium, contains sucrose to augment the density of the sample and proteinase inhibitors to slow protein degradation following cell disruption.

Materials and equipments

Trizma Base, Boric acid, EDTA–Na_2, bromophenol blue, phenylmethansulfonyl fluoride (PMSF), pepstatin, leupeptin, methanol, glass beads (425–600 µm in diameter), glass beakers, 2-mL glass potters, stirrer, filtration apparatus, 0.5 µm filters, Eppendorf tubes, Eppendorf-shaped, miniature homogenizers (Biomedix, UK), bench centrifuge, motorized drive, sonicator, Ultrafree®-MC centrifugal filter units Biomax-10 (Millipore Co., USA).

Reagents

The following products are dissolve in 100 mL distilled water: Trizma Base 12.1g (1 M), boric acid 5g (0.8 M), EDTA–Na_2 0.93g (25 mM). Proteinase inhibitor stock solutions: dissolve 10 mg PMSF in 1 mL methanol; 10 mg pepstatin in 10 ml methanol; 10 mg leupeptin in 10 mL distilled water. Extraction buffer (EB): in a 200 mL glass beaker add 10 mL of 1 M Trizma Base (0.1 M), 10 mL of 0.8 M boric acid (0.08 M), 10 mL of 25 mM EDTA–Na_2

(2.5 mM); add 20g sucrose (20 %; w/v) and dissolve by stirring; bring to 100 ml by distilled water; filter through a 0.5 μm filter. Final grinding buffer (fGB): take aliquots of 10 mL of EB and add 175 μL PMSF (1 mM), 7 μL pepstatin (1 μM), and 5 μL leupeptin (1 μM), from stock solutions; before homogenization, add 10 mg of the colorant bromophenol blue.

Procedure

i. Once we are ready to homogenize nematodes, we should transfer them in grinding devices specific for homogenization in 10–100 μL medium. Eppendorf tubes (1.5 mL) are the simplest tools to contain such minimal volumes. The problem is to arrange for pestles that perfectly fit the tubes and operate the crushing of nematodes. Models of these metal devices are shown in Fig 8, left. However, these pestles do not consent the proper handling of sample volumes as minimal as 10–20 μL that means a drop of liquid, which can be lost on the surface of the pestle. An innovative equipment has been proposed by Biomedix, UK, as Eppendorf-shaped, miniature homogenizers. Little plastic pestles fit into blunted-end Eppendorf-like tubes; a metal pole is inserted in the pestle for the connection to a motorized drive. The homogenization can be carried out leaving the grinding tube in ice (Fig 8, right).

ii. After homogenization, the pestle is maintained inside the tube to avoid loss of homogenate and discharged from the connector. Possible droplets of homogenate adhering on the pestles are collected at the bottom of the extraction tube by a centrifugation (12,000 rpm for few seconds is enough) in a bench centrifuge.

iii. Pestles are removed and the homogenate further centrifuged at 14,000g for 15 min at 4 °C. Supernatants are carefully drawn by a micropipette and placed in clean Eppendorfs; such samples are ready to be loaded on electrophoresis minigels.

iv. RKN females do not show any difficulty to be broken and chopped by plastic pestles, although part of many enzyme activities, which remain bound to nematode cuticle or heavy cellular organelles, are lost in such

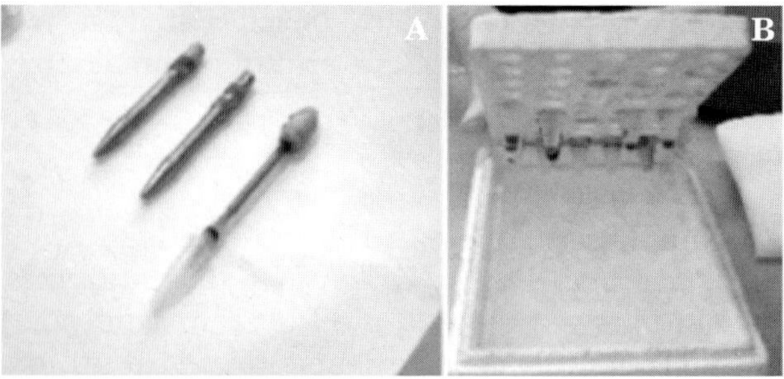

Fig 8. (A) Pestles and Eppendorf tubes to carry out homogenization in small volumes. (B) Device for maintaining samples in ice during homogenization.

a procedure. Cysts have a tougher surface and the contained eggs are difficult to be totally and completely broken: in this case, it is useful to add in the extraction mixture 0.1g of acid-washed glass beads (425–600 µm in diameter). The same addition is suggested when homogenization of the large longidorid vermiform individuals is to be done.

Homogenization of RKN J2

Suspensions coming from the concentration step (Experiment 7), which contains hundreds of thousands of specimens, cannot be smaller than 0.5–1 mL. Therefore, homogenization should be differently carried out:

i. Nematode suspensions are first homogenized, in the presence of glass beads, in a chilled 2-ml glass potter with a glass pestle connected to a motorized drive.

ii. Coarse homogenates are sonicated (for 5 min at 30 amplitude in a Vibra Cell, Sonics, Newtown CT USA) in a chilled water bath.

iii. Homogenates are centrifuged at 4000g for 15 min at 4 °C in a bench centrifuge.

iv. Supernatants are concentrated 4–5 fold in Ultrafree®-MC centrifugal filter units Biomax-10 (Millipore Co., USA).

v. 0.5 µg μL^{-1} bromophenol blue should be added just before runs.

Observations

A typical homogenization of 100–200 hundreds of thousands J2 gives a protein concentration of 2–3 mg mL^{-1}. Take a quote of 200 µL and concentrate 4-fold to have 20–30 µg of proteins in the loading samples (3 µL).

3.1. Miniaturized electrophoresis of nematode extracts

There are many thin-slab (0.6–1.5-mm thick gels) electrophoresis apparatus commercially available for studies on nematodes. However, preparing tank and gel buffers, casting polyacrylamide gels, and assembling an electrophoresis unit is a laborious task and requires specific expertise. Moreover, it may be dangerous for user safety because the powder of acrylamide is extremely toxic. Widely used procedures for electrophoresis of nematode extracts are fully described in Esbenshade and Triantaphyllou (An Advanced Treatise on *Meloidogyne,* Vol. 2, 1985) and the reader may refer to laboratory manuals of the firms that sell electrophoresis equipments and items for general procedures to set up an electrophoresis run. In the present experiment, the miniaturized and automated electrophoretic apparatus PhastSystem® (Amersham Biosciences, Piscataway, NJ, USA) (Fig 9) will be described, outlining its advantage with respect to previous techniques. PhastSystem is divided into two units, the first one is for

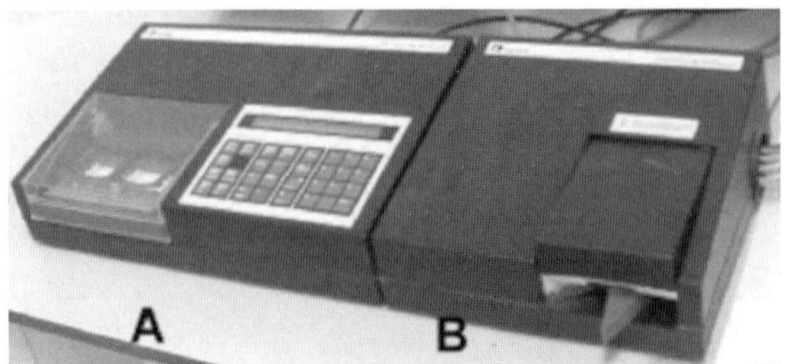

Fig 9. PhastSystem equipment for miniaturized and automated electrophoresis. (A) Electrophoresis unit and (B) Gel development unit.

electrophoresis runs (9A), the second one is for staining of the minigels (9B). As it concerns electrophoresis runs, it permits pre-programming of the chosen separation method, thus ensuring the exact reproduction of the method in each run by microprocessor control. There is no need to prepare or casting gels, as precast 0.45-mm-thick gels with a separation zone of only 3.8 × 3.3 cm (Fig 10) can be directly purchased. A 2000 V power supply and a cooled, thermostatically controlled, separation bed are built into give fields strength up to 500 V cm^{-1}, which allows high-speed and high-resolution separations.

As identification of nematodes is based on differential patterns of bands of enzyme activity, nondenaturizing conditions are required for maintenance of protein integrity. Different bands of enzyme activity reflect the polymorphism of different isozymes, which catalyze the same biochemical reaction but show differences in their shape, size, molecular weight, and net electric charge. The variations in shape, size, and molecular weight of isozymes can be detected by poly-acryl amide gel electrophoresis (PAGE) in which proteins move through a gel of acryl amide at the same (homogeneous) or different level of porosity (gradient). PAGE carried out in nondenaturizing conditions is named Native-PAGE. Five amino acids (glutamate, aspartate, arginine, lysine, and hystidine) carry ionizable lateral chains conferring an electric charge to the protein which depends on the pH of the medium. To detect this kind of variation, proteins can be separated according to their charge by the electrophoretic procedure of isoelectrofocusing (IEF), by which a determined pH gradient is established on the gel: after loading and application of an electric field, proteins move on the gel according to their charge and stop at that pH at which their net charge is zero: that point is called isoelectric point (p*I*).

Native-PAGE is carried out by using precast buffer strips made of 3 % agarose, with the buffer system consisting of 0.88 M l-alanine and 0.25 M Tris, pH 8.8. Homogenous and gradient polyacrylamide minigels show a 4.5 % acrylamide stacking zone and a separation zone with 12 % acrylamide or a gradient between 8 and 25 % with 2 % crosslinking, respectively. Gels are buffered with 0.112 M acetate and 0.112 M Tris, pH 6.4. The method consists of a pre-run of 10 Vh (volt hours) with the current set at 10 mM to initiate the moving boundary of the leading and trailing ions (acetate/l-alanine), a second step during which the samples (1–4 µL) are applied, while, the current is reduced to 1.0 mA, and a third step in which the current is raised again to 10 mM and the proteins are separated according to their size and charge. The separation bed is maintained at 15 °C throughout the run. Run is stopped when the marker dye reaches the bottom of the separating gel, which in normal conditions occurs about 30 min after the loading of samples. Therefore, runs are highly reproducible and quick. The IEF described in this procedure uses 5 % polyacrylamide homogenous minigels containing Pharmalyte® carrier ampholytes (pH 3–9). The separation method consists of three steps: a pre-focusing step, a sample application step and a focusing step. The first and last steps are done with the voltage set at 2000 V, but it was set at 200 V during the second step to avoid streaking caused by contaminants or poorly soluble proteins. Samples are applied in the middle position of gels. Gels are maintained at 15 °C and runs stopped at 500 Vh, which correspond to about 30 min.

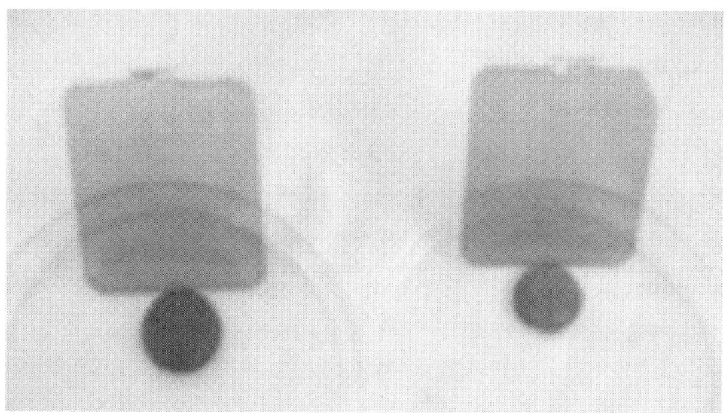

Fig 10. Pre-cast mini gels used with Phast System equipment

Experiment 11: Obtaining isozyme electrophoresis phenotypes of nematodes for identification of RKN species

Each genetically homogenous nematode population and possibly geographical isolate and pathotype – show a specific band pattern of a determined enzyme activity; such a band pattern characterizes a determined phenotype. Polymorphism of isozyme electrophoresis phenotypes (IEPs) is related to the difference existing at gene locus level among different nematode populations. Encodes of genes, after post-transcriptional arrangements, constitute the different isozymes which are detected on the patterns. The number and frequency of the different alleles which constitute a single or multiple genetic loci determine the level of polymorphism of a genus species and generally, a genetic group. A high number and low frequency of alleles correspond to a high polymorphism and major chances to discriminate among phenotypes of nematode populations. IEPs can be applied at species level for crop rotation, for diagnosis of Longidorid virus-vector species, advisory and quarantine services; at pathotype level for use of proper resistant cultivars; at geographical isolate level for determination of geographic origin and migration of nematode species. In the present experiment, only one major application of IEPs will be reported.

Materials and equipments

K-Phosphate buffers, α-naphtyl-acetate, β-naphtyl-acetate, acetone, Fast Blue RR Salt, glycerol, acetic acid, sodium carbonate, l-malic acid, Tris–HCl, NAD, nitroblutetrazolium (NBT), phenazine methasulphate (PMS), riboflavin, TEMED, filtration apparatus, 0.5 µm filters, PhastSystem® (Amersham Biosciences, Piscataway, NJ, USA), homogenous polyacrylamide (12.5 %) and gradient polyacrylamide (8–25 %) minigels (Amersham Biosciences, Piscataway, NJ, USA), pH meter.

Reagents

EST-staining solution: in 100 mL of 0.1 M K-phosphate buffer, pH 6.0, while stirring, add dropwise 40 mg α-naphtyl-acetate and 40 mg β-naphtyl-acetate dissolved in 2 mL acetone and 100 mg Fast Blue RR Salt. Preparation of stock solutions for MDH-staining solution: A = sodium carbonate 10.6g and l-malic acid 1.3g; B = Tris-HCl 6.1g dissolved in 100 mL distilled water, pH 7.1. MDH-staining solution: stock solution A 20 mL, stock solution B 30 mL, NAD 15 mg, NBT 10 mg, PMS 2 mg, distilled water up to 100 mL. SOD-staining solutions: 1. Reagent solution: dissolve 0.15g NBT in 500 ml distilled water (0.3 mg ml^{-1}); this solution can be stored in the refrigerator; 2. Colorant solution: dissolve 1 mg (a spot of powder) riboflavin (0.26 mM) and 200 µL TEMED (15 mM) in 50 mL K-phosphate buffer 0.1 M, pH 7.8; this solution must be used fresh and cannot be stored. Gel-preserving solution: 10 % glycerol + 10 % acetic acid.

Procedure

To differentiate among the most diffused tropical species *M. incognita*, *M. javanica*, and *M. arenaria*, band patterns obtained by the sole esterase (EST) activity are sufficient. If additional species are being investigated, IEPs of malate dehydrogenase (MDH) should be confronted to EST ones. If we want to have information about the intra-specific variability of a group of populations, superoxide dismutase (SOD) IEPs should be obtained.

Samples used: Adult, freshly collected females

Enzymes used: EST, MDH, SOD

Separation method: Native-PAGE on homogenous polyacrylamide gels (12.5 %)

EST staining:

i. Vacuum filter (0.5 µm filter) the staining solution to remove insoluble material. Use immediately after the preparation. Put the minigel in the development unit of PhastSystem.

ii. Let the gel stain at 37 °C for about 20 min. PhastSystem Development Unit makes the gel rotate to have a better staining.

iii. Extract the gel from the unit, wash with distilled water and put in the preserving solution (10 % glycerol + 10 % acetic acid) for 15 min.

iv. Analyze the band patterns corresponding to loaded samples that will appear black over a yellowish background.

IEPs are shown for identification of the most diffused species of *Meloidogyne* in Fig 11A. Fig 11B shows a typical minigel stained for EST.

Separation method: Native-PAGE on homogenous polyacrylamide gels (12.5 %)

MDH staining:

i. Put the minigel in the development unit of PhastSystem.

ii. Put the gel in the staining solution at 30 °C for about 20 min. PhastSystem Development Unit makes the gel rotate to have a better staining.

iii. Extract the gel from the unit, wash with distilled water and put in the preserving solution for 15 min.

iv. Analyze the band patterns corresponding to loaded samples that will appear blue over a white background.

MDH IEPs are specifically applied to distinguish *M. hapla* from the other most diffused species in that the unique band of *M. hapla* is much faster migrating than the corresponding bands of *M. incognita*, *M. javanica*, and *M. arenaria*. IEPs of EST and SOD are, conversely, barely distinguishable between *M. hapla* and *M. incognita*.

Separation method: Native-PAGE on gradient polyacrylamide gels (8–25 %). Gradient gels are used in this case because fast migrating SOD bands are numerous and very close to one another. Polyacrylamide gradients allow a high resolution and an adequate thinness of protein bands.

SOD staining:

i. Immerse the minigels in about 50 mL of NBT solution and incubate at 37 °C for 5 min.

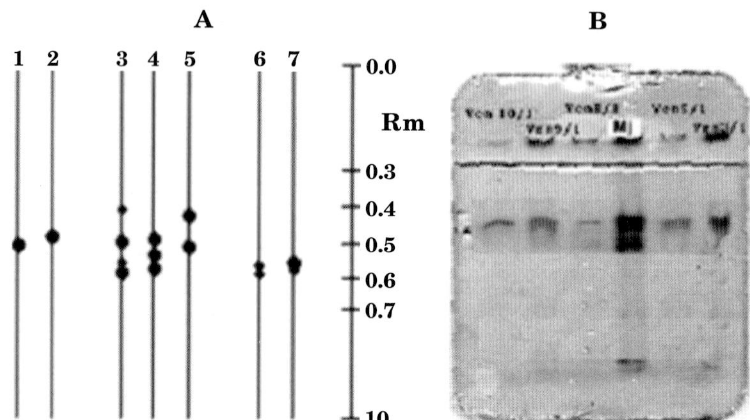

Fig 11. (A) Esterase phenotypes represented as isozyme electrophoresis patterns: 1. *M. hapla*; 2. *M. incognita*; 3–4. *M. javanica*; 5. *M. mayaguensis*; 6–7. *M. arenaria*; (B) Minigel loaded with female extracts from different *Meloidogyne* populations and stained for esterase.

ii. Rinse the minigels with distilled water and immerse in the colorant solution at room temperature and possibly, over a white light illuminator.

iii. Discolored bands will appear over a blue background.

iv. Put the minigels in the preserving solution for 15 min.

Minigels must be dried and saved as such for records. On the other hand, they can be directly scanned and arranged as digital images by computer's software.

4. A MODEL OF PLANT – NEMATODE INTERACTION: *MELOIDOGYNE* SPP. ATTACKING TOMATO

Reproduction of sedentary endoparasitic nematodes in the attacked plants depends on the successful initiation and establishment of highly specialized and intimate feeding relationships with their hosts. RKNs enter the roots as motile second stage juveniles (J2) and migrate intercellularly to the vascular cylinder, where they start to feed on living cells of the differentiation zone. In resistant plants, localized cell death of the root tissue surrounding the nematode occurs, thus avoiding the juvenile to develop into the enlarged, egg-laying adult female. The first nematode resistance gene to be cloned was $Hs1^{pro\text{-}1}$, a gene from a wild relative of sugar beet that confers resistance against *Heterodera schachtii*, the beet cyst nematode. *Gpa2*, a gene that confers resistance against some isolates of potato cyst nematode *Globodera pallida*, was more recently cloned as well. However, the best-studied of the cloned genes is the tomato gene *Mi*, which confers resistance against three species of root-knot nematodes. *Mi* is also active against some isolates of the potato aphid *Macrosiphum euphorbiae* and the white fly *Bemisia tabaci*.

According to the gene-for-gene theory, each dominant resistance gene in the plant is usually associated with a recessive avirulence gene in the pathogen. The encodes of such avirulence genes, directly or by means of "elicitors of resistance" derived from their action, are generally required for "recognition" of the pathogen; recognition leads to hypersensitive plant reaction and expression of resistance. Absence of avirulence genes causes loss of recognition and the chance for pathogens to induce the relative pathogenesis in plant tissues. The ability of some pathogens to develop on resistant plants is known as "virulence". To date, no avirulence genes have been conclusively isolated from nematodes. However, although most of the populations of the root-knot species, against which *Mi* is effective, induce a hypersensitive response, a high number of field populations have been isolated which break resistance. Besides these "natural" virulent populations often isolated from agricultural areas, which had never experienced *Mi*-carrying tomato crops, virulent isolates can be selected in greenhouse by repeated inoculation on *Mi*-carrying tomato, starting from wild standard populations.

Actually, the extremely high genetic variability of tomato germplasm and root-knot nematode populations, the marked effects on nematode–plant interaction of the environmental conditions, soil type, water regime and inoculation procedures make the quantification of the degree of infestation mandatory for attesting which kind of plant response we are dealing with when the interaction between specific tomato cultivars and nematode populations or isolates is investigated. More generally, plant breeding for nematode resistance must rely on an accurate rating of nematode–host suitability. Moreover, the effectiveness of a product, as it concerns its ability to "elicit" resistance or its nematocidal activity, can be assayed only after that a standard index of reproduction of nematodes on plants is established. The bioassays described in the next section are valid not only for *Meloidogyne*–tomato interaction but also for *Meloidogyne* spp. attacking most herbaceous hosts.

4.1. Rating of nematode–host suitability

An accurate rating of nematode–host suitability requires a strict standardization of experimental conditions. Experiments should be done in greenhouses provided with facilities to control the environmental conditions. Plants should be singly inoculated in pots, filled with soils of standard composition. The age of seedlings at inoculation should be fixed. The amount of inoculum should be evenly distributed to pots. Freshly hatched juveniles must be preferred to eggs because the rate of hatching can markedly vary according to a number of unpredictable factors. It should be noted that 50 J2/500 mL of soil is generally recognized as the parasite density able to cause the maximum damage to plants. However, the number of J2 needed for having a uniform diffuse infestation of the roots should be assayed in each plant–nematode interaction, according to the specific experimental conditions. Screening techniques must be devised to optimize the amount of inoculum for differentiating plant response among genotypes. Even the most resistant plant can be stressed if it is attacked by a large number of nematodes.

Generally, the evaluation of host suitability to *Meloidogyne* spp. has been based on root galling and nematode reproduction. Actually, the assessment of nematode reproduction seems more accurate as it can rely on the detectable numbers of eggs and/or egg masses per root system (for the techniques to measure such numbers see section 2.1). An index of egg masses (EI) has been associated with the degree of resistance (DR) of plants. Plants with a root system infested by up to 30 egg masses were considered showing various degrees of resistance, from high to slight resistance. Plants with roots affected by over 30 egg masses were considered susceptible. However, a method based only on egg mass counting does not adequately measure nematode reproduction as it does not quantify the eggs produced. The egg masses included in the counting can greatly vary in size and in number of contained eggs; the stain is active on gelatinous matrices which, sometimes, can even be empty or contain a negligible amount of eggs. It should be noted that only a fraction of the juveniles that establish a functional feeding site develops into mature females, and a fraction of mature females reproduces at their highest level of fecundity, in the limits of experimental time. Consequently, egg mass index gives more an idea of the intensity of nematode attack rather than the degree of reproduction of the inoculum. More juveniles are inoculated on a plant, higher is the competition for food to which those juveniles and developing adults are going to be subjected. Increase in competition leads to a smaller fraction of initial inoculum which is able to fully reproduce at the end of experimental time. Index of reproduction (IR) should be evaluated by the number of eggs that can be extracted from a root system.

If we establish the following standard system:

i. **Nematode populations:** Near-isogenic (a)-virulent clones selected from a standard population of *M. incognita.*

ii. **Amount of inoculum and method of inoculation:** 250 J2, inoculation carried out 1 week after transplanting.

iii. **Tomato cultivars:** Susceptible Roma VF, resistant Rossol.

iv. **Age of seedlings at inoculation:** 1 month.

v. **Pots:** Small 100 mL soil capable pots.

vi. **Temperature of soil:** 22–26 °C.

vii. **Harvest:** 2 months after inoculation.

Table 1. Egg masses and eggs per root system measured at the end of nematode cycle in different types of *Meloidogyne*–tomato interaction

***Meloidogyne*-tomato interaction**	**Eggs (range)**	**Egg masses (range)**
Avirulent J2/susceptible tomato	32000 – 46000	100 – 150
Avirulent J2/resistant tomato	0 – 3000	0 – 4
Virulent J2/resistant tomato	40000 – 50000	60 – 140
Virulent J2/susceptible tomato	23000 – 50000	70 – 160

The number of egg masses and eggs extracted per root system, according to different *Meloidogyne*–tomato interactions, are shown in Table 1.

If any of the standard factors previously appointed are changed, values shown in Table 1 may markedly differ. For instance, homozygous and heterozygous genotypes of tomato for the *Mi* gene influence the result of reproduction of virulent populations, thus suggesting a dosage effect of the gene. Moreover, a large span of reproduction degrees can be observed when virulent nematodes attack tomato cultivars showing the same resistant genotype. *M. mayaguensis*, which is spread in the Caribbean area, is considered a virulent species on *Mi*-carrying tomato. When, in my laboratory, Rossol tomato cultivar, which is heterozygous for the *Mi* gene, was repeatedly inoculated with 250 J2 of a Venezuelan population of this species, not more than 40 egg masses per root were detected. The same relatively low reproduction has been reported in literature for the interaction of Rossol cv. and another virulent species, *M. hapla*. Therefore, it is better to talk about increasing levels of virulence rather than fully avirulent and fully virulent populations or species. Practically, once established a standard system, interactions that produce more than 5000 eggs per root may generally be considered compatible, while interactions with less eggs produced may be indicative of some incompatibility between nematodes and plants. Obviously, as a control, an interaction that should produce full compatibility must be added to a screening as, sometimes, poor infestation can be the result of unfeasible environmental conditions or poor viability of inoculum, or other unpredictable problems, and not of plant resistance. Finally, attention must be paid to avoid contamination of the cultured nematode populations: tests on their initial characteristics should be regularly carried out.

5. BIOASSAYS ON NEMATODE CONTROL POTENTIAL OF EXOGENOUSLY ADDED ELICITORS OF RESISTANCE

All the chemicals used in practical agronomic procedures to control nematodes are toxic to different life stages of the parasites and are added to soil to kill the highest number of infective individuals. Organic nematicides are, among the pesticides commonly used in agriculture, the most dangerous to environment and risky for human and animal health. Residues of these chemicals represent a serious hazard for food safety, other than contributing to the general environmental pollution. Most countries and international organizations are addressing their agriculture policy toward an increasing banning of the use of such chemicals from agronomic practices. Consequently, the demand for research on alternative products, which may have a lower environmental impact as well as contribute to limit economical losses due to these parasites worldwide, is rapidly increasing.

One of the most promising researches in this field focuses on physiological metabolites involved as mediators in plant reactions due to genetic resistance or other naturally induced resistances, such as the most known Systemic Acquired Resistance (SAR). Necrosis-inducing pathogens, including nematodes, induce a hypersensitive response (HR) in incompatible plant tissues with the formation of necrotic areas, which, in the case of nematode attack, impede the formation of feeding sites and force nematodes to starve to death or leave the roots. It has been observed that after the formation of a necrotic lesion due to pathogens attacking leaves, either as a part of the hypersensitive response or a symptom of disease, the SAR is activated which results in the development of a broad-spectrum, systemic resistance. An interesting approach in disease control is the use of compounds able to activate the SAR response in plants. The physiological compound which mediates SAR in plants is salicylic acid, which, exogenously provided, has been proved to highly reduce damages provoked by some pathogens attacking leaves, such as fungi, bacteria, viruses. One of the analogues of salicylic acid, acilbenzolar-*S*-methyl (CGA 245704), has been commercially developed as Bion by Sygenta Co. On the contrary, research on the SAR response and external elicitors of resistance on root diseases, such as nematode infestation, is scarce.

In my laboratory, a rigorous study has been started on the effect of treatments with salicylic acid and its analogues on the fitness of plants as well as on nematode control. Considering that every compound, even those naturally produced by plants, has a toxic effect if provided at non-physiological dosages, the crucial issue of this study is to find the right balance between negative outcomes on plant fitness and acceptable control effectiveness.

5.1. Treatments with putative elicitors of resistance

Treatments must be carried out on inoculated and uninoculated plants; experiments must include negative controls, that is, untreated inoculated and uninoculated plants. Different dosages of the products must be tested. The following methods of application must be tried:

i. **Root dip in buffered solutions of the product for determined time intervals:** in this case, the uptake of the product by roots is direct; the dosages must be calculated as microgram of product per gram of fresh weight of adsorbing plant (ppm).

ii. **Soil-drench with product solutions:** in this case, root uptake is mediated by the interaction of the product with soil; the dosages must be calculated as microgram of product per gram of soil (ppm).

iii. **Foliar spray:** in this case, the availability of the product in the zone of pathogenesis is due to leaf absorption and transport to roots (or indirect systemic effect); the dosages must be calculated as microgram of product per gram of sprayed leaves.

5.2. Assays of plant fitness parameters

The quantitative parameters for testing plant fitness proposed here are the following:

i. Length of shoots and roots

ii. Number of branches or leaves

iii. Weight of shoots and roots

iv. Milligram of proteins per gram of fresh weight

Experiment 12: Detection of protein content by the enhanced alkaline (Lowry) protein assay

The enhanced alkaline copper (Lowry) protein assay is a quick method for measuring protein content in plant tissues. It detects proteins on the basis of their content in amino-acids provided with a phenolic residue. Consequently, protein content may be overestimated by the presence in the samples of free and bound phenols. Root extracts are particularly rich in phenols; therefore, it is suggested to do a previous dialysis or ultrafiltration of the samples to get rid of the low molecular weight free phenols. Phenols bound to particulate fractions may be discarded by a centrifugation at 9000g for 10–15 min. If there is an interest to detect the phenol content, the assay should be performed before and after dialysis or ultrafiltration.

Materials and equipments

Na_2CO_3, $CuSO_4{\cdot}5H_2O$, sodium potassium tartrate, sodium dodecyl sulfate (SDS), NaOH, Folin reagent, vortex, Eppendorf tubes, bovine serum albumin, spectrophotometer with a lamp for visible light.

Reagents

2× Lowry concentrate solution: Dissolve 20g Na_2CO_3 in 260 mL water, 0.4g $CuSO_4{\cdot}5H_2O$ in 20 mL water, and 0.2g sodium potassium tartrate in 20 mL water. Mix the solution to form the copper reagent; 1 % SDS solution: dissolve 1 g of SDS in 100 mL of water; 1 M NaOH: dissolve 4g NaOH in 100 mL of distilled water; basic reagent: immediately before use, mix 3 parts copper reagent with 1 part SDS and 1 part NaOH. This reagent is stable for

2–3 weeks: if a white precipitate forms warm the solution to 37 °C. If precipitate persists discard the solution; 0.2 N Folin reagent: mix 10 mL of 2 N Folin reagent with 90 mL of distilled water. This solution is stable for several months at room temperature if stored in an amber bottle.

Procedure

i. To a 400 µL sample add 400 µL 2× Lowry concentrate and incubate at room temperature for 10 min (if concentration of proteins in the samples is too high, take aliquots of samples and bring to 400 µL with protein buffer).

ii. Add 200 µL of the 0.2 N Folin reagent and vortex immediately after each addition. This rapid mixing is important since the reagent decomposes rapidly. Incubate 30 min at room temperature.

iii. Prepare standard protein solutions (bovine serum albumin is the most common standard) in the range 10–100 µg.

iv. Operate with a spectrophotometer provided with a visible lamp the absorbance at 750 nm or 500 nm; prepare a blank with each component except proteins, read its absorbance and, if present, activate autozero, or mark it up.

v. Prepare at least three repeats of standard and sample test solutions.

vi. Read the absorbance of the standard solutions at 750 and 500 nm and plot the standard curves.

vii. Read the absorbance of samples at 750 nm or, at 500 nm, if it is too high.

Calculations

Calculate the content in microgram protein from the equation of the standard curve [$A_{750} = m \times$ protein (µg)]:

µg protein (in 400 µL or in the aliquot chosen) = A_{750} / m

concentration of protein in mg mL^{-1} = µg protein (400 µL) × 2.5 / 1000

5.3. Assays of nematode development and reproduction

The effectiveness of a product in affecting nematode attack can be measured by comparing the following three values in untreated and treated inoculated plants:

i. Number of egg masses per root system

ii. Number of females per root system

iii. Number of eggs per root system

The first two measurements indicate the degree of infestation and the number of juveniles that have been able to develop into adult stages. A significant decrease in such values due to treatments with a natural product would indicate an interference of such a product, directly or through its metabolism in the plant, with the mechanisms by which juveniles establish functional feeding sites, or once established them, they rearrange plant gene expression to direct the whole metabolism toward a continuous sinking of nutrients, resulting in their development into fertile adult stages. The third amount is actually an index of the reproduction of the inoculum. A consistent decrease of such an amount in treated compared with untreated plants may practically be considered indicative of increased plant resistance and induce major investigations on the possible benefits of the product application in agricultural practices. It should be noted that the variations in the amount of egg masses and eggs may not be proportional. For instance, a decrease in egg mass amount, as far as it is commonly detected, in treated plants with respect to untreated controls, may match with an unchanged, or even augmented, amount of eggs. This is because the number of eggs contained in the gelatinous matrices stained by the colorant may greatly vary. An elicitor of resistance may act against a diffuse initial infestation by the juveniles, thus reducing developing individuals. The higher availability of food and the lesser competition to which these “selected” individuals are subjected may increase their fertility over the standards shown in control inoculations. This is particularly true when heavy inoculation is carried out. The ultimate target of this investigation is to find a suitable dosage of a natural product that, when provided to plants before inoculation, may be able to significantly and consistently decrease nematode reproduction without affecting plant fitness. Subsequent investigations should focus on testing, under the far more complicated growing crop conditions, such as a naturally infested field, the validity of the product(s) on crop yield. However, economic considerations should always be associated with the realization that such a pest management is fully sustainable and in agreement with the most up-to-date agricultural policies.

6. BIOCHEMICAL MARKERS OF RESISTANCE

Important biochemical markers of resistance may be represented by specific and reproducible changes in enzyme activity due to incompatible plant–nematode reaction, or levels of enzyme activity and derived metabolites, which can significantly differentiate between resistant and susceptible cultivars. The obtaining of such suitable biochemical markers for resistance relies on a sound understanding of pathogen biology, normal and deviate host–pathogen interactions, recognition process between plants and pathogens and bio-regulation of plants and pathogens. For instance, inhibition of catalase activity in the very first stages of *Meloidogyne*–tomato interaction is a specific characteristic of resistant cultivars attacked by avirulent populations of the parasites. Also peroxidases and many other enzymes have generally been involved in resistance response of plants to RKNs. In this section, the assay for detection of catalase activity in roots will be described. Assays for other key enzymes can be found in the literature (Molinari, 1995, 1999).

Experiment 13: Homogenization of roots

It is important to consider the amount of roots to be homogenated. The method described below refers to large quantities of roots (> 20g). All steps must be carried out on ice.

Materials and equipments

K-Phosphate buffer, glycerol, dithiothreitol, EDTA, ammonium sulphate, Polytron® PT–10–35 (Kinematica GmbH, Switzerland) or glass potters, centrifuge.

Reagents

Preparation of grinding buffer: 0.1 M, pH 7. Add 10 % glycerol, 1 mM dithiothreitol, as antioxidant, 0.1 mM EDTA and, if available, the protease inhibitors PMSF (1 mM), pepstatin (1 mM) and leupeptin (1 mM).

Procedure

i. Roots are weighed, placed in porcelain mortars and reduced to powder through immersion in liquid nitrogen.

ii. Powder is suspended in grinding buffer (1:5; w/v).

iii. Suspensions are ground further by a Polytron® PT–10–35 (Kinematica GmbH, Switzerland); if this kind of equipment is not available, suspensions can be ground in glass potters.

iv. Coarse homogenates are filtered through four layers of gauze and centrifuged at 9000g for 15 min.

v. Protein content is evaluated (see Experiment 11).

vi. Extracts as such can be used for assays of enzyme activity. However, dialysis or ultrafiltration of the extracts is useful to get rid of free phenols which lead to an overestimation of protein content and may interfere with some enzyme activity. Although, it should be noted that, for instance, dialysis causes a marked reduction of catalase activity.

vii. Further purification is possible through ammonium sulphate precipitation and chromatography.

Experiment 14: Assay of root enzymes. Catalase activity

Materials and equipments

Root extracts, Na–phosphate buffer, H_2O_2.

Procedure

i. Catalase activity of root extracts is measured as the initial rate of disappearance of hydrogen peroxide.

ii. Reaction mixture (0.5 mL final volume) consists of 0.1 M Na–phosphate, pH 7.0 and 20 mM H_2O_2.

iii. About 0.1 mg proteins should be added from crude extracts; about 0.01 mg proteins should be added from ultrafiltered extracts.

iv. Blanks must be prepared with all the reagents except H_2O_2.

v. Absorption at 240 nm is followed after the addition of H_2O_2.

vi. H_2O_2 disappearance is followed as decrease in the absorbance at 240 nm and oxidation of 1 μmole H_2O_2 min^{-1} ($\varepsilon = 0.038\ mM^{-1}\ cm^{-1}$) represents one unit of enzyme.

vii. The specific activity is referred to the amount of milligram protein present in the reaction mixture.

Calculations

Calculation of specific activity (A_s) is made as follows:

$$A_s \text{ (units mg}^{-1}\text{ protein)} = \frac{\Delta A_{240} \times \text{reaction volume (ml)} \times \text{chart rate (cm min}^{-1})}{h \text{ (cm)} \times \text{mg protein} \times 0.038}$$

Absorption decrease is monitored as a curve on the registration paper of the spectrophotometer. A straight line is drawn on the initial tract of the curve; this line must be considered as hypotenuse of a triangle whose vertical side represents ΔA_{240} (according to the scale set on the spectrophotometer), and horizontal side must be consider as height (h) in cm.

Suggested Readings

Abd-Elgawad, M.M. and Molinari, S. (2007). Markers of plant resistance to nematodes: Classical and molecular strategies. *Nematologia Mediterranea* (in press).

Acquaah, G. (1992). *Practical Protein Electrophoresis for Genetic Research*. Dioscorides Press, Oregon, USA.

Barker, K.R., Carter, C.C. and Sasser, J.N. (1985). An Advanced Treatise on Meloidogyne Vol. 1. *Biology and Control.* Vol. 2. *Methodology.* North Carolina State University, Raleigh, USA.

De Santos, M.S.N., Abrantes, I.M., Brown, D.J.F. and Lemos, R.M. (1997). *An Introduction to Virus Vector Nematodes and their Associated Viruses*. Instituto do Ambiente e Vida, Universidade de Coimbra, Portugal.

Eisenthal, R. and Danson, M.J. (1995). *Enzyme assays: A Practical Approach*. IRL Press, Oxford University, UK.

Molinari, S. (1995). Difference in isoperoxidase activities of tomato roots susceptible and resistant to root-knot nematodes. *Nematologia Mediterranea* **23**: 271–281.

Molinari, S. (1996). Molecular aspects of plant–nematode interaction. *Nematologia mediterranea* **24**: 139–154.

Molinari, S. (1999). Changes of catalase and SOD activities in the early response of tomato to *Meloidogyne* attack. *Nematologia Mediterranea* **27**: 167–172.

Molinari, S. and Loffredo, E. (2006). The role of salicylic acid in defense response of tomato to root-knot nematodes. *Physiological and Molecular Plant Pathology* **68**: 69–78.

Molinari, S., De Luca, F., Lamberti, F. and De Giorgi, C. (1997). Molecular methods for the identification of Longidorid nematodes. *Nematologia Mediterranea* **25**: 55–61.

Molinari, S., Lamberti, F., Duncan, L.W., Halbrendt, J., McKenry, M., Abawi, G.S., Magunacelaya, J.C., Crozzoli, R., Lemos, R.M., Nyczepir, A.P., Nagy, P., Robbins, R.T., Kotcon, J., Moens, M. and Brown, D.J.F. (2004). SOD polymorphism in *Xiphinema americanum*-group (Nematoda: Longidoridae). *Nematology* **6**: 867–876.

Sharma, S.B. (1998). *The Cyst Nematodes*. Kluwer Academic Publishers, The Netherlands.

Starr, J.L., Cook, R. and Bridge, J. (2002). *Plant Resistance to Parasitic Nematodes*. CABI Publishing, UK.

Appendices

Section IV. Appendices (i)

Abbreviations

ABA Abscisic acid

Ala Alanine

ANOVA Analysis of the variance

AOS Active oxygen species

ARA Acetylene-reducing activity (ARA)

Asp Aspartic acid

BSA Bovine serum albumin

BSMV Barley stripe mosaic virus

BSTFA Bis(trimethylsilyl) trifluoroacetamide

AP Alkaline phosphatase

ATCC American type culture collection

ATP Adenosine triphosphate

CAD Cadaverine

CAS Chrome azurol S

CE Capillary electrophoresis

CFU Colony forming units

DAB 3,3'-Diaminobenzidine

DEPC Diethylpyrocarbonate

DMSO Anhydrous dimethysulfoxide

DPPH 1,1-diphenyl-2-picrylhydrazyl

DR Degree of resistance

EC Electrical conductivity

EDTA Disodium ethylenediamine tetraacetic acid

ELISA Enzyme-linked immunosorbent assay

EtOAc Ethyl acetate

FID Fame ionization detection

FT-IR Fourier transformed IR

GA3 Gibberellic acid

GC Gas chromatography

Glu Glutamine

HPLC High-performance liquid chromatography

HPLC-F High performance liquid chromatography coupled to fluorescence detector

HR Hypersensitive response

HTD Heptadiamine

IAA Indole-3-acetic acid

IgG Immunoglobulin

J2 Second-stage juvenile

JA Jasmonic acid

LD_{50} Lethal dose 50

LMM Liquid malt medium

LPMA Low melting point agarose

LSC Liquid–solid chromatography

LSD Least significant difference

MBC Minimal bactericidal concentration

MDA Malondialdehyde

MDH Malate dehydrogenase

MES 2-[*N*-morpholino] ethanesulfonic acid

MFC Minimal fungicidal concentration

MHA Mueller Hinton agar

MHB Mueller Hinton broth

MIC Minimal inhibitory concentration

MS Mass spectrometry

MSTFA *N*-methyl-*N*-trimethylsilyltrifluoracetamide

MTT 3-{4,5-dimethylthiazol-2-yl}-2,5-diphenyltetraz-olium bromide

NADPH Nicotinamide adenine dinucleotide phosphate

NA Nutrient agar

NB Nutrient broth

NBT Nitro-blue tetrazolium

NCCLS National Committee for Clinical Laboratory Standards

NMPA Normal melting point agarose

OD Optical density

OPA *O*-Phthaldialdehyde

PAGE Poly-acrylamide gel electrophoresis

PAL Phenylalanine ammonia-lyase

PBS Phosphate buffered saline

PEG Poliethylen glycol

PGPR Plant growth promoting rhizobacteria

PGPB Plant growth promoting bacteria biocontrol

PMS Phenazine methasulphate

POD Peroxidase

PRs Pathogenesis-related proteins

PSHR Plant stress homeo-regulating rhizobacteria

PTFE Polytetrafluoroethylene

PVP Polyvinylpolypy-rrolidone

RKNs Root-knot nematodes

RP Reverse phase

Rt Retention time

SAR Systemic adquired resistance

SDA Sabouraud dextrose agar

SDS Sodium dodecyl sulfate

Ser Serine

SMM Solid malt medium

SOD Superoxide dismutase

SPE Solid–phase extraction

SPME Solid–phase microextraction

ssBHI Semi–solid Brain Heart Infusion

TBA 2-thiobarbituric acid

TCA Trichloroacetic acid

TLC Thin layer chromatography

TMV Tobacco mosaic Virus

TPA Trypticase phosphate agar

TSA Trypticase soya agar

TTC 2,3,5-Triphenyl-tetrazolium-chloride

UV–VIS Ultraviolet–visible

VOCs Volatile organic compounds

Z Zeatin

WMS Working microbial suspensions

Section IV. Appendices (ii)

Chemical formulae and molecular weight of solvents and reagents

Compound Name	Chemical Formulae	Molecular Weight
Acetic acid	CH_3CO_2H HOAc	60.1
Ammonia	NH_3	17
Ammonium chloride	NH_4Cl	53.5
Ammonium fluoride	NH_4F	37.04
Ammonium molybdate tetrahydrate	$(NH_4)_6Mo_7O_{24}\cdot 4H_2O$	1235.86
Ammonium dihydrogen ortophosphate	$NH_4H_2PO_4$	115.03
Ammonium sulfate	$(NH_4)_2SO_4$	132.1
Boric acid	H_3BO_3	61.8
Calcium chloride dihydrate	$CaCl_2\cdot 2H_2O$	147
Calcium nitrate	$Ca(NO_3)_2$	236.2
Carbon dioxide	CO_2	44
Chloroform	$CHCl_3$	119.4
Copper sulfate pentahydrate	$CuSO_4\cdot 5H_2O$	249.7
Cobalt chloride hexahydrate	$CoCl_2\cdot 6H_2O$	237.93
Dibasic sodium phosphate	Na_2HPO_4	141.9
Ethanol	EtOH	46.1
Ferric chloride	$FeCl_3$	162.2
Ferrous sulfate heptahydrate	$FeSO_4\cdot 7H_2O$	278.0
Hydrochloric acid	HCl	36.5 (d = 1.18 g/mL)
Hydrogen peroxide	H_2O_2	34
Magnesium chloride hexahydrate	$MgCl_2\cdot 6H_2O$	230.3
Magnesium chloride	$MgCl_2$	95.2

Appendices (ii). (Contd)

Compound Name	Chemical Formulae	Molecular Weight
Magnesium sulfate	$MgSO_4$	120.4
Manganese chloride tetrahydrate	$MnCl_2{\cdot}4H_2O$	197.8
Manganese sulfate hydrate	$MnSO_4{\cdot}H_2O$	169.01
Methanol	MeOH	32
Phosphorus pentoxide	P_2O_5	141.9
Potassium chloride	KCl	74.6
Potassium dihydrogen phosphate	KH_2PO_4	136.1
Potassium hydroxide	KOH	56.1
Potassium nitrate	KNO_3	101.1
Potassium oxide	K_2O	94.2
Sodium acetate trihydrate	$NaAcO{\cdot}3H_2O$	136.8
Sodium azide	NaN_3	65.01
Sodium bicarbonate	$NaHCO_3$	84.01
Sodium borate	$Na_2B_4O_7$	381.4
Sodium carbonate (anhydrous)	Na_2CO_3	105.9
Sodium chloride	NaCl	58.4
Sodium cyanide	NaCN	49.01
Sodium dihydrogen phosphate monohydrate	$NaH_2PO_4{\cdot}H_2O$	138
Sodium dodecyl sulfate	$NaC_{12}H_{25}SO_4$	288.38
Sodium hydrogen phosphate	Na_2HPO_4	142
Sodium hydroxide	NaOH	40
Sodium hypochlorite	NaOCl	74.45
Sodium sulfate (anhydrous)	Na_2SO_4	142
Stannous chloride dihydrate	$SnCl_2{\cdot}2H_2O$	225.63
Sulfuric acid	H_2SO_4	98.1 (d = 1.84 g/mL)
Tripotassium phosphate	K_3PO_4	212.27
Trisodium phosphate, dodecahydrate	$Na_3PO_4{\cdot}12H_2O$	380.12
Zinc sulfate heptahydrate	$ZnSO_4{\cdot}7H_2O$	287.6

Section IV. Appendices (iii)

Molecular weight of organic compounds

Compound Name	Molecular Weight
Abscisic Acid	264.32
Adenine	135.1
l-alanine	89.09
p-aminobenzoic acid	137.1
Ampicillin sodium salt	371.4
p-Anisaldehyde	136.15
l-arginine hydrochloride	210.7
l-ascorbic acid	176.1
l-asparagine	132.12
Aspartic acid	133.10
d-biotin	244.3
Calcium d-(+)-Pantothenate	476.54
β-Carotene	536.9
Cinnamic acid	148
trans-Cinnamic acid	148.2
Citric acid monohydrate	210.1
p-coumaric acid	164.2
Crystal violet	408
1-Deoxynojirimycin	164
2-Deoxyribose	134.1
Desacetyldoronine	417
2,6-Dichlorophenolindophenol	326.1
l-3,4-dihydroxyphenylalanine	197.2
N,N-Dimethylformamide	73.1
3-{4,5-Dimethylthiazol-2-yl}-2,5-diphenyltetrazolium bromide (MTT)	414.3
1,1-Diphenyl-2-picrylhydrazyl (DPPH)	394.3
Doronine	459
Ephedrine ($C_{10}H_{15}NO$)	165
Ethylenediaminetetraacetic acid disodium salt dihydrate (EDTA)	372.2
Evans blue	960.8

Appendices (iii). (Contd)

Compound Name	Molecular Weight
Ferulic acid	194.2
Fuchsin, basic	305.4
Gallic acid	188.14
Gibberellic acid	346.38
Glucose	180.2
d-Glucose-6-phosphate dipotassium salt	336.3
l-glutamic acid	147.1
Jasmonic acid	210.27
Florosenine	423
Heliotridine	155
l-Histidine monohydrochloride	209.6
p-Hydroxybenzoic acid	138.1
13α-hydroxylupanine	264
Indole-3-acetic acid	175.19
Isoretronecanol	157
Isovaleric acid	102
l-Leucine	131.2
Mannitol	182.17
l-Methionine	149.2
β-NADH disodium salt	709.4
β-NADP$^+$ sodium salt	765.4
Nicotinic acid	123.11
Nitro Blue Tetrazolium (NBT)	817.6
O-Phthaldialdehyde	134.13
Phenazine Methosulfate (PMS)	306.3
l-phenylalanine	165.2
Raffinose pentahydrate	594.52
Riboflavin	376.36
Salicylic acid	138.1
Serine	105.09
Sucrose	342.3
Thiamine hydrochloride	337.27
2-thiobarbituric acid	144.2
Tricloroacetic acid (TCA)	163.4
Tris	121.1
l-Tryptophan	204.2
Vanillic acid	168.1
trans-Zeatin	219.24

Subject Index

A

B

C

D

E

G

H

I

L

M

N

P

Author Index